Lecture Notes in Physics

T0211094

The Lecture Notes in Physics

The series Lecture Notes in Physics (LNP), founded in 1969, reports new developments in physics research and teaching – quickly and informally, but with a high quality and the explicit aim to summarize and communicate current knowledge in an accessible way. Books published in this series are conceived as bridging material between advanced graduate textbooks and the forefront of research and to serve three purposes:

- to be a compact and modern up-to-date source of reference on a well-defined topic

- to serve as an accessible introduction to the field to postgraduate students and nonspecialist researchers from related areas

- to be a source of advanced teaching material for specialized seminars, courses and schools

Both monographs and multi-author volumes will be considered for publication. Edited volumes should, however, consist of a very limited number of contributions only. Proceedings will not be considered for LNP.

Volumes published in LNP are disseminated both in print and in electronic formats, the electronic archive being available at springerlink.com. The series content is indexed, abstracted and referenced by many abstracting and information services, bibliographic networks, subscription agencies, library networks, and consortia.

Proposals should be sent to a member of the Editorial Board, or directly to the managing editor at Springer:

Christian Caron
Springer Heidelberg
Physics Editorial Department I
Tiergartenstrasse 17
69121 Heidelberg / Germany
christian.caron@springer.com

D.D. Schnack

Lectures
in Magnetohydrodynamics

With an Appendix on Extended MHD

 Springer

Dalton D. Schnack
Department Physics
University of Wisconsin
1150 University Avenue
Madison WI 53706
USA
schnack@wisc.edu

Dalton D. Schnack, *Lectures in Magnetohydrodynamics: With an Appendix on Extended MHD*, Lect. Notes Phys. 780 (Springer, Berlin Heidelberg 2009), DOI 10.1007/ 978-3-642-00688-3

ISBN 978-3-642-26921-9   e-ISBN 978-3-642-00688-3

DOI 10.1007/978-3-642-00688-3

Springer Dordrecht Heidelberg London New York

Lecture Notes in Physics ISSN 0075-8450   e-ISSN 1616-6361

Cover design: Integra Software Service Pvt. Ltd.

Printed on acid-free paper

Springer is part of Springer Science+Business Media (www.springer.com)

To my Teachers......

.... Reed Raines, Robert Zacur, Peter Havas, Herb Jeremias, Ed Greitzer, George Lamb, Art Mirin, Ken Marx, Bill Newcomb, John Killeen, Ed Caramana, Rick Nebel, Dan Barnes, Zoran Mikic, Jim Callen ....

......I hope to continue to learn.

And, of course .....

.... to C. K. Rowdyshrub for spiritual enlightenment.

About this Book

This book consists of lecture notes for an advanced graduate course in magnetohydrodynamics, or MHD, which was taught at the University of Wisconsin, Madison, in the Fall Semester of 2007. It is not a textbook, or a treatise, or a monograph, or an extended review paper. It is, as advertised, a set of lectures that were actually given in a classroom setting, presented as close to verbatim as possible while still being organized and grammatical. The course consisted of 26 class periods of an hour and a quarter each. However, there are 38 lectures, arranged by topic. Some are very short, others more lengthy. Clearly, any given class session might cover more than one topic, so the term "lecture" is somewhat arbitrary.

There was no final exam. The grades were based entirely on homework problems that were assigned every 2 weeks or so and graded. Devising these problems was the most difficult part of the course, and I was unhappy with the results. I have therefore not included any of them in the present volume. Almost anyone could make up better problems than I did. (Try it yourself!)

The course in MHD is part of a series of one semester core courses required of plasma physics graduate students at the University of Wisconsin. Others are Basic Plasma Physics, Fusion Fundamentals, Plasma Kinetic Theory, and Plasma Waves. With the exception of Basic Plasma Physics, the courses may be taken in any order, so the instructor cannot assume any specific advanced knowledge of these fields on the part of the students. Similarly, the students have a variety of backgrounds and interests. For example, in the Fall of 2007, the MHD class consisted of 18 students from the Physics, Engineering Physics, Electrical Engineering, and Astronomy/Astrophysics departments. Their interests varied accordingly. As a consequence, I tried to present broad a view of MHD, for the most part avoiding diversions into specialized topics and formalisms, such as tokamaks, that are difficult, time-consuming, and have limited general interest. Because the students' mathematical backgrounds were also varied, I spent some time reviewing certain mathematic topics, such as tensors, dyads, matched asymptotic expansions, and the calculus of variations that are essential to the development of MHD as a topic in theoretic physics. The goal was to make the course as self-contained as possible. These are included in the present notes.

Most of the material in the written lectures was actually covered in the classroom. I have not attempted to "flesh out" much of anything here. With a few exceptions, if it is in the notes, it was said in class. Conversely, the lectures intentionally omit some of the grinding details that accompany certain derivations. This is especially true with respect to some of the material regarding the energy principle, resistive instabilities, and MHD turbulence, and this is noted in the text. In almost all other cases I tried to present as many of the relevant details as possible. I strived for a reasonable balance given the time available for presentation. Because the scope of the material is tailored to the time constraint of a 13-week semester, the lectures can provide the basis for other courses in MHD. Of course, instructors are free to add or delete material; in particular, they might want to invent some problem sets!

The notes contained in the present volume have been slightly revised from the notes that were distributed to the students. I have added some footnotes referencing source material and corrected some errors that have been brought to my attention. I have also added a short quotation at the beginning of each lecture, which is meant to refer (hopefully in a pithy, entertaining way) to some aspect of the material to be presented. In doing this I was inspired by the famous book by Lord and Lady Jeffreys.[1] They were, however, true scholars who were well educated in classical literature. In contrast, I found most of my quotes on Google! I hope you find them at least interesting, and hopefully entertaining, because there's little enough entertainment value in MHD. Of course, you are invited to provide your own, alternative versions!

I have also included an Appendix on extended MHD. This material was contained in three lectures that I gave at UW in April, 2006. It talks about the kinetic theory underlying the MHD model, and how people have tried to extend the model to include additional physical effects. It is self-contained.

The Lectures on Turbulence (36) and Dynamos (38) have benefited from thorough and critical reading by Drs. Paul Terry, Fausto Cattaneo, and Ellen Zweibel. Their constructive comments and suggestions have greatly improved the presentation. As in the rest of the book, any errors that remain, either factual or interpretive, are my responsibility.

I also wish to acknowledge Dr. Christian Caron and Ms. Gabriele Hukuba of Springer-Verlag for all their assistance and support in the publication process.

I was partially supported by several grants from the U.S. Department of Energy during the preparation of this book.

I hope you enjoy the lectures, and maybe even get something from them!

Madison, WI   Dalton D. Schnack

---

[1] Harold Jeffreys and Bertha Jeffreys, *Methods of Mathematical Physics*, Cambridge University Press, Cambridge (1972).

Preface

Magnetohydrodynamics, or MHD, is a theoretical way of describing the statics and dynamics of electrically conducting fluids. The most important of these fluids occurring in both nature and the laboratory are ionized gases, called *plasmas*. These have the simultaneous properties of conducting electricity and being electrically charge neutral on almost all length scales. The study of these gases is called *plasma physics*.

MHD is the poor cousin of plasma physics. It is the simplest theory of plasma dynamics. In most introductory courses, it is usually afforded a short chapter or lecture at most: Alfvén waves, the kink mode, and that is it. (Now, on to Landau damping!) In advanced plasma courses, such as those dealing with waves or kinetic theory, it is given an even more cursory treatment, a brief mention on the way to things more profound and interesting. (It is *just* MHD! Besides, *real* plasma physicists do kinetic theory!)

Nonetheless, MHD is an indispensable tool in all applications of plasma physics. Even the simplest experiment will not be built unless it has first passed muster with MHD. The reason is that MHD deals with fundamental force balance and deviation from it, concepts that are surprisingly subtle and complex. MHD also provides the machinery for understanding the basic properties of global structure of magnetized plasmas, how they can sustain themselves, and why they share a small number of global properties. In this course we will look at many of these important issues in what I hope is sufficient detail as to reveal some of the elegance that underlies this immensely useful theory, and gain some appreciation for the skill of the pioneers of the field.

While MHD is the simplest mathematical model of a plasma, it is difficult to justify as a valid description of any interesting plasma. Surely plasmas know that they are made of individual ions and electrons (or at a minimum of separate ion and electron fluids) and that they are so hot that collisions between particles are relatively rare events! MHD completely ignores both of these issues. Nonetheless, it is a fact that MHD provides a remarkably accurate description of the low-frequency, long-wavelength dynamics of real plasmas. MHD seems to *work* (even when it shouldn't)!

The validity of MHD as a mathematical model for magnetized plasmas has been discussed in great detail elsewhere, particularly in the book by Freidberg.[2] It cannot be said better, so it will not be attempted here. Instead, from the beginning we adopt the point of view that MHD describes the dynamics of a continuum fluid that is capable of conducting an electric current, that this fluid can be characterized by a few parameters such as mass density, velocity, and pressure, and that the material properties of this fluid are independent of the physical size of the sample. That is, *the material looks exactly the same no matter how finely it is subdivided*, and behavior arising from the atomic structure of matter is not considered. This is precisely the approach taken with hydrodynamics, which is one of the most complex and difficult topics in classical physics. We will find that MHD is more complex, and even more difficult, primarily as a result of spatial anisotropy introduced by the magnetic field.

In this course, MHD will be treated as a topic in theoretical physics. There will be little attempt at experimental justification or motivation; these will be taken as self-evident. Mathematics is the language of theoretical physics, and a rudimentary knowledge of what is normally called "methods of theoretical physics" (e.g., linear algebra, ordinary differential equations, Fourier analysis, and some special functions) is assumed. However, the subject requires some additional emphasis on vectors, tensors and dyads, the calculus of variations, and what is called "matched asymptotic expansions," and these will be reviewed briefly as necessary. Of course, and as always, a certain amount of algebraic fortitude is useful.

The broad areas chosen for presentation are the derivation and properties of the fundamental equations, equilibrium, waves and instabilities, turbulence, self-organization, and dynamos. The latter topics require the inclusion of the effects of electrical resistivity and nonlinearity. Together, these span the range of MHD issues that have proven to be important for understanding magnetically confined plasmas, as they all describe either force balance, or some small deviation from it. Most have also been important in some space and astrophysical applications. It will take some effort to do them justice in the course of a single semester. Unfortunately, issues related to large deviations from force balance, such as strong flows and shocks, will not be covered, perhaps for no better reason than you cannot present what you do not know!

One of the areas where MHD has been useful is in the design and analysis of toroidally confined plasmas, in particular the tokamak configuration. The theory of these toroidal plasmas is extremely well developed and mathematically nuanced. It is unfortunate that this elegant theory has led to a jargon, a patois, that is difficult to penetrate while providing few additional physical insights. (Further, these theories of tokamak plasmas are applicable only in a vanishingly small volume of the universe, namely, the total volume of all the tokamaks!) Most of the important concepts can be illustrated in other, simpler geometries, and that is the plan here. (See also the last comment in the previous paragraph.) One exception is the equilibrium of an

---

[2] Jeffrey P. Freidberg, *Ideal Magnetohydrodynamics*, Plenum Press, New York (1987).

axially symmetric toroidal plasma, which provides an introduction to the necessary concepts and terminology. The plan is to provide sufficient background to enable further individual study.

Nonetheless, each of the topics to be presented in the course can be (and has been) pursued in significantly more (often mind-numbing) detail than can be presented here. These topics are somewhat analogous to a constellation of black holes (or perhaps rabbit holes); it is possible to disappear into any one of them for the rest of your life, never to be seen again. The best I can do is act as a guide around this universe and to try to provide you with a "star chart" so that you can avoid the most obvious traps and pitfalls in your future careers. Another goal is to provide sufficient background in MHD so as to make the published literature in these areas at least initially accessible.

In preparing this course, I have assumed that all students are at the level of advanced graduate education and that this may be one of their last formal courses before embarking on full-time research. These students have already proven that they are excellent at taking written exams. It is also conceivable, perhaps probable, that for each topic there will be some students in the class with more knowledge than the instructor. Therefore, it is useless to try to devise tricky exams or to try to be too didactic about any particular topic. The fundamental working assumption is that all students are sufficiently mature and motivated to accept and appreciate *guidance* rather than instruction and examination, and that is how I shall proceed.

Contents

Lecture 1
Introduction

> *You must start at the very beginning.*
>
> Julie Andrews, *The Sound of Music*

Magnetohydrodynamics: MHD. *Magneto*—having to do with electromagnetic fields; *hydro*—having to do with fluids; and *dynamics*—dealing with forces and the laws of motion. Magnetohydrodynamics, or MHD, is the mathematical model for the low-frequency interaction between electrically conducting fluids and electromagnetic fields.

So you will need to know about fluid dynamics, you will need to know about electromagnetism, and you will need to know some plasma physics.

An *isotropic medium* is one that has no preferred direction in space; it looks the same and behaves in the same way in all directions. Hydrodynamics, which generally deals with isotropic materials (fluids), is a very difficult subject. MHD is even more difficult because a magnetic field identifies a preferred direction in space. A magnetized fluid is said to be an *anisotropic medium*. Further, since there will be interaction between the fluid and the magnetic field, some form of Maxwell's equations must be solved simultaneously with the dynamical equations for the fluid. You will see that this can be a formidable task.

In this course, MHD will be treated as a topic in theoretical physics. We will make some fundamental assumptions regarding the properties of the material and the time and spatial scales of interest, and the rest will follow self-consistently (and hopefully logically). Little or no experimental motivation will be given, but will rather be taken as self-evident.

Mathematics is the language of theoretical physics. Specific topics that are essential to the theoretical analysis of the dynamics of a magnetized fluid are as follows:

- Vector and tensor analysis,
- ODEs,
- PDEs,
- Calculus of variations,
- Method of matched asymptotic expansions.

Short reviews of some of these topics will be provided at the appropriate time during the presentation.

From the beginning, we adopt the point of view that MHD describes the dynamics of a continuum fluid that is capable of conducting an electric current, that this

Schnack, D.D.: *Introduction*. Lect. Notes Phys. **780**, 1–4 (2009)
DOI 10.1007/978-3-642-00688-3_1

fluid can be characterized by a few parameters such as mass density, velocity, and pressure, and that the material properties of this fluid are independent of the physical size of the sample. As stated in Vol. 8 of Landau and Lifschitz,[1] "physical quantities are averaged over volumes that are 'physically infinitesimal', ignoring the variations that result from the molecular structure of matter." That is, the material looks exactly the same no matter how finely it is subdivided. This is our fundamental assumption.

This assumption implies an ordering of length scales. Some important distances are as follows:

- a_0, the atomic radius;
- λ, the "mean free path" between atomic collisions;
- δ, the "physically infinitesimal" distance;
- L, the smallest relevant macroscopic distance to be considered (e.g., $L \sim 1/k_{max}$, the shortest wavelength of interest).

Our fundamental assumption implies the ordering $\lambda \sim a_0 << \delta < L$. Whenever $\lambda \geq \delta, L$, the model breaks down, and it must be modified and further justified. This is the realm of *extended MHD*, and is beyond the scope of this course. (Except, see the Appendix.)

We will also assume an ordering of times scales. In particular, we will attempt to describe only low-frequency motions, those for which $V^2/c^2 << 1$, where V is a characteristic fluid velocity and c is the speed of light. With $V = \omega L$, where ω is a characteristic frequency, we have $\omega^2 << c^2/L^2$, or $\tau = 1/\omega >>>> L/c = \tau_c$. The characteristic time intervals for MHD are very much longer than the time it takes a light wave to transit the macroscopic system.

We further assume that the smallest subdivision of the medium is in local thermodynamic equilibrium, which is provided by inter-atomic collisions. It can be characterized by a temperature T. So, we also require that $\tau >> 1/\nu_c$, where ν_c is the frequency of inter-atomic collisions.

The usual definition of a *fluid* is *a substance that resists an applied compressive stress, but continually deforms, or flows, under an applied shear stress, regardless of the magnitude of the applied stress.*

This behavior is illustrated in Fig. 1.1, which shows the response of a fluid element to an applied force. Figure 1.1a shows a compressive stress. The fluid generates

Fig. 1.1 (a) A fluid element resisting a compressive force; (b) a fluid element cannot resist a shearing force. It deforms continuously

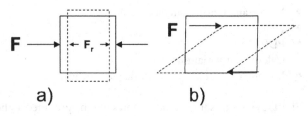

a)   b)

---

[1] L. D. Landau and E. M. Lifschitz, *Electrodynamics of Continuous Media*, Pergamon Press, Oxford, UK (1960).

a restoring force that opposes the applied force; it can *support* compressive stresses. Figure 1.1b shows a shearing stress. In this case, the fluid generates no restoring force, and the distortion will grow continually. The fluid *cannot* support a shearing stress.

We will see that the restoring force generated in response to an applied force leads to wave propagation. Thus a fluid is commonly said to allow the propagation of *compressional waves* (e.g., sound waves), but not *shear waves*.

Now consider the sheared deformation of a fluid element permeated by a magnetic field **B**. If the fluid is electrically conducting, we will see that the magnetic field is deformed along with the fluid. This situation is sketched in Fig. 1.2, where the shearing force is applied perpendicular to the direction of the magnetic field. The resulting bending of the field lines produces a restoring force that opposes the applied stress. Thus, *an electrically conducting fluid can support the propagation of shear waves*. This result was considered so counterintuitive and novel that at first it was not believed to be true. Subsequent astronomical observations and experiments showed the existence of these waves, and this led to the awarding in 1970 of the Nobel Prize in physics to Hannes Alfvén, whose name is attached to these waves. This remains the only Nobel awarded in plasma physics. (In contrast, a shearing force applied parallel to the magnetic field produces no such restoring force, and the fluid will act like it is not magnetized.)

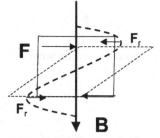

Fig. 1.2 A magnetized fluid element resists shearing stress because of the restoring force supplied by the magnetic field

At this point, it may be useful to define what we mean by a magnetic field line. The magnetic induction, or *magnetic field*, is a vector function, denoted as $\mathbf{B}(\mathbf{x}, t)$, that assigns a magnitude and direction to all points in space. (We will soon be more specific about what we mean by the term *vector*.) Physically, this field is produced by electric currents flowing somewhere in the universe. As with all vector fields, it is possible to define a set of three-dimensional curves that are everywhere tangent to the vector **B**. Sometimes these are called *streamlines*; here we call them *field lines*. The defining equations of these curves are

$$\frac{dx}{B_x} = \frac{dy}{B_y} = \frac{dz}{B_z} = \frac{dl}{B},$$

(1.1)

where (x, y, z) are Cartesian coordinates, $B = \sqrt{B_x^2 + B_y^2 + B_z^2}$ is the magnitude of the magnetic field (the *magnetic field strength*), and l is the distance along the field line. The trajectory of the field line $x(l)$, $y(l)$, and $z(l)$ passing through a point (x_0, y_0, z_0) can be found by integrating equations

$$\frac{dx}{dl} = \frac{B_x}{B},\tag{1.2}$$

$$\frac{dy}{dl} = \frac{B_y}{B},\tag{1.3}$$

and

$$\frac{dz}{dl} = \frac{B_z}{B},\tag{1.4}$$

beginning from the point (x_0, y_0, z_0).

We take it as an experimental and observational fact that magnetic field lines either close upon themselves or fill space *ergodically*. (An *ergodic trajectory* is one that passes arbitrarily close to all points in space *without* closing on itself.) In either case, the field line has neither beginning nor end. If we consider an arbitrarily shaped surface S enclosing a volume V, this means that as many field lines must leave the enclosed volume as enter it. This is expressed mathematically as

$$\oint_S \mathbf{B} \cdot d\mathbf{S} = 0,\tag{1.5}$$

or, using Gauss' theorem,

$$\int_V \nabla \cdot \mathbf{B} dV = 0.\tag{1.6}$$

Since this must hold for every volume V, we must have

$$\nabla \cdot \mathbf{B} = 0,\tag{1.7}$$

which expresses the "endless" nature of field lines. This is one of the fundamental equations of physics.

We will now proceed to derive the following:

1. The equations for the dynamics of the fluid.
2. The equations for the dynamics of the electromagnetic field.
3. Some properties of the combined equations.

These will require a familiarity with the language of scalars, vectors, tensors, and dyads. This is reviewed briefly in the next lecture.

Lecture 2
Review of Scalars, Vectors, Tensors, and Dyads

> *Our vice always lies in the direction of our virtues, and in*
> *their best estate are but plausible imitations of the latter.*
> Henry David Thoreau

In MHD, we will deal with relationships between quantities such as the magnetic field and the velocity that have both magnitude and direction. These quantities are examples of vectors (or, as we shall soon see, pseudovectors). The basic concepts of scalar and vector quantities are introduced early in any scientific education. However, to formulate the laws of MHD precisely, it will be necessary to generalize these ideas and to introduce the less familiar concepts of matrices, tensors, and dyads. The ability to understand and manipulate these abstract mathematical concepts is essential to learning MHD. Therefore, for the sake of both reference and completeness, this lecture is about the mathematical properties of scalars, vectors, matrices, tensors, and dyads. If you are already an expert, or think you are, please skip class and go on to Lecture 3. You can always refer back here if needed!

A *scalar* is a quantity that has *magnitude*. It can be written as

$$S \quad \alpha \quad 9 \tag{2.1}$$

It seems self-evident that such a quantity is independent of the coordinate system in which it is measured. However, we will see later in this lecture that this is somewhat naïve, and we will have to be more careful with definitions. For now, we say that the magnitude of a scalar is independent of coordinate transformations that involve translations or rotations.

A *vector* is a quantity that has both *magnitude* and *direction*. It is often printed with an arrow over it (as in $\vec{V}$) or in bold-face type (as in **V**, which is my preference). When handwritten, I use an underscore (as in $\underline{V}$, although many prefer the arrow notation here, too). It can be geometrically represented as an arrow. A vector has a tail and a head (where the arrowhead is). Its *magnitude* is represented by its length. We emphasize that the vector has an "absolute" orientation in space, i.e., it exists independent of any particular coordinate system. Vectors are therefore "coordinate-free" objects, and expressions involving vectors are true in any coordinate system. Conversely, if an expression involving vectors is true in one coordinate system, it is true in all coordinate systems. (As with the scalar, we will be more careful with our statements in this regard later in this lecture.)

Schnack, D.D.: *Review of Scalars, Vectors, Tensors, and Dyads.* Lect. Notes Phys. **780**,
5–18 (2009)
DOI 10.1007/978-3-642-00688-3_2

Vectors are added with the parallelogram rule. This is shown geometrically in Fig. 2.1.

Fig. 2.1 Illustration of the parallelogram rule for adding vectors

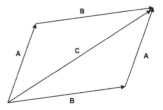

This is represented algebraically as $\mathbf{C} = \mathbf{A} + \mathbf{B}$.

We define the *scalar product* of two vectors $\mathbf{A}$ and $\mathbf{B}$ as

$$\mathbf{A} \cdot \mathbf{B} = AB \cos \theta, \tag{2.2}$$

where A and B are the magnitudes of $\mathbf{A}$ and $\mathbf{B}$ and θ is the angle (in radians) between them, as in Fig. 2.2:

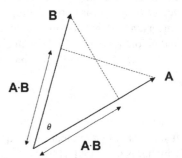

Fig. 2.2 Illustration of the scalar product of two vectors $\mathbf{A}$ and $\mathbf{B}$ as the projection of one on the other

The quantity $S = \mathbf{A} \cdot \mathbf{B}$ is the projection of $\mathbf{A}$ on $\mathbf{B}$, and vice versa. Note that it can be negative or zero. We will soon prove that S is a scalar.

It is sometimes useful to refer to a vector $\mathbf{V}$ with respect to some coordinate system (x_1, x_2, x_3), as shown in Fig. 2.3. Here the coordinate system is orthogonal. The vectors $\hat{\mathbf{e}}_1$, $\hat{\mathbf{e}}_2$, and $\hat{\mathbf{e}}_3$ have unit length and point in the directions of x_1, x_2, and x_3, respectively. They are called *unit basis vectors*. The *components* of $\mathbf{V}$ with respect to (x_1, x_2, x_3) are then defined as the scalar products

$$V_1 = \mathbf{V} \cdot \hat{\mathbf{e}}_1, \ V_2 = \mathbf{V} \cdot \hat{\mathbf{e}}_2, \ V_3 = \mathbf{V} \cdot \hat{\mathbf{e}}_3. \tag{2.3a,b,c}$$

The three numbers (V_1, V_2, V_3) also define the vector $\mathbf{V}$.

Fig. 2.3 A vector **V** and the
basis vectors in a
three-dimensional Cartesian
coordinate system

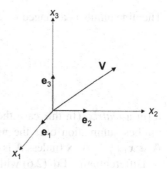

Of course, a vector can be referred to another coordinate system (x_1', x_2', x_3') by means of a *coordinate transformation*. This can be expressed as

$$
\begin{aligned}
x_1' &= a_{11}x_1 + a_{12}x_2 + a_{13}x_3, \\
x_2' &= a_{21}x_1 + a_{22}x_2 + a_{23}x_3, \\
x_3' &= a_{31}x_1 + a_{32}x_2 + a_{33}x_3,
\end{aligned}
\tag{2.4}
$$

where the nine numbers a_{ij} are independent of position; it is a *linear transformation*. Equation (2.4) can be written as

$$
x_i' = \sum_{j=1}^{3} a_{ij}x_j, \, i = 1, 2, 3.
\tag{2.5}
$$

We will often use the shorthand notation

$$
x_i' = a_{ij}x_j,
\tag{2.6}
$$

with an implied summation over the repeated index (in this case j). This is called the Einstein summation convention. Since the repeated index j does not appear in the result (the left-hand side), it can be replaced by any other symbol. It is called a *dummy index*. The economy of the notation of (2.6) over (2.4) is self-evident.

Equation (2.6) is often written as

$$
\mathbf{x}' = \mathbf{A} \cdot \mathbf{x},
\tag{2.7}
$$

where

$$
\mathbf{x} = \begin{pmatrix} x_1 \\ x_2 \\ x_3 \end{pmatrix}
\tag{2.8}
$$

is called a *column vector*. The transpose of **x** is the *row vector*

$$
\mathbf{x}^T = \begin{pmatrix} x_1 & x_2 & x_3 \end{pmatrix}.
\tag{2.9}
$$

The nine numbers arranged in the array

$$\mathbf{A} = \begin{pmatrix} a_{11} \ a_{12} \ a_{13} \\ a_{21} \ a_{22} \ a_{23} \\ a_{31} \ a_{32} \ a_{33} \end{pmatrix} \tag{2.10}$$

form a *matrix*. (In this case the matrix is 3×3.) The "dot product" in Eq. (2.7) implies summation over the neighboring indices, as in Eq. (2.6). Note that $\mathbf{x}^T \cdot \mathbf{A} \equiv x_j a_{ji} \neq \mathbf{A} \cdot \mathbf{x}$ (unless $\mathbf{A}$ is *symmetric*, i.e., $a_{ij} = a_{ji}$).

Differentiating Eq. (2.6) with respect to x_k, we find

$$\frac{\partial x_i'}{\partial x_k} = a_{ij} \frac{\partial x_j}{\partial x_k} = a_{ij} \delta_{jk} = a_{ik}, \tag{2.11}$$

which defines the transformation coefficients a_{ik}.

For reference, we give some matrix definitions and properties:

1. The *identity matrix* is defined as

$$\mathbf{I} = \begin{pmatrix} 1 \ 0 \ 0 \\ 0 \ 1 \ 0 \\ 0 \ 0 \ 1 \end{pmatrix} = \delta_{ij}. \tag{2.12}$$

2. The *inverse matrix* $\mathbf{A}^{-1}$, is defined by $\mathbf{A}^{-1} \cdot \mathbf{A} = \mathbf{I}$.
3. If a_{ij} are the components of $\mathbf{A}$, then a_{ji} are the components of $\mathbf{A}^T$, the *transpose* of $\mathbf{A}$. (If $\mathbf{A} = \mathbf{A}^T$, $\mathbf{A}$ is *symmetric*.)
4. The *adjoint matrix* is defined by $\mathbf{A}^\dagger = \mathbf{A}^{*T}$, where $(..)^*$ is the complex conjugate; i.e., $a_{ij}^\dagger = a_{ji}^*$.
5. If $\mathbf{A} = \mathbf{A}^\dagger$, then $\mathbf{A}$ is said to *self-adjoint*. (This is the generalization of a symmetric matrix to the case where the components are complex numbers.)
6. Matrix multiplication is defined by $\mathbf{A} \cdot \mathbf{B} = A_{ij} B_{jk}$.
7. $(\mathbf{A} \cdot \mathbf{B})^T = \mathbf{B}^T \cdot \mathbf{A}^T$.
8. $(\mathbf{A} \cdot \mathbf{B})^\dagger = \mathbf{B}^\dagger \cdot \mathbf{A}^\dagger$.

The prototypical vector is the *position vector*

$$\mathbf{r} = x_1 \hat{\mathbf{e}}_1 + x_2 \hat{\mathbf{e}}_2 + x_3 \hat{\mathbf{e}}_3 \equiv (x_1, x_2, x_3). \tag{2.13}$$

It represents a vector from the origin of coordinates to the point $P(x_1, x_2, x_3)$. We say that the three numbers (V_1, V_2, V_3) are the components of a vector if they transform like the components of the position vector $\mathbf{r}$ under *coordinate rotations*. Vectors are defined by their transformation properties.

We require that the *length* of the position vector, defined by $l^2 = \mathbf{x}^T \cdot \mathbf{x}$, be invariant under coordinate rotations, i.e., $l^2 = \mathbf{x}^T \cdot \mathbf{x} = \mathbf{x}'^T \cdot \mathbf{x}'$. Then

$$l^2 = \mathbf{x}^T \cdot \mathbf{x} = \mathbf{x}'^T \cdot \mathbf{x}'$$
$$= (\mathbf{A} \cdot \mathbf{x})^T \cdot (\mathbf{A} \cdot \mathbf{x})$$
$$= \left(\mathbf{x}^T \cdot \mathbf{A}^T\right) \cdot (\mathbf{A} \cdot \mathbf{x})$$
$$= \mathbf{x}^T \cdot \left(\mathbf{A}^T \cdot \mathbf{A}\right) \cdot \mathbf{x},$$

so that $\mathbf{A}^T \cdot \mathbf{A} = \mathbf{I}$, or $\mathbf{A}^T = \mathbf{A}^{-1}$. Matrices with this property are called *orthogonal matrices*, and the rotation matrix $\mathbf{A}$ is an orthogonal matrix, i.e.,

$$a_{ij}^{-1} = a_{ji}, \tag{2.14}$$

or, since $\mathbf{x} = \mathbf{A}^{-1} \cdot \mathbf{x}'$,

$$\frac{\partial x_i'}{\partial x_j} = \frac{\partial x_j}{\partial x_i'}. \tag{2.15}$$

Then the components of the rotation matrix $\mathbf{A}$ have the property

$$a_{ij}a_{ik} = \frac{\partial x_i'}{\partial x_j}\frac{\partial x_i'}{\partial x_k} = \frac{\partial x_i'}{\partial x_j}\frac{\partial x_k}{\partial x_i'} = \frac{\partial x_k}{\partial x_j} = \delta_{jk}. \tag{2.16}$$

We now say that *the three numbers V_1, V_2, and V_3 are the components of a vector if they transform like the position vector $\mathbf{r}$ under coordinate rotations*, i.e.,

$$V_i' = a_{ij}V_j, \tag{2.17}$$

where the a_{ij} are the components of an orthogonal matrix. (Note that *not* all triplets are components of vectors.)

Suppose that $\mathbf{A}$ and $\mathbf{B}$ are vectors. As an illustration of the algebraic manipulation of vector quantities, we now prove that the product defined in Eq. (2.2) is a scalar. To do this, we must show that $S' = \mathbf{A}' \cdot \mathbf{B}'$, the value of the product in the primed coordinate system, is the same as $S = \mathbf{A} \cdot \mathbf{B}$, the value in the unprimed system:

$$S' = \mathbf{A}' \cdot \mathbf{B}'$$
$$= a_{ij}A_j a_{ik}B_k$$
$$= a_{ij}a_{ik}A_j B_k$$
$$= \delta_{jk}A_j B_k$$
$$= A_j B_j = S,$$

where the property of orthogonal matrices defined in Eq. (2.16) has been used. Further, if $S = \mathbf{A} \cdot \mathbf{B}$ is a scalar, and $\mathbf{B}$ is a vector, then $\mathbf{A}$ is also a vector.

In addition to the scalar product of two vectors, we can also define the *vector product* of two vectors. The result is another vector. This operation is written symbolically as $\mathbf{C} = \mathbf{A} \times \mathbf{B}$. The magnitude of $\mathbf{C}$, C, is given by $C = AB\sin\theta$, in analogy with Eq. (2.2). By definition, the direction of **C** is perpendicular to the plane defined by **A** and **B** along with the "right-hand rule." Note that it can point either "above" or "below" the plane, and may be zero.

In index notation, the vector product is written as

$$C_i = \varepsilon_{ijk} A_j B_k. \tag{2.18}$$

The quantity ε_{ijk} is called the *Levi-Civita tensor density*. It will prove to be quite important and useful in later analysis. It has 27 components, most of which vanish. These are defined as

$$\left.\begin{array}{l} \varepsilon_{123} = \varepsilon_{231} = \varepsilon_{312} = 1 \text{ (even permutation of the indices)} \\ \varepsilon_{132} = \varepsilon_{213} = \varepsilon_{321} = -1 \text{ (odd permutation of the indices)} \\ \varepsilon_{ijk} = 0 \text{ if } i = j, \text{ or } i = k, \text{ or } j = k \end{array}\right\}. \tag{2.19}$$

We should really prove that **C** so defined is a vector, but I will leave it as an exercise for the student.

The Levi-Civita symbol satisfies the very useful identity

$$\varepsilon_{ijk}\varepsilon_{lmk} = \delta_{il}\delta_{jm} - \delta_{im}\delta_{jl}. \tag{2.20}$$

The expression can be used to derive a wide variety of formulas and identities involving vectors and tensors. For example, consider the "double cross product" **A** × (**B** × **C**). We write this in a Cartesian coordinate system as

$$\mathbf{A} \times (\mathbf{B} \times \mathbf{C}) \Rightarrow \varepsilon_{ijk} A_j (\mathbf{B} \times \mathbf{C})_k = \varepsilon_{ijk} A_j (\varepsilon_{klm} B_l C_m)$$

$$= \varepsilon_{ijk}\varepsilon_{klm} A_j B_l C_m = \varepsilon_{ijk} \underbrace{\varepsilon_{lmk}}_{\text{Even permutation of indices}} A_j B_l C_m$$

$$= \underbrace{(\delta_{il}\delta_{jm} - \delta_{im}\delta_{jl})}_{\text{From Equation (2.20)}} A_j B_l C_m = \delta_{il} A_j B_l C_j - \delta_{im} A_j B_j C_m$$

$$= A_j B_i C_j - A_j B_j C_i = B_i (A_j C_j) - C_i (A_j B_j)$$

$$\Rightarrow \mathbf{B} (\mathbf{A} \cdot \mathbf{C}) - \mathbf{C} (\mathbf{A} \cdot \mathbf{B}), \tag{2.21}$$

which is commonly known as the "BAC-CAB rule." The first step in the derivation was to translate the vector formula into components in some convenient Cartesian coordinate system, then turn the crank. We recognized the formula in the line preceding Eq. (2.21) as another vector formula expressed in the same Cartesian system. However, *if a vector formula is true on one system, it is true in all systems* (even generalized, non-orthogonal, curvilinear coordinates), so we are free to translate it back into vector notation. This is a very powerful technique for simplifying and manipulating vector expressions.

We define the *tensor product* of two vectors **B** and **C** as **A** = **BC** or

$$A_{ij} = B_i C_j. \tag{2.22}$$

How do the nine numbers A_{ij} transform under rotations? Since $\mathbf{B}$ and $\mathbf{C}$ are vectors, we have

$$A'_{ij} = B'_i C'_j = (a_{ik} B_k)(a_{jl} C_l) = a_{ik} a_{jl} B_k C_l,$$

or

$$A'_{ij} = a_{ik} a_{jl} A_{kl}. \tag{2.23}$$

Equation (2.23) is the *tensor transformation law*. Any set of nine numbers that transform like this under rotations form the components of a tensor.

The *rank* of the tensor is the number of indices. We notice that a scalar is a tensor of rank zero, a vector is a first-rank tensor, the 3×3 array just defined is a second-rank tensor, etc. In general, a tensor transforms according to

$$A'_{ijkl.....} = a_{ip} a_{jq} a_{kr} a_{ls} \ldots \ldots A_{pqrs.....}. \tag{2.24}$$

We can also write $\mathbf{A}$ in *dyadic notation*:

$$\begin{aligned}
\mathbf{A} = \mathbf{BC} &= (B_1\hat{\mathbf{e}}_1 + B_2\hat{\mathbf{e}}_2 + B_3\hat{\mathbf{e}}_3)(C_1\hat{\mathbf{e}}_1 + C_2\hat{\mathbf{e}}_2 + C_3\hat{\mathbf{e}}_3) \\
&= B_1 C_1 \hat{\mathbf{e}}_1\hat{\mathbf{e}}_1 + B_1 C_2 \hat{\mathbf{e}}_1\hat{\mathbf{e}}_2 + B_1 C_3 \hat{\mathbf{e}}_1\hat{\mathbf{e}}_3 \\
&\quad + B_2 C_1 \hat{\mathbf{e}}_2\hat{\mathbf{e}}_1 + B_2 C_2 \hat{\mathbf{e}}_2\hat{\mathbf{e}}_2 + B_2 C_3 \hat{\mathbf{e}}_2\hat{\mathbf{e}}_3 \\
&\quad + B_3 C_1 \hat{\mathbf{e}}_3\hat{\mathbf{e}}_1 + B_3 C_2 \hat{\mathbf{e}}_3\hat{\mathbf{e}}_2 + B_3 C_3 \hat{\mathbf{e}}_3\hat{\mathbf{e}}_3.
\end{aligned} \tag{2.25}$$

The quantities $\hat{\mathbf{e}}_i\hat{\mathbf{e}}_j$ are called *unit dyads*. Note that

$$\hat{\mathbf{e}}_1 \cdot \mathbf{A} = B_1 C_1 \hat{\mathbf{e}}_1 + B_1 C_2 \hat{\mathbf{e}}_2 + B_1 C_3 \hat{\mathbf{e}}_3 \tag{2.26}$$

is a vector, while

$$\mathbf{A} \cdot \hat{\mathbf{e}}_1 = B_1 C_1 \hat{\mathbf{e}}_1 + B_2 C_1 \hat{\mathbf{e}}_2 + B_3 C_1 \hat{\mathbf{e}}_3 \tag{2.27}$$

is a *different* vector. In general, $\mathbf{BC} \neq \mathbf{CB}$.

We could similarly define higher-rank tensors and dyads as $\mathbf{D} = \mathbf{AE}$, or $D_{ijk} = A_{ij} E_k$, etc.

Contraction is defined as summation over a pair of indices, e.g., $D_i = A_{ij} E_j$. Contraction reduces the rank by 2. We have also used the notation $\mathbf{D} = \mathbf{A} \cdot \mathbf{E}$ to indicate contraction over "neighboring" indices. (Note that $\mathbf{A} \cdot \mathbf{E} \neq \mathbf{E} \cdot \mathbf{A}$.) The "double-dot" notation $(\mathbf{ab}) : (\mathbf{cd})$ is often used, but is ambiguous. We define $\mathbf{A} : \mathbf{B} \equiv A_{ij} B_{ij}$, a scalar.

We now define a *differential operator* in our Cartesian coordinate system[1]

$$\nabla \equiv \hat{\mathbf{e}}_1 \frac{\partial}{\partial x_1} + \hat{\mathbf{e}}_2 \frac{\partial}{\partial x_2} + \hat{\mathbf{e}}_3 \frac{\partial}{\partial x_3} \equiv \hat{\mathbf{e}}_i \frac{\partial}{\partial x_i} \equiv \hat{\mathbf{e}}_i \partial_i. \qquad (2.28)$$

The symbol ∇ is sometimes called "nabla," and more commonly, "grad," which is short for "gradient." So far, it is just a linear combination of partial derivatives; it needs something more. What happens when we let it "operate" on a scalar function $f(x_1, x_2, x_3)$? We have

$$\nabla f = \hat{\mathbf{e}}_1 \frac{\partial f}{\partial x_1} + \hat{\mathbf{e}}_2 \frac{\partial f}{\partial x_2} + \hat{\mathbf{e}}_3 \frac{\partial f}{\partial x_3}. \qquad (2.29)$$

What kind of a "thing" is ∇f? Consider the quantity $g = d\mathbf{x} \cdot \nabla f$, where $d\mathbf{x} = \hat{\mathbf{e}}_1 dx_1 + \hat{\mathbf{e}}_2 dx_2 + \hat{\mathbf{e}}_3 dx_3$ is a vector defining the differential change in the position vector:

$$g = d\mathbf{x} \cdot \nabla f,$$
$$= dx_1 \frac{\partial f}{\partial x_1} + dx_2 \frac{\partial f}{\partial x_2} + dx_3 \frac{\partial f}{\partial x_3} = df,$$

which we recognize as the differential change in f, and therefore a scalar. Therefore, by the argument given previously, since $d\mathbf{x} \cdot \nabla f$ is a scalar, and $d\mathbf{x}$ is a vector, the three quantities $\partial f/\partial x_1$, $\partial f/\partial x_2$, and $\partial f/\partial x_3$ form the components of a vector, so ∇f is a vector. It measures the magnitude and direction of the rate of change of the function f at any point in space.

Now form the dyad $\mathbf{D} = \nabla \mathbf{V}$, where $\mathbf{V}$ is a vector. Then the nine quantities $D_{ij} = \partial_i V_j$ are the components of a second-rank tensor. If we contract over the indices i and j we have

$$D = \partial_i V_i \equiv \nabla \cdot \mathbf{V}, \qquad (2.30)$$

which is a scalar. It is called the *divergence* of $\mathbf{V}$.

We can take the vector product of ∇ and $\mathbf{V}$, $\mathbf{D} = \nabla \times \mathbf{V}$, or

$$D_i = \varepsilon_{ijk} \partial_j V_k. \qquad (2.31)$$

This is called the *curl* of $\mathbf{V}$. For example, in Cartesian coordinates, the x_1 component is

---

[1] The discussion of the vector nature of the gradient operator follows that of the Feynman's lectures: R. P. Feynman, R. B. Leighton, and M. Sands, *The Feynman Lectures on Physics* Vol. 1, Addison-Wesley, Reading, MA (1963).

$$D_1 = \varepsilon_{123}\frac{\partial V_2}{\partial x_3} + \varepsilon_{132}\frac{\partial V_3}{\partial x_2} = \frac{\partial V_2}{\partial x_3} - \frac{\partial V_3}{\partial x_2},$$

by the properties of ε_{ijk}.

We could also have ∇ operate on a tensor or dyad: $\nabla \mathbf{A} \Rightarrow \partial_i A_{jk}$, which is a third-rank tensor. A common notation for this is $A_{jk,i}$ (the comma denotes differentiation with respect to x_i). Contracting over i and j,

$$D_k = \partial_j A_{jk} = \nabla \cdot \mathbf{A}, \tag{2.32}$$

or $A_{jk,j}$, which is the *divergence of a tensor* (it is a vector). In principle we could define the curl of a tensor, etc.

So far we have worked in Cartesian coordinates. This is because they are easy to work with, and if a vector expression is true in Cartesian coordinates it is true in *any* coordinate system.

We will now talk about *curvilinear coordinates*. Curvilinear coordinates are still orthogonal but the unit vectors $\hat{\mathbf{e}}_i$ are functions of $\mathbf{x}$, and this complicates the computation of derivatives. Examples of orthogonal curvilinear coordinates are cylindrical and spherical coordinates.

The gradient operator is

$$\nabla = \sum_i \hat{\mathbf{e}}_i \frac{\partial}{\partial x_i} \tag{2.33}$$

(the order of the unit vector and the derivative is now important), and any vector $\mathbf{V}$ is

$$\mathbf{V} = \sum_j V_j \hat{\mathbf{e}}_j, \tag{2.34}$$

where now $\hat{\mathbf{e}}_i = \hat{\mathbf{e}}_i(\mathbf{x})$. Then the tensor (or dyad) $\nabla \mathbf{V}$ is

$$\nabla \mathbf{V} = \sum_i \hat{\mathbf{e}}_i \frac{\partial}{\partial x_i} \sum_j V_j \hat{\mathbf{e}}_j$$

$$= \sum_i \hat{\mathbf{e}}_i \sum_j \frac{\partial}{\partial x_j} V_j \hat{\mathbf{e}}_j$$

$$= \sum_i \sum_j \left(\underbrace{\hat{\mathbf{e}}_i \hat{\mathbf{e}}_j \frac{\partial V_j}{\partial x_i}}_{\text{"Cartesian" part}} + \underbrace{\hat{\mathbf{e}}_i V_j \frac{\partial \hat{\mathbf{e}}_j}{\partial x_i}}_{\text{Extra terms if } \hat{\mathbf{e}}_j = \hat{\mathbf{e}}_j(x_i)} \right). \tag{2.35}$$

The first term is just the usual Cartesian derivative. The remaining terms arise in curvilinear coordinates. They must always be accounted for.

For example, in familiar cylindrical (r, θ, z) coordinates, the unit vectors $\hat{\mathbf{e}}_r$ and $\hat{\mathbf{e}}_\theta$ are functions of space with the properties

$$\frac{\partial \hat{\mathbf{e}}_r}{\partial \theta} = \hat{\mathbf{e}}_\theta \tag{2.36}$$

and

$$\frac{\partial \hat{\mathbf{e}}_\theta}{\partial \theta} = -\hat{\mathbf{e}}_r. \tag{2.37}$$

Then in these polar coordinates,

$$
\begin{aligned}
\nabla \mathbf{V} &= \left(\hat{\mathbf{e}}_r \frac{\partial}{\partial r} + \hat{\mathbf{e}}_\theta \frac{1}{r} \frac{\partial}{\partial \theta} \right) (\hat{\mathbf{e}}_r V_r + \hat{\mathbf{e}}_\theta V_\theta) \\
&= \hat{\mathbf{e}}_r \hat{\mathbf{e}}_r \frac{\partial V_r}{\partial r} + \hat{\mathbf{e}}_r \hat{\mathbf{e}}_\theta \frac{\partial V_\theta}{\partial r} \\
&\quad + \hat{\mathbf{e}}_\theta \hat{\mathbf{e}}_r \frac{1}{r} \frac{\partial V_r}{\partial \theta} + \hat{\mathbf{e}}_\theta \frac{V_r}{r} \frac{\partial \hat{\mathbf{e}}_r}{\partial \theta} \\
&\quad + \hat{\mathbf{e}}_\theta \hat{\mathbf{e}}_\theta \frac{1}{r} \frac{\partial V_\theta}{\partial \theta} + \hat{\mathbf{e}}_\theta \frac{V_\theta}{r} \frac{\partial \hat{\mathbf{e}}_\theta}{\partial \theta} \\
&= \hat{\mathbf{e}}_r \hat{\mathbf{e}}_r \frac{\partial V_r}{\partial r} + \hat{\mathbf{e}}_r \hat{\mathbf{e}}_\theta \frac{\partial V_\theta}{\partial r} \\
&\quad + \hat{\mathbf{e}}_\theta \hat{\mathbf{e}}_r \left(\frac{1}{r} \frac{\partial V_r}{\partial \theta} - \frac{V_\theta}{r} \right) + \hat{\mathbf{e}}_\theta \hat{\mathbf{e}}_\theta \left(\frac{1}{r} \frac{\partial V_\theta}{\partial \theta} + \frac{V_r}{r} \right).
\end{aligned}
\tag{2.38}
$$

Expressions of the form $\mathbf{U} \cdot \nabla \mathbf{V}$ appear often in MHD. It is a vector that expresses the rate of change of $\mathbf{V}$ in the direction of $\mathbf{U}$. Then, for polar coordinates,

$$\mathbf{U} \cdot \nabla \mathbf{V} = \hat{\mathbf{e}}_r \left(U_r \frac{\partial V_r}{\partial r} + \frac{U_\theta}{r} \frac{\partial V_r}{\partial \theta} - \frac{U_\theta V_\theta}{r} \right) + \hat{\mathbf{e}}_\theta \left(U_r \frac{\partial V_\theta}{\partial r} + \frac{U_\theta}{r} \frac{\partial V_\theta}{\partial \theta} + \frac{U_\theta V_r}{r} \right). \tag{2.39}$$

The third term in each of the brackets is the new terms that arise from the differentiation of the unit vectors in curvilinear coordinates.

Of course, there is no need to insist that the bases $\hat{\mathbf{e}}_i(\mathbf{x})$ even be orthogonal. (An orthogonal system has $\hat{\mathbf{e}}_i \cdot \hat{\mathbf{e}}_j = \delta_{ij}$.) Such systems are called *generalized curvilinear coordinates*.[2] Then the bases $(\hat{\mathbf{e}}_1, \hat{\mathbf{e}}_2, \hat{\mathbf{e}}_3)$ are not unique, because it is always possible to define equivalent, *reciprocal* basis vectors $(\hat{\mathbf{e}}^1, \hat{\mathbf{e}}^2, \hat{\mathbf{e}}^3)$ at each point in space by the process

$$\hat{\mathbf{e}}^3 = \hat{\mathbf{e}}_1 \times \hat{\mathbf{e}}_2 / J, \quad \hat{\mathbf{e}}^2 = \hat{\mathbf{e}}_3 \times \hat{\mathbf{e}}_1 / J, \quad \hat{\mathbf{e}}^1 = \hat{\mathbf{e}}_2 \times \hat{\mathbf{e}}_3 / J, \tag{2.40}$$

---

[2] This discussion follows that of Donald H. Menzel, *Mathematic Physics*, Dover Publications, New York (1961).

where $J = \hat{\mathbf{e}}_1 \cdot \hat{\mathbf{e}}_2 \times \hat{\mathbf{e}}_3$ is called the *Jacobian*. A vector $\mathbf{V}$ can be equivalently expressed as

$$\mathbf{V} = V^i \hat{\mathbf{e}}_i, \tag{2.41}$$

or

$$\mathbf{V} = V_i \hat{\mathbf{e}}^i. \tag{2.42}$$

The V^i are called the *contravariant* components of $\mathbf{V}$, and the V_i are called the *covariant* components. (Of course, the vector $\mathbf{V}$, which is invariant by definition, is neither contravariant or covariant.)

Our previous discussion of vectors, tensors, and dyads can be generalized to these non-orthogonal coordinates, *as long as extreme care is taken in keeping track of the contravariant and covariant components.* Of particular interest is the generalization of vector differentiation, previously discussed for the special case of polar coordinates. The tensor $\nabla \mathbf{V}$ can be written as

$$\nabla \mathbf{V} = \hat{\mathbf{e}}^i \hat{\mathbf{e}}_j D_i V^j, \tag{2.43}$$

where

$$D_i V^j = \partial_i V^j + V^k \Gamma^j_{ik} \tag{2.44}$$

is called the *covariant derivative*. The quantities Γ^j_{ik} are called the *Christoffel symbols* and are defined by

$$\partial_i \hat{\mathbf{e}}^k = -\Gamma^j_{ik} \hat{\mathbf{e}}^k. \tag{2.45}$$

They are the generalization of Eqs. (2.36) and (2.37). [We remark that the Γ^j_{ik} are *not* tensors, as they do not obey the transformation law, Eq. (2.24).] Expressions for the Γ^j_{ik} in any particular coordinate system are given in terms of the metric tensor components $g_{ij} = \hat{\mathbf{e}}_i \cdot \hat{\mathbf{e}}_j$, and $g^{ij} = \hat{\mathbf{e}}^i \cdot \hat{\mathbf{e}}^j$, as

$$\Gamma^k_{ij} = g^{kl} \left(\partial_i g_{il} + \partial_j g_{li} - \partial_l g_{ij} \right). \tag{2.46}$$

We stated previously that if an expression is true in Cartesian coordinates, it is true in all coordinate systems. In particular, expressions for generalized curvilinear coordinates can be obtained by replacing everywhere the derivative $\partial_i V^j$ with the covariant derivative $D_i V^j$, defined by Eq. (2.44). In analogy with the discussion preceding Eq. (2.32), covariant differentiation is often expressed in the shorthand notation $D_i A^{jk} \equiv A^{jk}_{..;i}$. Misner, Thorne and Wheeler[3] call this the "comma goes to semi-colon rule" for obtaining tensor expressions in generalized curvilinear coordinates: first get an expression in orthogonal coordinates, and then change all commas to semi-colons!

Generalized curvilinear coordinates play an essential role in the theoretical description of tokamak plasmas. The topic is so detailed and complex (and, frankly, difficult) that it will not be covered further here. I hope this short introduction will allow you to learn more about this on your own.

We now return to Cartesian coordinates.

The divergence of a tensor **T** has been defined as

$$\nabla \cdot \mathbf{T} = \partial_i T_{ij}. \tag{2.47}$$

It is a vector whose jth component is

$$(\nabla \cdot \mathbf{T})_j = \frac{\partial T_{1j}}{\partial x_1} + \frac{\partial T_{2j}}{\partial x_2} + \frac{\partial T_{3j}}{\partial x_3}. \tag{2.48}$$

Integrate this expression over all space:

$$
\begin{aligned}
\int (\nabla \cdot \mathbf{T})_j \, d^3x &= \int dx_1 dx_2 dx_3 \left(\frac{\partial T_{1j}}{\partial x_1} + \frac{\partial T_{2j}}{\partial x_2} + \frac{\partial T_{3j}}{\partial x_3} \right) \\
&= \int dx_2 dx_3 T_{1j} + \int dx_1 dx_3 T_{2j} + \int dx_1 dx_2 T_{3j} \\
&= \int dS_1 T_{1j} + \int dS_2 T_{2j} + \int dS_3 T_{3j} \\
&= \int (d\mathbf{S} \cdot \mathbf{T})_j,
\end{aligned}
$$

or

$$\int \nabla \cdot \mathbf{T} d^3x = \oint d\mathbf{S} \cdot \mathbf{T}. \tag{2.49}$$

This is the *generalized Gauss' theorem*.

It is also possible to derive the following integral theorems:

$$\int \nabla \mathbf{V} d^3x = \oint d\mathbf{S}\mathbf{V} \tag{2.50}$$

$$\int \nabla f d^3x = \oint d\mathbf{S} f \tag{2.51}$$

---

[3] C. W. Misner, K. S. Thorn, and J. A. Wheeler, *Gravitation*, W. H. Freeman and Company, San Francisco (1973).

$$\int \nabla \times \mathbf{V} d^3 x = \oint d\mathbf{S} \times \mathbf{V} \tag{2.52}$$

$$\int_S d\mathbf{S} \times \nabla f d^3 x = \oint_C d\mathbf{l} f \tag{2.53}$$

$$\int_S d\mathbf{S} \cdot \nabla \times \mathbf{A} d^3 x = \oint_C \mathbf{A} \cdot d\mathbf{l} \tag{2.54}$$

It seems intuitive that a physically measurable quantity should not care what coordinate system it is referred to. We have shown that scalars, vectors, tensors, etc., things that we can associate with physical quantities, are invariant under rotations. This is good!

There is another important type of coordinate transformation called an *inversion*. It is also known as *parity transformation*. Mathematically, this is given by

$$x_i' = -x_i. \tag{2.55}$$

An inversion is shown in Fig. 2.4.

The first coordinate system is "right-handed"; the second coordinate system is "left-handed." Consider the position vector $\mathbf{r}$. In the unprimed coordinate system, it is given by $\mathbf{r} = x_i \hat{\mathbf{e}}_i$. In the primed (inverted) coordinate system, it is given by

$$\mathbf{r}' = x_i' \hat{\mathbf{e}}_i' = (-x_i)(-\hat{\mathbf{e}}_i) = x_i \hat{\mathbf{e}}_i = \mathbf{r}, \tag{2.56}$$

so it is invariant under inversions. Such a vector is called a *polar vector*. (It is sometimes called a *true* vector.) We remark the gradient operator $\nabla = \hat{\mathbf{e}}_i \partial_i$ transforms like a polar vector, since $\partial_i' = -\partial_i$.

Now consider the vector $\mathbf{C}$, defined by $\mathbf{C} = \mathbf{A} \times \mathbf{B}$ (or $C_i = \varepsilon_{ijk} A_j B_k$), where $\mathbf{A}$ and $\mathbf{B}$ are polar vectors, i.e., $\mathbf{A}$ and $\mathbf{B}$ transform according to $A_i' = -A_i$ and $B_i' = -B_i$. Then under inversion, the components of $\mathbf{C}$ transform according to

$$C_i' = \varepsilon_{ijk} A_j' B_k' = \varepsilon_{ijk} (-A_j)(-B_k) = \varepsilon_{ijk} A_j B_k = +C_i. \tag{2.57}$$

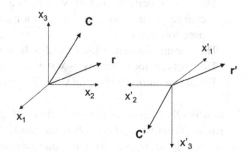

Fig. 2.4 An inversion of the Cartesian coordinate system

Then

$$\mathbf{C}' = C_i'\hat{\mathbf{e}}_i' = C_i\left(-\hat{\mathbf{e}}_i\right) = -\mathbf{C}, \tag{2.58}$$

so that $\mathbf{C}$ does *not* transform like a true vector under coordinate inversions. Such a vector (that changes direction in space under coordinate inversion; see Fig. 2.4) is called an *axial vector* (or *pseudovector*). Since it is defined as a vector (or cross) product, it usually describes some process involving rotation. For example, the vector area $d\mathbf{S}_k = d\mathbf{x}_i \times d\mathbf{x}_j$ is a pseudovector. However, notice that if $\mathbf{A}$ is a polar vector and $\mathbf{B}$ is an axial vector, then $C_i' = -C_i$, and $\mathbf{C}$ is a polar vector.

The elementary volume is defined as $dV = d\mathbf{x}_1 \cdot d\mathbf{x}_2 \times d\mathbf{x}_3 = \varepsilon_{ijk}dx_1dx_2dx_3$. It is easy to see that under inversions, $dV' = -dV$; the volume changes sign! Such quantities are called *pseudoscalars*: they are invariant under rotations, but change sign under inversions.

Again, it is intuitive that physical quantities should exist independent from coordinate systems. How then to account for the volume? Apparently it should be considered a "derived quantity," not directly measurable. For example, one can measure directly the true vectors (i.e., lengths) $d\mathbf{x}_i$, but one has to *compute* the volume. This renders as a pseudoscalar any quantity that expresses an amount of a scalar quantity per unit volume; these are not directly measurable. This includes the mass density ρ (mass/volume) and the pressure (internal energy/volume). (An exception is the electric charge density, which is a true scalar.) Apparently, one can measure directly mass and length (both true scalars), but must then infer the mass density.

The following is a list of some physical variables that appear in MHD, and their transformation properties:

  Time is a scalar.
  Temperature, which has units of energy, is a scalar.
  Mass density, $\rho = M/V$, is a pseudoscalar.
  Pressure, $p = \rho k_B T$, is a pseudoscalar.
  Velocity, $V_i = dx_i/dt$, is a vector.
  The vector potential $\mathbf{A}$ is a vector.
  The magnetic flux, $\int \mathbf{A} \cdot d\mathbf{x}$, is a scalar.
  The magnetic field, $\mathbf{B} = \nabla \times \mathbf{A}$, is a pseudovector.
  The current density, $\mu_0\mathbf{J} = \nabla \times \mathbf{B}$, is a vector. (Note that, since J is electric charge per unit area, and area is a pseudoscalar, electric charge must be a pseudoscalar.)
  The Lorentz force density, $\mathbf{f}_L = \mathbf{J} \times \mathbf{B}$, is a pseudovector.
  The pressure force density, $-\nabla p$, is a pseudovector.
  The acceleration density, $\rho d\mathbf{V}/dt$, is a pseudovector.

Now, it is OK to express physical relationships by using pseudovectors and pseudoscalars. What is required is that the resulting expressions be consistent, i.e., we do not end up adding scalars and pseudoscalars, or vectors and pseudovectors.

Lecture 3
Mass Conservation and the Equation of Continuity

You are my density.

Crispin Glover, *Back to the Future*

We now begin the derivation of the equations governing the behavior of the fluid. We will start by looking at the mass flowing into and out of a physically infinitesimal volume element.

There are two "viewpoints," and they are equivalent:

1. *Eulerian*: A volume element is fixed in space in the "laboratory" frame of reference.
2. *Lagrangian*: The surface of the volume element is co-moving with the fluid, in the "fluid" frame of reference.

We will use whichever is most convenient.

First consider the Eulerian picture. The volume element dV is shown in Fig. 3.1.

Fig. 3.1 An Eulerian volume element. Its boundaries are fixed in space

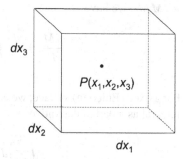

dx_3

$P(x_1, x_2, x_3)$

dx_2

dx_1

$P(x_1, x_2, x_3)$ is the centroid of the volume element. The sides of the volume element are fixed in space. Fluid can flow into and out of the volume element through the sides.

Let the mass density at $P(x_1, x_2, x_3)$ be $\rho(x_1, x_2, x_3)$ (mass/volume). It is the average (and nearly uniform) mass density throughout dV. The total mass contained within dV is

$$M = \int \rho \, dV = \int \rho \, dx_1 dx_2 dx_3. \tag{3.1}$$

Schnack, D.D.: *Mass Conservation and the Equation of Continuity*. Lect. Notes Phys. **780**, 19–24 (2009)
DOI 10.1007/978-3-642-00688-3_3

Assume that there are no sources or sinks of mass within dV. Then dM/dt is the rate at which mass enters or leaves through the surface dS.

A surface element dS is shown in Fig. 3.2.

Fig. 3.2 Eulerian surface element, showing the velocity vector $\mathbf{V}$ and the surface area element vector $d\mathbf{S}$

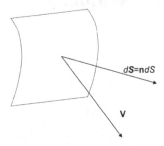

The surface area is dS, and $\hat{\mathbf{n}}$ is a unit vector normal (perpendicular) to the surface (in an average sense). When dS is a side of a volume element dV, $\hat{\mathbf{n}}$ is assumed to point *out* of the volume element (i.e., from inside to outside). The flux of mass (mass/unit area/unit time) passing through a surface is $\rho\mathbf{V}$, where $\mathbf{V}$ is the fluid velocity. It is a vector quantity (actually, a pseudovector, because of the presence of ρ). Then the mass per unit time flowing through dS is $\rho\mathbf{V} \cdot d\mathbf{S} = \rho\mathbf{V} \cdot \hat{\mathbf{n}} dS$, and the total rate of flow of mass *out* of the volume dV is

$$\sum_{Faces} \rho\mathbf{V} \cdot d\mathbf{S} \Rightarrow \oint_S \rho\mathbf{V} \cdot d\mathbf{S} = \oint_S \rho\mathbf{V} \cdot \hat{\mathbf{n}} dS, \qquad (3.2)$$

where the integral is over the surface enclosing dV. Since this must be equal to $-dM/dt$, we have

$$\frac{dM}{dt} = \frac{d}{dt} \int_V \rho dV = - \oint_S \rho\mathbf{V} \cdot \hat{\mathbf{n}} dS. \qquad (3.3)$$

For a fixed (Eulerian) surface, we can take the total time derivative inside the volume integral as a partial derivative:

$$\int_V \frac{\partial \rho}{\partial t} dV = - \oint_S \rho\mathbf{V} \cdot \hat{\mathbf{n}} dS. \qquad (3.4)$$

By Gauss' theorem,

$$\oint_S \rho\mathbf{V} \cdot \hat{\mathbf{n}} dS = \int_V \nabla \cdot (\rho\mathbf{V}) dV, \qquad (3.5)$$

so that

$$\int_V \left[\frac{\partial \rho}{\partial t} + \nabla \cdot (\rho \mathbf{V}) \right] dV = 0. \tag{3.6}$$

This expression must hold for every arbitrarily shaped volume; the only way that it can be satisfied is if the integrand vanishes identically, or

$$\frac{\partial \rho}{\partial t} = -\nabla \cdot (\rho \mathbf{V}). \tag{3.7}$$

This is called the *continuity equation*. It expresses *conservation of mass* in the Eulerian frame of reference.

We remark that Eq. (3.7) is a partial differential equation with four dependent variables: ρ and the three components of $\mathbf{V}$. If the velocity were known a priori, the system would be closed and we could solve Eq. (3.7) for the evolution of ρ. Problems in which the velocity field is fixed, or specified in advance, are called *kinematic*. Problems where $\mathbf{V}$ is determined from other physical principles are called *dynamic*, and the latter is the case of interest here. We therefore have three more unknowns than we have equations; the problem is not *closed*. This problem of *closure* is of fundamental importance in MHD, and we will discuss it in more detail later in this course.

We now describe conservation of mass in the Lagrangian picture. Here the volume element dV is co-moving with the fluid, as sketched in Fig. 3.3.

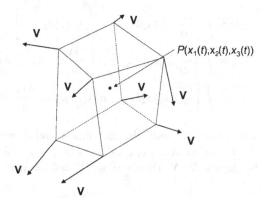

Fig. 3.3 A Lagrangian fluid element. Its boundaries move with the local fluid velocity

Every point on the surface and within the volume is moving with the local velocity $\mathbf{V} = d\mathbf{x}/dt$; the coordinates of each "bit" of the volume element are thus time dependent: $x_i = x_i(t)$. The shape of the volume element can distort with time. However, since each point on the boundary S moves with the fluid, *no fluid can flow across the surface*, so that *the total mass within the volume element is fixed in time*: $dM/dt = 0$, and mass is automatically conserved.

However, things are still complicated. As the volume element moves through space, its total mass, as given by Eq. (3.1), remains constant, but since the total volume of the element can change as it distorts due to fluid motions, the mass density must be considered to be a function of time. Conservation of mass is then stated as

$$\frac{dM}{dt} = 0 = \frac{d}{dt} \int_V \rho(t) dx_1(t) dx_2(t) dx_3(t). \tag{3.8}$$

So, we not only need to calculate the change in ρ, but we need to account for the change in the volume dV as it moves through space.

To this end, we introduce a "new" infinitesimal, δx_i. The infinitesimal operator δ is taken to operate only on the spatial coordinates x_i; the notation d is reserved for time. In all other respects, d and δ are the same. Equation (3.8) is written as

$$\frac{d}{dt} \int_V \rho(t) \delta x_1(t) \delta x_2(t) \delta x_3(t) = 0, \tag{3.9}$$

and the time dependence of the coordinates is given by

$$\frac{dx_i}{dt} = V_i. \tag{3.10}$$

Now time is the only independent variable. Differentiating under the integral sign in Eq. (3.9), we have

$$
\begin{aligned}
0 &= \int \left[\frac{d\rho}{dt} \delta x_1 \delta x_2 \delta x_3 + \rho \frac{d}{dt} (\delta x_1 \delta x_2 \delta x_3) \right], \\
&= \int \left\{ \frac{d\rho}{dt} \delta x_1 \delta x_2 \delta x_3 + \rho \left[\frac{d\delta x_1}{dt} \delta x_2 \delta x_3 + \delta x_1 \frac{d\delta x_2}{dt} \delta x_3 + \delta x_1 \delta x_2 \frac{d\delta x_3}{dt} \right] \right\}, \\
&= \int \left[\frac{d\rho}{dt} \delta x_1 \delta x_2 \delta x_3 + \rho \delta x_1 \delta x_2 \delta x_3 \left(\frac{\delta V_1}{\delta x_1} + \frac{\delta V_2}{\delta x_2} + \frac{\delta V_3}{\delta x_3} \right) \right], \tag{3.11}
\end{aligned}
$$

where we have used the fact that δ and d are both infinitesimals, along with Eq. (3.10), to write $d(\delta x_i)/dt = \delta(dx_i/dt) = \delta V_i$. We recognize the last term in brackets as $\nabla \cdot \mathbf{V}$. Then writing $dV = \delta x_1 \delta x_2 \delta x_3$,

$$0 = \int \left[\frac{d\rho}{dt} + \rho \nabla \cdot \mathbf{V} \right] dV. \tag{3.12}$$

As with Eq. (3.6), since this must hold for arbitrary volume elements, we require

$$\frac{d\rho}{dt} + \rho \nabla \cdot \mathbf{V} = 0. \tag{3.13}$$

This is the expression for conservation of mass in the Lagrangian frame of reference.

Equation (3.13), the Lagrangian expression, appears to be different from Eq. (3.7), the Eulerian expression. In particular, how are we to interpret the time derivative

d/dt that appears in Eq. (3.13)? Since these equations each express the law of conservation of mass, they must be consistent. Note that we can write Eq. (3.7) as

$$\frac{\partial \rho}{\partial t} + \nabla \cdot (\rho \mathbf{V}) = \frac{\partial \rho}{\partial t} + \mathbf{V} \cdot \nabla \rho + \rho \nabla \cdot \mathbf{V} = 0, \tag{3.14}$$

which will be consistent with Eq. (3.13) if we identify

$$\frac{d\rho}{dt} = \frac{\partial \rho}{\partial t} + \mathbf{V} \cdot \nabla \rho. \tag{3.15}$$

Generally, the operator $d/dt = \partial/\partial t + \mathbf{V} \cdot \nabla$ is called the *total time derivative* or the *Lagrangian derivative*. It measures the total change in a quantity associated with a fluid element as it moves about in space. This can be seen as follows. The Lagrangian change in the density in time dt consists of two parts, $d\rho = d\rho_1 + d\rho_2$, where $d\rho_1$ is the change in ρ during dt at a *fixed point in space*,

$$d\rho_1 = \frac{\partial \rho}{\partial t} dt, \tag{3.16}$$

and $d\rho_2$ is the difference between densities separated by a distance $d\mathbf{x}$, at the same time t,

$$d\rho_2 = d\mathbf{x} \cdot \nabla \rho. \tag{3.17}$$

The total change in ρ is therefore

$$d\rho = \frac{\partial \rho}{\partial t} dt + d\mathbf{x} \cdot \nabla \rho, \tag{3.18}$$

so that

$$\frac{d\rho}{dt} = \frac{\partial \rho}{\partial t} + \mathbf{V} \cdot \nabla \rho. \tag{3.19}$$

Of course, this result can also be obtained formally by applying the chain rule to $\rho = \rho[x(t), t]$, i.e.,

$$\frac{d\rho}{dt} = \frac{\partial \rho}{\partial t} + \frac{\partial \rho}{\partial x} \cdot \frac{d\mathbf{x}}{dt} = \frac{\partial \rho}{\partial t} + \mathbf{V} \cdot \nabla \rho,$$

but this provides little physical insight.

The term $\mathbf{V} \cdot \nabla \rho$ is called the *advective derivative*. It measures the change of ρ in the direction of $\mathbf{V}$. The terminology originated in weather and climate modeling, where *convection* refers to vertical uplift driven by buoyancy and thermal forces and *advection* refers to wind-driven horizontal transport. The terminology has been carried over to MHD, where it refers to all velocity-driven transport.

The term $-\rho \nabla \cdot \mathbf{V}$ measures the change in ρ due to *compression* or *dilation* of the fluid element. This is illustrated in Fig. 3.4.

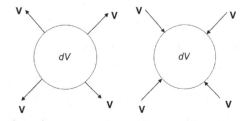

Fig. 3.4 *Left*: An expanding flow for which $\nabla \cdot \mathbf{V} > 0$. *Right*: A contracting, or compressing, fluid element for which $\nabla \cdot \mathbf{V} < 0$

In the figure on the left, $\nabla \cdot \mathbf{V} > 0$, the flow is *diverging*, there is net flow *out* of the volume element, and the mass within the volume element is *decreasing*. In the figure on the right, $\nabla \cdot \mathbf{V} < 0$, the flow is *converging*, there is net flow *into* the volume element, and the mass within the volume element is *increasing*.

The continuity equation and conservation of mass are exactly the same in hydrodynamics and MHD.

Lecture 4
The Equation of Motion

> *The Force is with you, young Skywalker, but you are not a Jedi yet.*
>
> Darth Vader, *The Return of the Jedi*

Newton's law for a fluid element is

$$\rho \frac{d\mathbf{V}}{dt} = \mathbf{F},\tag{4.1}$$

where ρ is the mass density (mass per unit volume), $\mathbf{V}$ is the velocity of the fluid element, and $\mathbf{F}$ is the force per unit volume acting on the element. The latter is composed of two types: *volumetric* forces and *surface* forces.

Volumetric forces act throughout the volume of the fluid element. They can be thought of as acting at the centroid. Examples are:

1. Gravity: $\mathbf{F}_g = \rho\mathbf{g}$, where $\mathbf{g}$ is the *gravitational acceleration*. If the force is a *central force*, then $\mathbf{F}_g = -\nabla\Phi$, where Φ is the *gravitational potential*.
2. Electromagnetic forces: Since the fluid can conduct electricity, it can have a current density $\mathbf{J} = \sum_\alpha n_\alpha q_\alpha \mathbf{V}_\alpha$ (with the sum being over all species of ions and electrons, and n_α, q_α, and $\mathbf{V}_\alpha$ the number density, electric charge, and velocity of species α, respectively), and, in principle, a net electric charge per unit volume, ρ_q. The electromagnetic force (per unit volume) are then the *electric force*, $\mathbf{F}_q = \rho_q\mathbf{E}$ ($\mathbf{E}$ is the electric field), and the *Lorentz force*, $\mathbf{F}_L = \mathbf{J} \times \mathbf{B}$ ($\mathbf{B}$ is the magnetic field).

Surface forces are more complicated.[1] Consider the forces acting on a surface $\mathbf{S}$. We assume the convention that the material in *front* of $\mathbf{S}$ exerts a force on the material *behind* $\mathbf{S}$ that is given by

$$\mathbf{F} = \mathbf{S} \cdot \mathbf{P}\tag{4.2}$$

or

$$F_i = S_i P_{ij}.\tag{4.3}$$

---

[1] This discussion follows lecture notes by W. A. Newcomb (unpublished), 1975. The author was privileged to attend these lectures.

Schnack, D.D.: *The Equation of Motion*. Lect. Notes Phys. **780**, 25–29 (2009)
DOI 10.1007/978-3-642-00688-3_4

We consider three orientations for $\mathbf{S}$, along each of the three coordinate directions. If $\mathbf{S} = \hat{\mathbf{e}}_1$,

$$\mathbf{F} = P_{11}\hat{\mathbf{e}}_1 + P_{12}\hat{\mathbf{e}}_2 + P_{13}\hat{\mathbf{e}}_3, \tag{4.4}$$

which is a vector. Similarly, if $\mathbf{S} = \hat{\mathbf{e}}_2$, then

$$\mathbf{F} = P_{21}\hat{\mathbf{e}}_1 + P_{22}\hat{\mathbf{e}}_2 + P_{23}\hat{\mathbf{e}}_3, \tag{4.5}$$

and if $\mathbf{S} = \hat{\mathbf{e}}_3$, then

$$\mathbf{F} = P_{31}\hat{\mathbf{e}}_1 + P_{32}\hat{\mathbf{e}}_2 + P_{33}\hat{\mathbf{e}}_3. \tag{4.6}$$

It therefore takes nine numbers to define the force on the surface $\mathbf{S}$. These are the components of the *stress tensor*, P_{ij}.

The total surface force acting on a fluid element is the sum of the forces on its faces. We want the total force acting *on* the volume. Since $\mathbf{F}$ has been defined as the force exerted by the material in front of $\mathbf{S}$ acting on the material behind $\mathbf{S}$, all of the material within the element is *behind* the faces, and the net force is the *negative* of the surface forces, i.e.,

$$\mathbf{F} = - \oint_S d\mathbf{S} \cdot \mathbf{P} \tag{4.7}$$

or, by Gauss' theorem,

$$\mathbf{F} = - \int_V \nabla \cdot \mathbf{P} dV. \tag{4.8}$$

As $V \to 0$, we obtain the net force per unit volume as

$$\mathbf{f} = -\nabla \cdot \mathbf{P}. \tag{4.9}$$

This is the *volumetric equivalent* of the surface forces. The equation of motion considering only surface forces is then

$$\rho \frac{d\mathbf{V}}{dt} = -\nabla \cdot \mathbf{P}. \tag{4.10}$$

We will now prove that *the stress tensor is symmetric*, i.e., $P_{ij} = P_{ji}$. Consider the angular momentum per unit volume of fluid,

$$\mathbf{L} = \mathbf{r} \times \rho \mathbf{V}. \tag{4.11}$$

Using Eq. (4.10), the total (i.e., Lagrangian) time rate of change of $\mathbf{L}$ for a fluid with fixed volume V_0 is

$$\dot{\mathbf{L}} = \int_{V_0} \rho \mathbf{r} \times \frac{d\mathbf{V}}{dt} dV = -\int_{V_0} \mathbf{r} \times \nabla \cdot \mathbf{P} dV, \tag{4.12}$$

or, in Cartesian tensor notation,

$$\dot{L}_i = -\int_{V_0} \varepsilon_{ijk} r_j \left(\partial_l P_{lk} \right) dV. \tag{4.13}$$

Since $\partial_l r_j = \delta_{lj}$, we can write

$$r_j \left(\partial_l P_{lk} \right) = \partial_l \left(r_j P_{lk} \right) - P_{jk}, \tag{4.14}$$

so that

$$\begin{aligned}
\dot{L}_i &= -\int_{V_0} \varepsilon_{ijk} \left[\partial_l \left(r_j P_{lk} \right) - P_{jk} \right] dV \\
&= -\int_{V_0} \partial_l \left(\varepsilon_{ijk} r_j P_{lk} \right) dV + \int_{V_0} \varepsilon_{ijk} P_{lk} dV \\
&= -\oint_{S_0} dS_l \varepsilon_{ijk} r_j P_{lk} + \int_{V_0} \varepsilon_{ijk} P_{lk} dV,
\end{aligned} \tag{4.15}$$

where we have used Gauss' theorem and S_0 is the surface bounding V_0. We recognize the first term on the right-hand side of Eq. (4.15) as the total external torque applied to the surface of the volume. The remaining term is the rate of change of internal angular momentum of the fluid. In the absence of applied torque, we require $\dot{L}_i = 0$ or

$$\int_{V_0} \varepsilon_{ijk} P_{lk} dV = 0, \tag{4.16}$$

which can be written as

$$\int_{V_0} \frac{1}{2} \left(\varepsilon_{ijk} P_{jk} + \varepsilon_{ikj} P_{kj} \right) dV = \int_{V_0} \frac{1}{2} \varepsilon_{ijk} \left(P_{jk} - P_{kj} \right) dV = 0. \tag{4.17}$$

(The first expression comes from interchanging the dummy indices j and k; the second follows from the properties of ε_{ijk}.) Since Eq. (4.17) must hold for an arbitrary volume, we have $P_{jk} = P_{kj}$, which is the desired result.

The symmetry of the stress tensor is a very general result; it can be considered a general principle of physics.[2] It is independent of the properties of the medium, which can be solid, liquid, or gas (or even plasma). It is required to prevent the internal angular momentum of the system from increasing without bound.

The stress tensor (indeed, any tensor) can always be decomposed as

$$\mathbf{P} = p\mathbf{I} - \mathbf{\Pi} \tag{4.18}$$

or

$$P_{ij} = p\delta_{ij} - \Pi_{ij}. \tag{4.19}$$

The first term on the right-hand side is called the *scalar pressure*. The second term is called the *viscous stress tensor*.

Including all the volumetric and equivalent volumetric forces, Eq. (4.1) becomes

$$\rho\frac{d\mathbf{V}}{dt} = \rho_q\mathbf{E} + \mathbf{J} \times \mathbf{B} - \nabla p + \nabla \cdot \mathbf{\Pi}, \tag{4.20}$$

which is the Lagrangian form of the *equation of motion*. The Eulerian form is

$$\rho\left(\frac{\partial\mathbf{V}}{\partial t} + \mathbf{V} \cdot \nabla\mathbf{V}\right) = \rho_q\mathbf{E} + \mathbf{J} \times \mathbf{B} - \nabla p + \nabla \cdot \mathbf{\Pi}. \tag{4.21}$$

As always, they are equivalent.

We now compute $W_V = \mathbf{V} \cdot \mathbf{F}_V$, the work done *on* a volume element *by* the viscous force $\mathbf{F}_V = \nabla \cdot \mathbf{\Pi}$:

$$W_V = V_i\partial_j\Pi_{ji} = \partial_j\left(V_i\Pi_{ji}\right) - \Pi_{ji}\partial_j V_i$$

or

$$W_V = \nabla \cdot (\mathbf{\Pi} \cdot \mathbf{V}) - \mathbf{\Pi} : \nabla\mathbf{V}. \tag{4.22}$$

Then

$$\int_{V_0} W_V dV = \int_{V_0} \nabla \cdot (\mathbf{\Pi} \cdot \mathbf{V}) dV - \int_{V_0} \mathbf{\Pi} : \nabla\mathbf{V} dV$$

$$= \oint_{S_0} d\mathbf{S} \cdot (\mathbf{\Pi} \cdot \mathbf{V}) - \int_{V_0} \mathbf{\Pi} : \nabla\mathbf{V} dV. \tag{4.23}$$

---

[2] This discussion follows that of L. D. Landau and E. M. Lifschitz, *Fluid Mechanics*, Pergamon Press, London, UK (1959).

The first term on the right-hand side is the work done on the surface S; the second term is the work done throughout the volume. In the absence of the surface term, kinetic energy is *lost* from the fluid if $\mathbf{\Pi} : \nabla \mathbf{V} > 0$; it must show up as internal (thermal) energy. We therefore identify the *volumetric viscous heating rate* as

$$Q_\mathrm{V} = \mathbf{\Pi} : \nabla \mathbf{V} \tag{4.24}$$

We remark that the equation of motion has introduced six new dependent variables; i.e., the six independent components of the stress tensor. This is a further example of the closure problem first mentioned in Lecture 3.

Lecture 5
Energy Flow

Energy falls just short of being joy.

Mason Cooley, *City Aphorisms*

The enumeration of the various sources and flows of energy is enabled by working in the Lagrangian frame of reference; i.e., we consider a volume element that is co-moving with the fluid. The heat input rate per unit volume is equal to the rate of energy flow across the surface plus the volumetric heating rate, i.e.,

$$\rho \frac{dQ}{dt} = -\nabla \cdot \mathbf{q} + R_V, \qquad (5.1)$$

where Q is the *heat per unit mass*, $\mathbf{q}$ is the *heat flux through the boundary*, and

$$R_V = \mathbf{\Pi} : \nabla \mathbf{V} + \eta J^2 \qquad (5.2)$$

is the *volumetric heating rate*. The first term is the viscous heating rate, defined in Lecture 4, and the second term is the Ohmic heating rate ($\mathbf{J} \cdot \mathbf{E}$, where $\mathbf{E}$ is the electric field, $\mathbf{J}$ is the current density, η is the electrical resistivity of the fluid, and $\mathbf{E} = \eta \mathbf{J}$ in the Lagrangian frame; Ohm's law and the transformation properties of the electric field will be discussed in Lecture 6).

We write the first law of thermodynamics (conservation of energy) as

$$dQ = pd \left(\frac{1}{\rho} \right) + de, \qquad (5.3)$$

where dQ is the change in heat per unit mass, $pd(1/\rho)$ is the PV work per unit mass, and de is the change in energy per unit mass. Substituting this into Eq. (5.2), we have

$$p\rho \frac{d}{dt} \left(\frac{1}{\rho} \right) + \rho \frac{de}{dt} = -\nabla \cdot \mathbf{q} + \mathbf{\Pi} : \nabla \mathbf{V} + \eta J^2. \qquad (5.4)$$

Now,

$$\frac{d}{dt} \left(\frac{1}{\rho} \right) = -\frac{1}{\rho^2} \frac{d\rho}{dt} = \frac{1}{\rho} \nabla \cdot \mathbf{V}, \qquad (5.5)$$

Schnack, D.D.: *Energy Flow*. Lect. Notes Phys. **780**, 31–33 (2009)
DOI 10.1007/978-3-642-00688-3_5

where we have used the Lagrangian form of the continuity equation. Then Eq. (5.4) becomes

$$\rho \frac{de}{dt} = -p\nabla \cdot \mathbf{V} - \nabla \cdot \mathbf{q} + \mathbf{\Pi} : \nabla \mathbf{V} + \eta J^2. \tag{5.6}$$

The term on the left-hand side is the rate of change of energy per unit volume. It is equal to the sum of the work done to expand or compress the fluid element, the rate of heat flow through the surface, and the volumetric heating rate due to viscous and resistive processes. We can again use the continuity equation to write

$$\rho \frac{de}{dt} = \frac{d}{dt}(\rho e) + \rho e \nabla \cdot \mathbf{V} \tag{5.7}$$

or, in the Eulerian frame,

$$\rho \frac{de}{dt} = \frac{\partial}{\partial t}(\rho e) + \nabla \cdot (\rho e \mathbf{V}). \tag{5.8}$$

Then the equation describing energy and heat flow in the Eulerian frame of reference is

$$\frac{\partial}{\partial t}(\rho e) = -\nabla \cdot (\rho e \mathbf{V}) - p\nabla \cdot \mathbf{V} - \nabla \cdot \mathbf{q} + \mathbf{\Pi} : \nabla \mathbf{V} + \eta J^2. \tag{5.9}$$

Not surprisingly (and certainly not very originally), Eq. (5.9) is called the *energy equation*.

To an excellent approximation, a plasma behaves as an *ideal gas*, i.e., the energy depends only on the pressure. This relationship is written as

$$\rho e = \frac{p}{\Gamma - 1}, \tag{5.10}$$

where Γ is the adiabatic index. For a plasma $\Gamma = 5/3$. Using Eq. (5.10), and the identity $\nabla \cdot (p\mathbf{V}) + (\Gamma - 1)\, p\nabla \cdot \mathbf{V} = \Gamma p\nabla \cdot \mathbf{V} + \mathbf{V} \cdot \nabla p$, we obtain evolutionary equations for the pressure in both the Eulerian frame,

$$\frac{\partial p}{\partial t} + \mathbf{V} \cdot \nabla p = -\Gamma p\nabla \cdot \mathbf{V} + (\Gamma - 1)\left[-\nabla \cdot \mathbf{q} + \mathbf{\Pi} : \nabla \mathbf{V} + \eta J^2\right], \tag{5.11}$$

and the Lagrangian frame

$$\frac{dp}{dt} = -\Gamma p\nabla \cdot \mathbf{V} + (\Gamma - 1)\left[-\nabla \cdot \mathbf{q} + \mathbf{\Pi} : \nabla \mathbf{V} + \eta J^2\right]. \tag{5.12}$$

The first term on the right-hand side of Eq. (5.12) represents reversible PV work. The second term represents irreversible heating processes. If these are absent, we say that the fluid is *ideal*. In that case, we can again use the continuity equation to eliminate $\nabla \cdot \mathbf{V}$ and obtain

$$\frac{d}{dt}\left(\frac{p}{\rho^{\Gamma}}\right) = 0, \tag{5.13}$$

so that as the fluid element moves about in space it obeys the so-called adiabatic law $p/\rho^{\Gamma} = $ constant.

We can then summarize the results of Lectures 3, 4, and 5 by stating the *fluid equations* in Eulerian form:

$$\frac{\partial \rho}{\partial t} + \nabla \cdot \rho \mathbf{V} = 0, \tag{5.14}$$

$$\rho\left(\frac{\partial \mathbf{V}}{\partial t} + \mathbf{V} \cdot \nabla \mathbf{V}\right) = \rho_q \mathbf{E} - \nabla p + \mathbf{J} \times \mathbf{B} + \nabla \cdot \mathbf{\Pi}, \tag{5.15}$$

and

$$\frac{\partial p}{\partial t} + \mathbf{V} \cdot \nabla p = -\Gamma p \nabla \cdot \mathbf{V} + (\Gamma - 1)\left[-\nabla \cdot \mathbf{q} + \mathbf{\Pi} : \nabla \mathbf{V} + \eta J^2\right]. \tag{5.16}$$

Note that the advective derivative $\mathbf{V} \cdot \nabla$ appears prominently in all the equations. It will also appear in the equations that describe the dynamics of the electromagnetic fields.

We again remark that Eqs. (5.14, 5.15, 5.16) are not closed, i.e., there are more unknowns than there are equations. In particular, we will have to say something about $\mathbf{J}$, $\mathbf{B}$, $\mathbf{\Pi}$, and $\mathbf{q}$. The first will come from electrodynamics. The rest require a further discussion of closures. These are the next two topics.

Lecture 6
The Electromagnetic Field

> *Mysterious affair, electricity.*
>
> Samuel Beckett, *Ends and Odds*

The electromagnetic fields are **E**, the electric field, and **B**, the magnetic flux density or magnetic field. The sources of these fields are the electric charge density, ρ_q, and the electric current density, **J**. Together, these must satisfy Maxwell's equations:

Faraday's law:

$$\frac{\partial \mathbf{B}}{\partial t} = -\nabla \times \mathbf{E}. \tag{6.1}$$

Ampére's law:

$$\mu_0 \mathbf{J} = \nabla \times \mathbf{B} - \frac{1}{c^2} \frac{\partial \mathbf{E}}{\partial t}. \tag{6.2}$$

Gauss' law:

$$\nabla \cdot \mathbf{E} = \frac{\rho_q}{\varepsilon_0}. \tag{6.3}$$

Absence of magnetic monopoles:

$$\nabla \cdot \mathbf{B} = 0. \tag{6.4}$$

These equations are written in MKS units. This convention will be used throughout. In these units, the *square of the speed of light* is

$$c^2 = \frac{1}{\varepsilon_0 \mu_0}. \tag{6.5}$$

The constant ε_0 is called the *permittivity of free space* and the constant μ_0 is called the *permeability of free space*.

The dynamics of the electromagnetic fields and the fluid are coupled through *Ohm's law*,

$$\mathbf{E}' = \eta \mathbf{J}, \tag{6.6}$$

Schnack, D.D.: *The Electromagnetic Field*. Lect. Notes Phys. **780**, 35–38 (2009)
DOI 10.1007/978-3-642-00688-3_6   © Springer-Verlag Berlin Heidelberg 2009

where η is the *electrical resistivity*, which is to be considered a material property of the fluid, and $\mathbf{E}'$ is the electric field as seen by a conductor moving with velocity $\mathbf{V}$. According to the theory of relativity, this is given by

$$\mathbf{E}' = \frac{\mathbf{E} + \mathbf{V} \times \mathbf{B}}{\sqrt{1 - \frac{V^2}{c^2}}}, \tag{6.7}$$

where $\mathbf{E}$ is the electric field in the stationary frame.

Maxwell's equations and Ohm's law are Lorentz invariant, i.e., they are physically accurate to all orders of V^2/c^2. However, the fluid equations (5.14, 5.15, 5.16) are Gallilean invariant; they are physically accurate only to $O(V/c)$. The two systems of equations are incompatible as presently formulated. So, we either need to make the fluid equations relativistic (not a savory task!) or need to render Maxwell's equations' Gallilean invariant. As stated in the Introduction (Lecture 1), in MHD we will consider only low frequencies, i.e., $V^2/c^2 = (\omega L/c)^2 << 1$. We therefore choose the latter course and seek a form of Maxwell's equations that is only accurate through $O(V/c)$.

Consider Ohm's law, Eq. (6.6). From Eq. (6.7), when $V^2/c^2 << 1$ we can write the electric field in the moving frame as

$$\mathbf{E}' = (\mathbf{E} + \mathbf{V} \times \mathbf{B}) \left(1 - \frac{1}{2} \frac{V^2}{c^2} + \ldots \right)$$

$$= \mathbf{E} + \mathbf{V} \times \mathbf{B} + O\left(\frac{V^2}{c^2} \right). \tag{6.8}$$

Ohm's law then becomes

$$\mathbf{E} + \mathbf{V} \times \mathbf{B} = \eta \mathbf{J}, \tag{6.9}$$

which is the proper MHD form. It is sometimes called the *resistive Ohm's law*. When $\eta = 0$, it is called the *ideal* MHD Ohm's law. Note that, for this ideal MHD, the electric field scales like $E_0 \sim V_0 B_0$ or $V_0 \sim E_0/B_0$. We will find these useful in a moment.

Now consider Ampére's law, Eq. (6.2). The ratio of the two terms on the right-hand side is, approximately,

$$\frac{\left| \frac{1}{c^2} \frac{\partial \mathbf{E}}{\partial t} \right|}{|\nabla \times \mathbf{B}|} \sim \frac{E_0 \omega/c^2}{B_0/L} \sim \frac{V_0 \omega L}{c^2} \sim \frac{V_0^2}{c^2} << 1, \tag{6.10}$$

where we have set $V_0 \sim \omega L$. We can therefore ignore the second term (the displacement current) compared with the first, and the low-frequency version of Ampére's law is

$$\mu_0 \mathbf{J} = \nabla \times \mathbf{B}. \tag{6.11}$$

In MHD, this equation *defines* the current density.

Next consider Gauss' law, Eq. (6.3). When combined with the ideal MHD Ohm's law, we have

$$\rho_q = -\varepsilon_0 \nabla \cdot (\mathbf{V} \times \mathbf{B}) \neq 0, \tag{6.12}$$

so that MHD allows for a non-vanishing charge density. This net charge must arise from a difference Δn between the local number densities of positive and negative charges. Then we can write $\Delta n/n_0 \sim E_0 \varepsilon_0/(n_0 L e) \sim \varepsilon_0 V_0 B_0/n_0 L e$, where n_0 is the average number density of positive and negative charges. Then using Eq. (6.5) to eliminate ε_0, we find

$$\frac{\Delta n}{n_0} \sim \frac{V_0}{c^2} \frac{B_0^2}{\mu_0 n_0 M} \frac{M}{e B_0} \frac{1}{L}, \tag{6.13}$$

where M is the mass of the individual positively charged particles (ions). We anticipate future results and identify $V_A^2 \equiv B_0^2/(\mu_0 n_0 M)$ as the *Alfvén speed*. (This turns out to be the propagation speed of the shear waves briefly described in Lecture 1.) We also identify $\Omega = e B_0/M$ as the ion gyro-frequency (the frequency at which the individual ions orbit the magnetic field lines). Then setting $V_0 \sim V_A$, we have

$$\frac{\Delta n}{n_0} \sim \frac{V_0^2}{c^2} \frac{V_0}{\Omega L}. \tag{6.14}$$

Finally, identifying $V_A/\Omega L \equiv d_i/L$ (where $d_i = c/\omega_{pi}$ is the ion skin depth and $\omega_{pi}^2 = n_0 e^2/\varepsilon_0 M$ is the square of the plasma frequency), we can estimate the size of the excess electric charge as

$$\frac{\Delta n}{n_0} \sim \frac{d_i}{L} \frac{V_0^2}{c^2}, \tag{6.15}$$

which is $<< V_0^2/c^2$ since $d_i/L << 1$. This result is called *quasi-neutrality*; it is a consequence of the low-frequency assumption.

However, the charge density cannot be ignored if the parameters are such that $\Delta n/n_0 \sim V_0/c$. This can occur if $(d_i/L)(V_0/c) \sim 1$ or, on length scales, $L \sim (V_0/c)d_i$. If we estimate $V \sim V_{thi} \sim \sqrt{T/M}$, then $L \sim \sqrt{\varepsilon_0 T/n_0 e^2} = \lambda_D$, the *Debye length*. This is assumed to be much smaller than any macroscopic scale length.

The virtual vanishing of the electric charge density does *not* imply that the electrostatic field vanishes. In steady state ($\partial/\partial t = 0$), Faraday's law requires $\nabla \times \mathbf{E} = 0$ or $\mathbf{E} = -\nabla \phi$, and so the electric field is *completely* electrostatic, and can be large. Instead, regions of smooth field (where $\nabla \cdot \mathbf{E} \sim 0$) are "patched together" across layers with finite charge density and thickness that is vanishingly small, i.e., $O(\lambda_D)$.

This is reminiscent of (although not completely analogous to) the role of shock waves in hydrodynamics.

Finally, it can be shown that the ratio of the electric force to the Lorentz force is

$$\frac{|\rho_q \mathbf{E}|}{|\mathbf{J} \times \mathbf{B}|} \sim \frac{V^2}{c^2} \ll 1, \tag{6.16}$$

so that it can be dropped from the equation of motion. The charge density therefore never enters the MHD equations. However, if you ever want to know what it is, all you have to do is compute $\rho_q = -\varepsilon_0 \nabla \cdot (\mathbf{V} \times \mathbf{B})$ (at least in ideal MHD).

In Eulerian form, the final equations of the MHD model are

Equations for the fluid:

$$\frac{\partial \rho}{\partial t} + \nabla \cdot \rho \mathbf{V} = 0, \tag{6.17}$$

$$\rho \left(\frac{\partial \mathbf{V}}{\partial t} + \mathbf{V} \cdot \nabla \mathbf{V} \right) = -\nabla p + \mathbf{J} \times \mathbf{B} + \nabla \cdot \mathbf{\Pi}, \tag{6.18}$$

$$\frac{\partial p}{\partial t} + \mathbf{V} \cdot \nabla p = -\Gamma p \nabla \cdot \mathbf{V} + (\Gamma - 1) \left[-\nabla \cdot \mathbf{q} + \mathbf{\Pi} : \nabla \mathbf{V} + \eta J^2 \right]. \tag{6.19}$$

Equations for the electromagnetic fields:

$$\frac{\partial \mathbf{B}}{\partial t} = -\nabla \times \mathbf{E}, \tag{6.20}$$

$$\mu_0 \mathbf{J} = \nabla \times \mathbf{B}. \tag{6.21}$$

Ohm's law, which couples the fluid and the fields:

$$\mathbf{E} + \mathbf{V} \times \mathbf{B} = \eta \mathbf{J}. \tag{6.22}$$

Equations (6.20) and (6.21) are sometimes called the "pre-Maxwell equations," because they represent the state of knowledge of the electromagnetic field before Maxwell's introduction of the displacement current.

Equations (6.17, 6.18, 6.19, 6.20, 6.21, 6.22) are 14 equations in 27 unknowns: ρ (1 unknown), $\mathbf{V}$ (3), p (1), $\mathbf{\Pi}$ (9), $\mathbf{J}$ (3), $\mathbf{B}$ (3), $\mathbf{E}$ (3), $\mathbf{q}$ (3), and η (1). The conditions $\nabla \cdot \mathbf{B} = 0$ implied by Faraday's law, and $\nabla \cdot \mathbf{J} = 0$ implied by Ampére's law, either increase the number of equations by two or decrease the number of unknowns by two, depending on your point of view. So, we need expressions for 13 of the variables in terms of the other 14. This is the problem of *closure*. It will be discussed next.

Lecture 7
Closures

A closed mouth catches no flies.

Miguel De Cervantes, *Don Quixote*

We have just seen that the MHD equations constitute 14 equations in 27 unknowns. These can be reduced to eight equations by substituting for $\mathbf{J}$ from Ampére's law [Eq. (6.21)] and $\mathbf{E}$ from Ohm's law [Eq. (6.22)]. The result is

$$\frac{\partial \rho}{\partial t} + \nabla \cdot \rho \mathbf{V} = 0, \tag{7.1}$$

$$\rho \left(\frac{\partial \mathbf{V}}{\partial t} + \mathbf{V} \cdot \nabla \mathbf{V} \right) = -\nabla p + \frac{1}{\mu_0} (\nabla \times \mathbf{B}) \times \mathbf{B} + \nabla \cdot \mathbf{\Pi}, \tag{7.2}$$

$$\frac{\partial p}{\partial t} + \mathbf{V} \cdot \nabla p = -\Gamma p \nabla \cdot \mathbf{V} + (\Gamma - 1) \left[-\nabla \cdot \mathbf{q} + \mathbf{\Pi} : \nabla \mathbf{V} + \eta J^2 \right], \tag{7.3}$$

and

$$\frac{\partial \mathbf{B}}{\partial t} = \nabla \times \left(\mathbf{V} \times \mathbf{B} - \frac{\eta}{\mu_0} \nabla \times \mathbf{B} \right). \tag{7.4}$$

We take ρ, p, $\mathbf{V}$, and $\mathbf{B}$ to be the primary dependent variables. Then in order to close the equations (i.e., have as many unknowns as equations), we require expressions for the stress tensor, the heat flux, and the resistivity. These are usually expressed as the functional relations $\mathbf{\Pi}(\rho, \mathbf{V}, \mathbf{B})$ (9 variables), $\mathbf{q}(p, \rho, \mathbf{B})$ (3 variables), and $\eta(p, \rho)$ (1 variable), so that we require 13 additional closure relations. The symmetry of the stress tensor, $\Pi_{ij} = \Pi_{ji}$, eliminates three unknowns and reduces the number of required closure relations to ten. These generally are obtained from a knowledge of the material properties of the fluid, and must therefore come from outside the framework of the fluid model.

However, it is possible to say some more general things about the form of the stress tensor.[1] First, we consider the case of an *unmagnetized* fluid, i.e., hydrodynamics. We know that this stress must arise from internal friction between different

---

[1] This discussion follows that of L. D. Landau and E. M. Lifschitz, *Fluid Mechanics*, Pergamon Press, London, UK (1959).

Schnack, D.D.: *Closures*. Lect. Notes Phys. **780**, 39–42 (2009)
DOI 10.1007/978-3-642-00688-3_7 © Springer-Verlag Berlin Heidelberg 2009

parts of the fluid that are in relative motion. For an *isotropic* medium, a general form for the components of $\mathbf{\Pi}$ can be deduced from the following considerations:

1. $\mathbf{\Pi} = 0$ if there is no relative motion between the different parts of the fluid. This implies that $\Pi_{ij} \sim \left(\partial_i V_j\right)^\alpha$ or $\mathbf{\Pi} \sim (\nabla \mathbf{V})^\alpha$. If $\nabla \mathbf{V}$ is "small," we expect a linear relationship ($\alpha = 1$), so $\mathbf{\Pi} \sim \nabla \mathbf{V}$.
2. $\mathbf{\Pi} = 0$ when $\nabla \mathbf{V} = 0$, so there can be no terms that are independent of $\nabla \mathbf{V}$.
3. $\mathbf{\Pi} = 0$ for rigid body rotation, i.e., when $\mathbf{V} = \mathbf{\Omega} \times \mathbf{r}$ with constant $\mathbf{\Omega}$, since then there is then no relative motion of the fluid elements.
4. $\mathbf{\Pi}$ must be symmetric (stated previously).

It turns out that the tensor (dyadic) $\mathbf{e} = \nabla \mathbf{V} + \nabla \mathbf{V}^T$ satisfies these constraints, so that $\Pi_{ij} \sim \partial_i V_j + \partial_j V_i$. The most general tensor that satisfies this condition is

$$\Pi_{ij} = a \left(\partial_i V_j + \partial_j V_i\right) + b \partial_l V_l \delta_{ij} \tag{7.5}$$

or, in dyadic notation, $\mathbf{\Pi} = a(\nabla \mathbf{V} + \nabla \mathbf{V}^T) + b(\nabla \cdot \mathbf{V})\mathbf{I}$. It is conventional to write this as $\mathbf{\Pi} = \mu \mathbf{W}$, where the *dynamic viscosity* $\mu(\rho, p)$ is a property of the medium, and

$$\mathbf{W} = \nabla \mathbf{V} + \nabla \mathbf{V}^T - \frac{2}{3} \mathbf{I} \nabla \cdot \mathbf{V}, \tag{7.6}$$

is called the *rate of strain tensor*. A relationship of the form stress $\sim$ strain (such as $\mathbf{\Pi} = \mu \mathbf{W}$) is called *Hooke's law*. A fluid is therefore an example of an elastic medium.

For the special case $\mu = $ constant, the viscous force density $\mathbf{F}_\mu = \nabla \cdot \mathbf{\Pi}$ can be written as

$$F_{\mu i} = \mu \partial_j \left(\partial_i V_j + \partial_j V_i - \frac{2}{3} \partial_l V_l \delta_{ij}\right)$$

$$= \mu \left(\partial_j \partial_j V_i - \frac{2}{3} \partial_i \partial_l V_l\right)$$

or

$$\mathbf{F}_\mu = \mu \left[\nabla^2 \mathbf{V} - \frac{2}{3} \nabla (\nabla \cdot \mathbf{V})\right]. \tag{7.7}$$

For the further special case of incompressible flow, $\nabla \cdot \mathbf{V} = 0$ and we obtain the especially simple form

$$\mathbf{F}_\mu = \mu \nabla^2 \mathbf{V}. \tag{7.8}$$

Under these circumstances, it is common to write the viscous force per unit mass $(\mathbf{F}_\mu/\rho)$ as $\nu \nabla^2 \mathbf{V}$, where $\nu = \mu/\rho$ is the *kinematic viscosity*. It has the units of a diffusion coefficient, L^2/T. The form of Eq. (7.8) is widely used in theory and computation, often when $\mu \neq$ constant or $\nabla \cdot \mathbf{V} \neq 0$. This is not physically justified.

When the fluid is magnetized ($\mathbf{B} \neq 0$), it is not isotropic and the preceding discussion must be generalized. For an *anisotropic* medium, the general comments relating to the form of the stress tensor still apply (i.e., stress $\sim$ strain), except that the simple relationship $\mathbf{\Pi} = \mu\mathbf{W}$ is modified according to

$$\Pi_{ij} = E_{ijkl}W_{jl}, \tag{7.9}$$

which is called the *generalized* Hooke's law. The quantity E_{ijkl} is a fourth-rank tensor called the *elastic constant tensor*, whose specific form depends on the fluid properties and must be obtained from considerations outside of the fluid model. So, in MHD, things are much more complicated than they are in hydrodynamics.

Expressions for $\mathbf{\Pi}$ have been derived by Braginskii.[2] For the case when the field is strong, the collisionality is large, and the mean free path is small; specifically, $\nu_c/\Omega << 1$ (with ν_c the collision frequency) and $\lambda_f/L << 1$ (with λ_f the mean free path). This involves calculations that are heroic in scale, and we will state the results here without proof. The viscous stress is decomposed into three "components": parallel, $\mathbf{\Pi}_{\|} = \hat{\mathbf{b}}\hat{\mathbf{b}} \cdot \mathbf{\Pi}$ (with $\hat{\mathbf{b}} = \mathbf{B}/B$); "cross," $\mathbf{\Pi}_{\wedge} = (\hat{\mathbf{b}} \times \mathbf{I}) \cdot \mathbf{\Pi}$; and perpendicular, $\mathbf{\Pi}_{\perp} = \hat{\mathbf{b}} \times (\hat{\mathbf{b}} \times \mathbf{I}) \cdot \mathbf{\Pi}$. The resulting expressions are complicated, and we give them here for reference only.

For the parallel component,

$$\mathbf{\Pi}_{\|} = \frac{3}{2}\eta_0 \left(\hat{\mathbf{b}} \cdot \mathbf{W} \cdot \hat{\mathbf{b}}\right) \left(\hat{\mathbf{b}}\hat{\mathbf{b}} - \frac{1}{3}\mathbf{I}\right), \tag{7.10}$$

where $\eta_0 \sim p/\nu_c$. This coefficient diverges when $\nu_c \to 0$, indicating that the fluid model is no longer strictly valid in the limit of low collisionality. (Note that when there is no magnetic field this expression reduces to $\mathbf{\Pi}_{\|} = \eta_0\mathbf{W}$, since then the fluid is isotropic and $\hat{\mathbf{b}}\hat{\mathbf{b}} \to \mathbf{I}$.)

For the "cross" component,

$$\mathbf{\Pi}_{\wedge} = \frac{p}{4\Omega} \left[(\hat{\mathbf{b}} \times \mathbf{W}) \cdot (\mathbf{I} + 3\hat{\mathbf{b}}\hat{\mathbf{b}}) - (\mathbf{I} + 3\hat{\mathbf{b}}\hat{\mathbf{b}}) \cdot (\mathbf{W} \times \hat{\mathbf{b}})\right], \tag{7.11}$$

This is called the *gyro-viscous stress*. The second term in the square brackets is just the transpose of the first term, as is required for the symmetry of $\mathbf{\Pi}_{\wedge}$. Note that Eq. (7.11) is independent of the collision frequency ν_c. It is therefore *non-dissipative*. The gyro-viscous stress becomes important when the Larmor radius of the ions becomes "finite" (but still small compared with the size of a fluid element); it is called an "FLR" (for finite Larmor-, or gyro-, radius) effect. It represents a completely reversible flux of momentum due to the gyro-motion of the individual particles.

---

[2] The original reference is S. I. Braginskii, "Transport Processes in Plasmas", in *Reviews of Plasma Physics*, M. A. Leontovich (ed.), Consultants Bureau, New York (1965). For further discussion and some more recent developments, see Per Helander and Dieter J. Sigmar, *Collisional Transport in Magnetized Plasmas*, Cambridge University Press, Cambridge (2002).

For the perpendicular component,

$$\mathbf{\Pi}_\perp = \eta_1 \left\{ (\mathbf{I} - \hat{\mathbf{b}}\hat{\mathbf{b}}) \cdot \mathbf{W} \cdot (\mathbf{I} - \hat{\mathbf{b}}\hat{\mathbf{b}}) - \frac{1}{2} (\mathbf{I} - \hat{\mathbf{b}}\hat{\mathbf{b}}) (\mathbf{I} - \hat{\mathbf{b}}\hat{\mathbf{b}}) : \mathbf{W} \right.$$
$$\left. + 4 \left[(\mathbf{I} - \hat{\mathbf{b}}\hat{\mathbf{b}}) \cdot \mathbf{W} \cdot \hat{\mathbf{b}}\hat{\mathbf{b}} + \hat{\mathbf{b}}\hat{\mathbf{b}} \cdot \mathbf{W} \cdot (\mathbf{I} - \hat{\mathbf{b}}\hat{\mathbf{b}}) \right] \right\}, \tag{7.12}$$

where $\eta_1 \sim p v_c / \Omega^2$. The coefficient vanishes in the limit of low collisionality, so Eq. (7.12) is generally ignored in the analysis of hot fusion plasmas. However, it survives in the fluid limit. (The use of the symbol η for the dynamic viscosity is common in hydrodynamics. In MHD we reserve this symbol for the electrical resistivity.)

We require a closure for the heat flux, $\mathbf{q}$. As is the case for the stress tensor, in a strongly magnetized plasma it can be decomposed as $\mathbf{q} = \mathbf{q}_\| + \mathbf{q}_\wedge + \mathbf{q}_\perp$, where

$$\mathbf{q}_\| = -\kappa_\| \hat{\mathbf{b}}\hat{\mathbf{b}} \cdot \nabla T, \tag{7.13}$$

$$\mathbf{q}_\wedge = -\kappa_\wedge \hat{\mathbf{b}} \times \nabla T, \tag{7.14}$$

and

$$\mathbf{q}_\perp = -\kappa_\perp (\mathbf{I} - \hat{\mathbf{b}}\hat{\mathbf{b}}) \cdot \nabla T, \tag{7.15}$$

where the temperature is given by the equation of state $T = p/\rho$. The coefficients are called the *thermal conductivities*. As was the case with the stress tensor, $\kappa_\| \sim 1/v_c$, $\kappa_\wedge$ is independent of v_c, and $\kappa_\perp \sim v_c$. The ratio of parallel to perpendicular conductivity diverges like $1/v_c^2$, so the heat flux can become highly anisotropic in the low-collisionality limit. (This proves true in spite of the fact that the expression for $\mathbf{q}_\|$ is no longer valid.) As with $\mathbf{\Pi}_\wedge$, the "cross" heat flux $\mathbf{q}_\wedge$ is an FLR effect and is non-dissipative. It represents a reversible flux of heat due to the gyro-motion of the individual particles.

We still require a closure for the electrical resistivity η. This is generally taken to be a known function of the temperature T. Its particular form is not important for MHD.

We emphasize that the closures presented here are valid only in the regime of large collisionality and strong magnetic field. The derivation of closure expressions for the case of low collisionality and long mean free path is an important topic of current plasma physics research, and there is no consensus regarding the correct result.

With closure expressions such as those presented in this lecture, Eqs. (7.1, 7.2, 7.3, 7.4) are eight partial differential equations in the eight unknowns ρ, p, $\mathbf{V}$, and $\mathbf{B}$, and together constitute the *MHD equations*. If we set $\mathbf{q} = 0$ and $\mathbf{\Pi} = 0$, we obtain the *resistive MHD model*. If we further set $\eta = 0$ we obtain the *ideal MHD model*. Ideal and resistive MHD will be the topics of the remainder of this course.

Lecture 8
Conservation Laws

Conservation must come before recreation.
Prince Charles of England, July 5, 1989

The general form of a conservation law is

$$\frac{\partial U_{ijk...}}{\partial t} = -\frac{\partial}{\partial x_m} F_{mijk...},$$

(8.1)

where $U_{ijk...}$ is a tensor of rank N and $F_{mijk...}$ is a tensor or rank $N+1$. Integrating over the volume and applying Gauss' theorem, we have

$$\frac{\partial}{\partial t} \int_{V_0} U_{ijk...} dV = \oint_{S_0} dS_m F_{mijk...},$$

(8.2)

which expresses the conservation of the volume integral of $U_{ijk...}$. The tensor $F_{mijk...}$ is the flux of $U_{ijk...}$ in the direction of x_m; the surface integral expresses the total flux through the bounding surface.

For example, if $N = 0$, we have the *scalar conservation law*

$$\frac{\partial U}{\partial t} = -\nabla \cdot \mathbf{F},$$

(8.3)

where U is a scalar and $\mathbf{F}$ is a vector. We have already encountered one of these in the continuity equation, which expresses the conservation of mass. If $N = 1$, we have the vector conservation law

$$\frac{\partial \mathbf{U}}{\partial t} = -\nabla \cdot \mathbf{F},$$

(8.4)

where $\mathbf{U}$ is a vector and $\mathbf{F}$ is a second-rank tensor (or, equivalently, dyad). The quantity F_{ij} is the flux of U_i in the x_j direction. This can be continued ad infinitum.

The MHD equations can be written in the form of conservation laws that express the physical principles of conservation of mass, momentum, and energy.

As mentioned above, we already have the law of *conservation of mass*:

$$\frac{\partial \rho}{\partial t} = -\nabla \cdot \rho \mathbf{V}.$$

(8.5)

Schnack, D.D.: *Conservation Laws*. Lect. Notes Phys. **780**, 43–47 (2009)
DOI 10.1007/978-3-642-00688-3_8

The quantity $\rho \mathbf{V}$ is called the *mass flux*.

We now obtain the law of *conservation of momentum*. This will be of the form

$$\frac{\partial \rho \mathbf{V}}{\partial t} = -\nabla \cdot \mathbf{T}, \tag{8.6}$$

where $\rho \mathbf{V}$ is the momentum per unit volume and $\mathbf{T}$ is called the *total stress tensor*. We have seen that the equation of motion (Newton's law) is

$$\rho \left(\frac{\partial \mathbf{V}}{\partial t} + \mathbf{V} \cdot \nabla \mathbf{V} \right) = -\nabla \cdot (p\mathbf{I} - \mathbf{\Pi}) + \mathbf{J} \times \mathbf{B}. \tag{8.7}$$

The first term on the right-hand side is already in conservation form. We transform the left-hand side according to

$$\rho \frac{\partial \mathbf{V}}{\partial t} = \frac{\partial \rho \mathbf{V}}{\partial t} - \mathbf{V} \frac{\partial \rho}{\partial t} = \frac{\partial \rho \mathbf{V}}{\partial t} + \mathbf{V} \nabla \cdot \rho \mathbf{V} \tag{8.8}$$

and

$$\nabla \cdot \rho \mathbf{V} \mathbf{V} = \mathbf{V} \nabla \cdot \rho \mathbf{V} + \rho \mathbf{V} \cdot \nabla \mathbf{V}. \tag{8.9}$$

We have used the continuity equation to obtain the second equality in Eq. (8.8). Then we have

$$\frac{\partial \rho \mathbf{V}}{\partial t} + \nabla \cdot \rho \mathbf{V} \mathbf{V} = -\nabla \cdot (p\mathbf{I} - \mathbf{\Pi}) + \mathbf{J} \times \mathbf{B}. \tag{8.10}$$

Consider the second term on the right-hand side. Using Cartesian tensor notation,

$$(\mathbf{J} \times \mathbf{B})_i = \frac{1}{\mu_0} \varepsilon_{jki} \varepsilon_{jlm} B_k \partial_l B_m$$

$$= \frac{1}{\mu_0} (B_l \partial_l B_i - B_m \partial_i B_m). \tag{8.11}$$

Now $B_l \partial_l B_i = \partial_l (B_l B_i)$, since $\partial_l B_l = 0$ ($\nabla \cdot \mathbf{B} = 0$) and $B_m \partial_i B_m = \partial_l (B^2 \delta_{li}/2)$, so we have

$$(\mathbf{J} \times \mathbf{B})_i = \frac{1}{\mu_0} \partial_l \left(B_l B_i - \frac{1}{2} B^2 \delta_{li} \right) \tag{8.12}$$

or, in coordinate-free vector notation,

$$\mathbf{J} \times \mathbf{B} = \frac{1}{\mu_0} \nabla \cdot \left(\mathbf{B} \mathbf{B} - \frac{1}{2} B^2 \mathbf{I} \right). \tag{8.13}$$

The final result is in the conservation form of Eq. (8.6), with

$$\mathbf{T} = \rho\mathbf{VV} - \frac{1}{\mu_0}\mathbf{BB} + \left(p + \frac{1}{2\mu_0}B^2\right)\mathbf{I} - \mathbf{\Pi}. \tag{8.14}$$

Note that $\mathbf{T}$ is symmetric, as is required of a proper stress tensor. The component T_{ij} is the total flux of the jth component of momentum in the direction x_i.

The tensor $\rho\mathbf{VV}$ is called the *Reynolds' stress*; $\rho V_i V_j$ is the rate at which the momentum component ρV_j is carried through surface element dS_i by velocity component V_i. We recognize $\mathbf{P} = p\mathbf{I} - \mathbf{\Pi}$ from our previous discussion as the *hydrodynamic stress tensor*. The tensor $\mathbf{T}_M = \frac{1}{\mu_0}\left(\frac{1}{2}B^2\mathbf{I} - \mathbf{BB}\right)$ is called the *Maxwell stress tensor*. Note that the magnetic field transports (or carries) momentum. The quantity $p + B^2/2\mu_0$ is the *total pressure*. The second contribution is called the *magnetic pressure*; the magnetic field resists compression, just like the fluid pressure. The tensor $\mathbf{BB}/\mu_0$ is called the *hoop stress*. We will see that it resists shearing motions.

Conservation of angular momentum, $\mathbf{L} = \mathbf{x} \times \rho\mathbf{V}$, follows directly from the form of Eq. (8.6):

$$\frac{\partial\mathbf{L}}{\partial t} = -\mathbf{x} \times \nabla \cdot \mathbf{T}. \tag{8.15}$$

Using Cartesian tensor notation, the right-hand side of Eq. (8.15) can be written as

$$(\mathbf{x} \times \nabla \cdot \mathbf{T})_i = \varepsilon_{ijk}x_j\partial_m T_{mk} = \varepsilon_{ijk}\left[\partial_m\left(x_j T_{mk}\right) - \delta_{mj}T_{mk}\right]$$
$$= \partial_m\left(\varepsilon_{ijk}x_j T_{mk}\right) - \varepsilon_{imk}T_{mk}. \tag{8.16}$$

The last term in Eq. (8.16) vanishes because $\mathbf{T}$ is a symmetric tensor. Then $\partial L_i/\partial t = -\partial_m \mathcal{L}_{mi}$, where

$$\mathcal{L}_{mi} = T_{mk}\varepsilon_{ijk}x_j = \varepsilon_{ijk}x_j T_{mk} = \varepsilon_{ijk}x_j T_{km} = (\mathbf{x} \times \mathbf{T})_{im}$$
$$= (\mathbf{x} \times \mathbf{T})_{mi}^T , \tag{8.17}$$

and we have again used the symmetry of $\mathbf{T}$. Conservation of angular momentum is therefore expressed as

$$\frac{\partial\mathbf{L}}{\partial t} = -\nabla \cdot \left[(\mathbf{x} \times \mathbf{T})^T\right]. \tag{8.18}$$

The total energy is the sum of the kinetic, magnetic, and internal energies:

$$u = \frac{1}{2}\rho V^2 + \frac{B^2}{2\mu_0} + \rho e. \tag{8.19}$$

To obtain the law of conservation of energy, we work in the Lagrangian frame. We first consider the rate of change of kinetic energy. The scalar product of $\mathbf{V}$ with the equation of motion is

$$\rho \frac{d}{dt} \left(\frac{1}{2} V^2 \right) = -\mathbf{V} \cdot \nabla \mathbf{P} + \mathbf{V} \cdot \mathbf{J} \times \mathbf{B}. \tag{8.20}$$

Using the Lagrangian form of the continuity equation, the identity $\mathbf{V} \cdot \nabla \mathbf{P} = \nabla \cdot (\mathbf{P} \cdot \mathbf{V}) - \mathbf{P} : \nabla \mathbf{V}$, and interchanging terms in the second term on the right-hand side, this becomes

$$\frac{d}{dt} \left(\frac{1}{2} \rho V^2 \right) + \frac{1}{2} \rho V^2 \nabla \cdot \mathbf{V} = -\nabla \cdot (\mathbf{P} \cdot \mathbf{V}) + \mathbf{P} : \nabla \mathbf{V} - \mathbf{J} \cdot \mathbf{V} \times \mathbf{B}. \tag{8.21}$$

We can use the resistive form of Ohm's law to eliminate $\mathbf{V} \times \mathbf{B}$ in favor of $\mathbf{J}$ and $\mathbf{E}$. Then writing $\rho V^2 \nabla \cdot \mathbf{V} = \nabla \cdot (\rho V^2 \mathbf{V}) - \mathbf{V} \cdot \nabla (\rho V^2)$ and using the definition of $\mathbf{P}$,

$$\frac{d}{dt} \left(\frac{1}{2} \rho V^2 \right) + \nabla \cdot \left(\frac{1}{2} \rho V^2 \mathbf{V} \right) - \mathbf{V} \cdot \nabla \left(\frac{1}{2} \rho V^2 \right)$$
$$= -\nabla \cdot (\mathbf{P} \cdot \mathbf{V}) + p \nabla \cdot \mathbf{V} - \mathbf{\Pi} : \nabla \mathbf{V} + \mathbf{J} \cdot \mathbf{E} - \eta J^2. \tag{8.22}$$

If we now transform to the Eulerian frame, the last term on the left-hand side cancels, and we obtain the final expression for the rate of change of kinetic energy:

$$\frac{\partial}{\partial t} \left(\frac{1}{2} \rho V^2 \right) = -\nabla \cdot \underbrace{\left[\left(\frac{1}{2} \rho V^2 \mathbf{I} + \mathbf{P} \right) \cdot \mathbf{V} \right]}_{\text{flux through surface}}$$
$$+ \underbrace{p \nabla \cdot \mathbf{V}}_{\text{PV work}} + \underbrace{\mathbf{J} \cdot \mathbf{E}}_{\text{EM work}} - \underbrace{\mathbf{\Pi} : \nabla \mathbf{V}}_{\text{viscous dissipation}} - \underbrace{\eta J^2}_{\text{resistive dissipation}}. \tag{8.23}$$

Note that this is not yet a conservation law. The last four terms act as sources and sinks of kinetic energy. They must show up elsewhere.

The calculation of the rate of change of magnetic energy is relatively straightforward. Taking $\mathbf{B}$ the induction equation, and using $\mathbf{B} \cdot \nabla \times \mathbf{E} = \nabla \cdot (\mathbf{E} \times \mathbf{B}) + \mathbf{E} \cdot \nabla \times \mathbf{B}$ along with Ampére's law, we find

$$\frac{\partial}{\partial t} \left(\frac{B^2}{2\mu_0} \right) = -\nabla \cdot \underbrace{\left[\frac{1}{\mu_0} \mathbf{E} \times \mathbf{B} \right]}_{\text{Poynting flux}} - \underbrace{\mathbf{J} \cdot \mathbf{E}}_{\text{EM work}}. \tag{8.24}$$

We already have an expression for the rate of change of internal energy:

$$\frac{\partial}{\partial t}(\rho e) = -\nabla \cdot (\rho e \mathbf{V}) - \underbrace{p\nabla \cdot \mathbf{V}}_{\text{PV work}} - \underbrace{\nabla \cdot \mathbf{q}}_{\text{heat flow}} + \underbrace{\boldsymbol{\Pi} : \nabla \mathbf{V}}_{\text{viscous heating}} + \underbrace{\eta J^2}_{\text{resistive heating}} \quad . \tag{8.25}$$

The law of *conservation of energy* is obtained by adding Eqs. (8.23), (8.24), and (8.25). All the sources and sinks cancel and the result is

$$\frac{\partial u}{\partial t} = -\nabla \cdot \left\{ \left[\left(\rho e + \frac{1}{2}\rho V^2 \right) \mathbf{I} + \mathbf{P} \right] \cdot \mathbf{V} + \frac{1}{\mu_0} \mathbf{E} \times \mathbf{B} + \mathbf{q} \right\}. \tag{8.26}$$

Another conserved quantity is the magnetic flux density,

$$\frac{\partial \mathbf{B}}{\partial t} = -\nabla \times \mathbf{E}. \tag{8.27}$$

We remark that this can be written in the form of Eq. (8.1) as

$$\frac{\partial B_i}{\partial t} = -\varepsilon_{ijk}\partial_j E_k = -\partial_j \left(\varepsilon_{ijk} E_k \right) = -\partial_j \left(\varepsilon_{jki} E_k \right) = -\partial_j F_{ji}. \tag{8.28}$$

The tensor F_{ji} is formally the flux of B_i in the direction of x_j. Note that F_{ji} is completely antisymmetric, i.e., $F_{ij} = \varepsilon_{ikj}E_k = -\varepsilon_{jki}E_k = -F_{ji}$ by the properties of ε_{jki}. For the case of ideal MHD, then $E_k = -\varepsilon_{kmn}V_m B_n$, so $F_{ij} = B_i V_j - B_j V_i$ and

$$\frac{\partial \mathbf{B}}{\partial t} = -\nabla \cdot (\mathbf{BV} - \mathbf{VB}). \tag{8.29}$$

Recall that $\mathbf{B}$ is a pseudovector; Eq. (8.29) is an example of a *pseudovector conservation law*. The conserved quantity is the magnetic flux through an area element, $d\Phi = \mathbf{B} \cdot d\mathbf{S}$, which is a true scalar (since both $\mathbf{B}$ and $d\mathbf{S} = d\mathbf{l}_1 \times d\mathbf{l}_2$ are pseudovectors). The appropriate integral theorems are

$$\frac{d\Phi}{dt} = -\int_S \nabla \times \mathbf{E} \cdot d\mathbf{S} = -\oint_C \mathbf{E} \cdot d\mathbf{l} \tag{8.30}$$

and

$$\frac{d}{dt}\int_V \mathbf{B}dV = -\oint_S d\mathbf{S} \times \mathbf{E}. \tag{8.31}$$

Note that the evolution of $\bar{\mathbf{B}}$, the volume average of $\mathbf{B}$, is completely determined by the tangential component of $\mathbf{E}$ in the bounding surface.

Lecture 9
Ideal MHD and the Frozen Flux Theorem

And moveth all together, if it moves at all.
William Wordsworth, *Resolution and Independence*

We have just derived the "complete" MHD fluid model. We now consider briefly a simpler model that is obtained by ignoring Π and $\mathbf{q}$ and setting $\eta = 0$. (Note that this is a form of closure.) The medium is thus a perfect electrical conductor and has no viscosity or thermal conductivity. This is a highly idealized situation, not attainable in nature; it is called ideal MHD. However, it turns out that ideal MHD describes to a remarkably good approximation many of the dynamical properties of hot, strongly magnetized plasmas. This is primarily because most hot plasmas are excellent (although not perfect) conductors of electricity. Ideal MHD is thus of considerable interest. Under these circumstances, the equations of the model reduce to

$$\frac{\partial \rho}{\partial t} = -\nabla \cdot \rho \mathbf{V}, \tag{9.1}$$

$$\rho \left(\frac{\partial \mathbf{V}}{\partial t} + \mathbf{V} \cdot \nabla \mathbf{V} \right) = -\nabla p + \frac{1}{\mu_0} \left(\nabla \times \mathbf{B} \right) \times \mathbf{B}, \tag{9.2}$$

$$\frac{\partial p}{\partial t} = -\mathbf{V} \cdot \nabla p - \Gamma p \nabla \cdot \mathbf{V}, \tag{9.3}$$

and

$$\frac{\partial \mathbf{B}}{\partial t} = -\nabla \times (\mathbf{V} \times \mathbf{B}). \tag{9.4}$$

These are eight equations in the eight unknowns ρ, p, $\mathbf{V}$, and $\mathbf{B}$. We remark that, in this case, Ohm's law is

$$\mathbf{E} = -\mathbf{V} \times \mathbf{B}. \tag{9.5}$$

We now derive the most important property of ideal MHD. Consider a closed curve C within the fluid, and let every pointon the curve be moving with the local

Schnack, D.D.: *Ideal MHD and the Frozen Flux Theorem.* Lect. Notes Phys. **780**, 49–53 (2009)
DOI 10.1007/978-3-642-00688-3_9 © Springer-Verlag Berlin Heidelberg 2009

fluid velocity. We say that C is co-moving with the fluid, in the Lagrangian sense. Let S be a surface bounded by C. Define

$$\Psi = \int_S \mathbf{B} \cdot d\mathbf{S} \tag{9.6}$$

as the flux through S. We ask how Ψ changes as C moves with the fluid. As we discussed regarding the Lagrangian change in mass density, $d\Psi$ consists of two parts:

1. $d\Psi_1$, due to the changes in $\mathbf{B}$ with C (and S) held fixed, i.e.,

$$\left(\frac{d\Psi}{dt}\right)_1 = \int_S \frac{\partial \mathbf{B}}{\partial t} \cdot d\mathbf{S} = -\int_S \nabla \times \mathbf{E} \cdot d\mathbf{S} = -\oint_C \mathbf{E} \cdot d\mathbf{l}; \tag{9.7}$$

2. $d\Psi_2$, the amount of magnetic flux swept out by C as it moves with the fluid. This is calculated as follows. As S moves about, each line element comprising it moves a distance $\mathbf{V}dt$, and sweeps out a lateral area $d\mathbf{S} = \mathbf{V}dt \times d\mathbf{l}$. This is shown in Fig. 9.1.

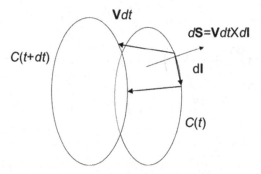

Fig. 9.1 Computing the magnetic flux through a volume element swept out by a surface moving with the fluid

The flux through this area is $d\Psi_2 = \mathbf{B} \cdot d\mathbf{S} = \mathbf{B} \cdot \mathbf{V} \times d\mathbf{l}dt$, so that

$$\left(\frac{d\Psi}{dt}\right)_2 = \oint_C \mathbf{B} \cdot \mathbf{V} \times d\mathbf{l} = -\oint_C \mathbf{V} \times \mathbf{B} \cdot d\mathbf{l}, \tag{9.8}$$

where we have used the properties of the triple vector product.

The total rate of change of flux through C is then

$$
\begin{aligned}
\frac{d\Psi}{dt} &= \left(\frac{d\Psi}{dt}\right)_1 + \left(\frac{d\Psi}{dt}\right)_2 \\
&= -\oint_C \mathbf{E} \cdot d\mathbf{l} - \oint_C \mathbf{V} \times \mathbf{B} \cdot d\mathbf{l} \\
&= -\oint_C (\mathbf{E} + \mathbf{V} \times \mathbf{B}) \cdot d\mathbf{l}.
\end{aligned}
\tag{9.9}
$$

However, in ideal MHD, $\mathbf{E} + \mathbf{V} \times \mathbf{B} = 0$ [see Eq. (9.5)], so that $d\Psi/dt = 0$. We conclude that in ideal MHD, *the magnetic flux through any co-moving closed circuit remains constant*. This important result is called the *frozen flux condition*. It means that the field lines can be thought of as being *attached* to the fluid (and *vice versa*); the fluid cannot move *across* the magnetic field. (However, the fluid is free to slide *along* **B**.) A perpendicular velocity will induce an electric field through $\mathbf{E} = -\mathbf{V} \times \mathbf{B}$. This will cause a change in **B** through Faraday's law that is sufficient to make the field lines appear to move with the fluid. If there are both electric and magnetic fields, there will be a perpendicular velocity given by

$$
\mathbf{V}_\perp = \frac{\mathbf{E} \times \mathbf{B}}{B^2}.
\tag{9.10}
$$

This is sometimes called the *MHD velocity*.

Now consider a volume whose lateral sides are everywhere parallel to the magnetic field, as shown in Fig. 9.2.

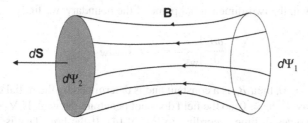

Fig. 9.2 A flux tube whose lateral sides are everywhere parallel to the magnetic field

By construction, $d\Psi_1 = d\Psi_2 = d\Psi$. If we now consider the case where the volume is long and thin, and make the cross-sectional area infinitesimal, then the volume is called a *flux tube*. Since in ideal MHD the magnetic field is co-moving with the fluid, the flux tubes also move with the fluid and the flux through every flux tube in the fluid remains constant as the tube moves about in space. This in turn implies that *the magnetic field lines cannot change their topology*, or connectivity, since doing so would require violation of the integrity of fluid elements.

If at $t = 0$ we start at a point $\mathbf{x}_0(0) = (x_0(0), y_0(0), z_0(0))$, and integrate the field line equations

$$\frac{dx}{B_x} = \frac{dy}{B_y} = \frac{dz}{B_z} = \frac{dl}{B} \tag{9.11}$$

for a distance L, we arrive at a point $\mathbf{x}_1(0) = (x_1(0), y_1(0), z_1(0))$; the field line can be thought of as a *mapping* $\mathbf{x}_0 \Rightarrow \mathbf{x}_1$. During a time Δt, these two points will have moved to $\mathbf{x}_0(\Delta t)$ and $\mathbf{x}_1(\Delta t)$ according to $d\mathbf{x}_i/dt = \mathbf{V}_i$. If we then integrate Eq. (9.11) starting from the point $\mathbf{x}_0(\Delta t)$, in ideal MHD the trajectory of this field line must also pass through $\mathbf{x}_1(\Delta t)$, since each point on the field line can be thought of as moving with the local fluid velocity. Thus, in ideal MHD, the mapping $\mathbf{x}_0(t) \Rightarrow \mathbf{x}_1(t)$ is valid at all times. We say that a field line (or flux tube) retains its identity for all times, i.e., again, its topology cannot change.

We comment on the *boundary conditions* for ideal MHD. We consider the boundary of the domain as a "wall" with a unit normal vector $\hat{\mathbf{n}}$. From electromagnetic theory, we know that the tangential component of $\mathbf{E}$ and the normal component of $\mathbf{B}$ must be continuous across an interface. Applying these across the boundary, we have

$$\mathbf{E}_{\text{tan}}|_W \equiv \hat{\mathbf{n}} \times \mathbf{E}|_W \text{ is continuous} \tag{9.12}$$

and

$$B_n|_W \equiv \mathbf{B} \cdot \hat{\mathbf{n}}|_W \text{ is continuous.} \tag{9.13}$$

If these values are known just outside the boundary, they can be applied just inside the boundary as boundary conditions on the ideal MHD equations. By applying Faraday's law in the two-dimensional plane of the boundary, we find

$$\left. \frac{\partial B_n}{\partial t} \right|_W = - \nabla \times \mathbf{E}_{\text{tan}}|_W . \tag{9.14}$$

If $\nabla \times \mathbf{E}_{\text{tan}}|_W = 0$, then B_n is a constant and is determined by the initial conditions. A common case is $B_n = 0$, so the field does not penetrate the wall. If $\nabla \times \mathbf{E}_{\text{tan}}|_W \neq 0$, then B_n evolves in time according to Eq. (9.14). If the boundary is a *perfectly conducting wall*, it cannot support an electric field, so $\hat{\mathbf{n}} \times \mathbf{E}|_W = 0$. From the ideal MHD Ohm's law $\mathbf{E} = -\mathbf{V} \times \mathbf{B}$, we have

$$\hat{\mathbf{n}} \times \mathbf{E} = \mathbf{B}(\hat{\mathbf{n}} \cdot \mathbf{V}) - \mathbf{V}(\hat{\mathbf{n}} \cdot \mathbf{B}), \tag{9.15}$$

so that $\mathbf{V}$ must satisfy the condition

$$\hat{\mathbf{n}} \cdot \mathbf{V}|_W = \hat{\mathbf{n}} \cdot \left. \frac{\mathbf{E} \times \mathbf{B}}{B^2} \right|_W + \left. \frac{(\hat{\mathbf{n}} \cdot \mathbf{B})(\mathbf{B} \cdot \mathbf{V})}{B^2} \right|_W . \tag{9.16}$$

The first term on the right-hand side is just the Poynting flux arising from an externally applied electric field. The second term introduces complications when $B_n \neq 0$, for it allows coupling between the tangential and normal components of the velocity. (It is usually ignored, but this can lead to complications because the boundary conditions are then not compatible with Ohm's law.) We have seen that the density and the total energy satisfy conservation equations, so it is necessary to specify their fluxes at the wall. When $\hat{\mathbf{n}} \cdot \mathbf{V}|_W = 0$ both of these fluxes vanish. When $\hat{\mathbf{n}} \cdot \mathbf{V}|_W \neq 0$ (as, for example, when there is an electric field at the wall), then, again, the situation is more complicated.

We summarize the *ideal MHD boundary conditions* for the commonly seen case of a *perfectly conducting wall with no normal magnetic field*:

$$\mathbf{B} \cdot \hat{\mathbf{n}}|_W = 0, \tag{9.17}$$

$$\hat{\mathbf{n}} \times \mathbf{E}|_W = 0, \tag{9.18}$$

and

$$\hat{\mathbf{n}} \cdot \mathbf{V}|_W = 0. \tag{9.19}$$

These are sufficient to determine the solution of the ideal MHD equations.

Lecture 10
Resistivity and Viscosity

Resistance is futile.

The Borg, *Star Trek: The Next Generation*

In this lecture, we briefly discuss the effect of resistivity and viscosity on the dynamics of a magnetized fluid.

We just proved that the change in magnetic flux passing through a co-moving closed circuit is

$$\frac{d\Psi}{dt} = -\oint_C (\mathbf{E} + \mathbf{V} \times \mathbf{B}) \cdot d\mathbf{l}. \tag{10.1}$$

Since in ideal MHD $\mathbf{E} + \mathbf{V} \times \mathbf{B} = 0$, we have $d\Psi/dt = 0$, and we say that the flux is "frozen in" the fluid.

However, in the more general MHD case when the fluid is no longer a perfect electrical conductor, $\mathbf{E} + \mathbf{V} \times \mathbf{B} = \eta\mathbf{J}$ and

$$\frac{d\Psi}{dt} = -\oint_C \eta\mathbf{J} \cdot d\mathbf{l} \neq 0, \tag{10.2}$$

so that the frozen flux condition no longer applies. This is called *resistive MHD*. In this case, the fluid can "move" separately from the field and the field lines can "slip across" the fluid. We will eventually see that this can be an important effect, even when the resistivity is very small.

In resistive MHD, the combination of Faraday's law and Ohm's law becomes

$$\frac{\partial \mathbf{B}}{\partial t} = \underbrace{\nabla \times (\mathbf{V} \times \mathbf{B})}_{\text{Ideal MHD}} - \underbrace{\nabla \times \left(\frac{\eta}{\mu_0} \nabla \times \mathbf{B} \right)}_{\text{Resistive modification}}. \tag{10.3}$$

The first term is just ideal MHD. The second term is a modification introduced when the electrical conductivity $\sigma = 1/\eta$ is finite (rather than infinite). When $\eta = $ constant, the last term can be written as

Schnack, D.D.: *Resistivity and Viscosity.* Lect. Notes Phys. **780**, 55–64 (2009)
DOI 10.1007/978-3-642-00688-3_10   © Springer-Verlag Berlin Heidelberg 2009

$$\nabla \times \left(\frac{\eta}{\mu_0} \nabla \times \mathbf{B} \right) = \frac{\eta}{\mu_0} \nabla \times \nabla \times \mathbf{B}$$

$$= \frac{\eta}{\mu_0} \left[\nabla \left(\nabla \cdot \mathbf{B} \right) - \nabla^2 \mathbf{B} \right]$$

$$= -\frac{\eta}{\mu_0} \nabla^2 \mathbf{B},$$

so that Eq. (10.3) becomes

$$\frac{\partial \mathbf{B}}{\partial t} = \nabla \times (\mathbf{V} \times \mathbf{B}) + \frac{\eta}{\mu_0} \nabla^2 \mathbf{B}. \tag{10.4}$$

The effect of resistivity is to introduce *diffusion* of the magnetic field, with a diffusion coefficient $D_\eta = \eta/\mu_0$ (m^2/s). The characteristic time scale for the diffusion of structures with length scale L is

$$\tau_R = L^2 D_\eta = \mu_0 L^2 / \eta. \tag{10.5}$$

This is called the *resistive diffusion time*.

We will soon see that the characteristic time scale associated with *ideal* ($\eta = 0$) MHD processes is the *Alfvén time*

$$\tau_A = \frac{L}{V_A}, \tag{10.6}$$

where $V_A^2 = B^2/\mu_0 \rho$ is the square of the *Alfvén velocity*. The ratio of the resistive and ideal MHD time scales is called the *Lundquist number*

$$S = \frac{\tau_R}{\tau_A} = \mu_0 \frac{L V_A}{\eta}. \tag{10.7}$$

It turns out that for many (but not all) MHD situations, $S \gg 1$. The Lundquist number plays an important role in describing the dynamics of hot magnetized plasmas. We will return to the Lundquist number and its importance when we discuss magnetic reconnection later in this course.

We now inquire as to the overall effect of electrical resistivity on plasma confinement. We will discuss confinement in more detail when we discuss MHD equilibrium states. For now, we assume that we can attain a state of quasi-force balance in which the plasma is contained (or confined) within a magnetic field, with more plasma on the "inside" and more field on the "outside," as sketched in Fig. 10.1

By quasi-force balance, we mean that the time on which the system evolves is more longer than the time required for wave propagation across the system (all concepts to be defined later), so that inertia ($\rho d\mathbf{V}/dt$) can be neglected in the equation of motion. This approach is difficult (but possible) to justify theoretically, but it is useful and we will adopt it here without justification.

Fig. 10.1 Sketch of a plasma confined by a magnetic field

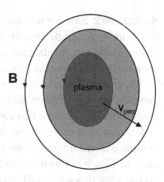

With this *ansatz*, the configuration evolves according to the continuity equation

$$\frac{\partial n}{\partial t} = -\nabla \cdot n\mathbf{V} = -\nabla \cdot n\mathbf{V}_\perp, \tag{10.8}$$

where n is defined by $\rho = Mn$ (M is the ion mass); the equation of motion neglecting inertia,

$$\nabla p = \mathbf{J} \times \mathbf{B}; \tag{10.9}$$

and the steady-state Ohm's law

$$\mathbf{E} = -\mathbf{V} \times \mathbf{B} + \eta\mathbf{J} = \nabla\phi, \tag{10.10}$$

where ϕ is the scalar potential. Here $\mathbf{E}$ can be thought of as an applied electric field. (The last equality assures that $\nabla \times \mathbf{E} = 0$, as is required for steady state.) Taking $\mathbf{B} \times$ Eq. (10.10) and using Eq. (10.9), we find the perpendicular velocity to be

$$\mathbf{V}_\perp = \frac{\mathbf{E} \times \mathbf{B}}{B^2} - \frac{\eta}{B^2}\nabla p. \tag{10.11}$$

Substituting this into Eq. (10.8), we have

$$\frac{\partial n}{\partial t} = \nabla \cdot \left(\frac{n\eta}{B^2}\nabla p\right) - \nabla \cdot \left(n\frac{\nabla\phi \times \mathbf{B}}{B^2}\right). \tag{10.12}$$

The primary effects are illustrated for the isothermal case $p = nT$, with $T =$ constant. Then the density evolves according to

$$\frac{\partial n}{\partial t} = \nabla \cdot (D_n \nabla n) - \nabla \cdot (n\mathbf{V}_E), \tag{10.13}$$

where $D_n = \eta nT/B^2$ is a diffusion coefficient and $\mathbf{V}_E = \nabla\phi \times \mathbf{B}/B^2$ is the MHD velocity resulting from the applied electric field. The first term represents diffusion across the confining field lines and the second is a generally inward convection. The characteristic time for diffusion of the plasma across the size of the system is $\tau_n = B^2 L^2/\eta n T$; we might expect the plasma to be substantially lost from the system on this time scale in the absence of an applied electric field. Therefore, resistivity has a negative effect on plasma confinement.

The concepts illustrated above form the basis of what are called "1 1/2-dimensional transport models." When this calculation is done in a torus with all the geometric complexity that ensues, it is called *Pfirsch–Schlüter transport*.

We now turn to viscosity. In hydrodynamics, a primary effect of viscosity is to introduce internal momentum transport between parts of the fluid that are in relative motion. This was discussed in Lecture 7. Because of this friction, the part of a fluid in contact with a solid wall must assume the velocity of the wall. This introduces the requirement $\mathbf{V}_\| = \mathbf{V}_{W\|}$, where "$\|$" represents the directions parallel to the wall. (If the wall is not moving, the condition on the fluid velocity is $\mathbf{V}_\| = 0$. This is sometimes called the *no-slip* boundary condition.) This additional boundary condition is allowed mathematically because the viscous equations are one order higher in the derivatives of the velocity than the inviscid equations. Differences between the flow far from the wall and the flow at the wall are taken up in a transition region called a *boundary layer*.

A similar situation occurs in a magnetized fluid, and the resistivity plays a role in determining the thickness of the boundary layer. We consider the case of one-dimensional, incompressible, steady flow between two parallel plates located at $x = \pm L$. The plates are prefect electrical conductors. The flow between the plates is $\mathbf{V} = V_z(x)\hat{\mathbf{e}}_z$, which is maintained by a constant imposed pressure gradient $dp/dz = p_0'$. There is an externally produced magnetic field in the x-direction, perpendicular to both the flow and the plates, which penetrates through the plates with normal component B_n. We allow for an induced z-component of the field, so that $\mathbf{B} = B_x(x)\hat{\mathbf{e}}_x + B_z(x)\hat{\mathbf{e}}_x$. The fluid is assumed to have a constant and uniform viscosity and resistivity. The geometry is sketched in Fig. 10.2. Then from $\nabla \cdot \mathbf{V} = 0$ we have $\rho = $ constant, and the governing steady-state equations are

$$-\frac{1}{\rho}\nabla p + \frac{1}{\rho}\mathbf{J} \times \mathbf{B} + \nu\nabla^2\mathbf{V} = 0 \tag{10.14}$$

and

$$\nabla \times (\mathbf{V} \times \mathbf{B}) + \frac{\eta}{\mu_0}\nabla^2\mathbf{B} = 0, \tag{10.15}$$

where $\nu = \mu/\rho$ is the kinematic viscosity (see Lecture 7). For the flow, the no-slip boundary condition is $V_z(\pm L) = 0$. For the field, we are allowed to specify the normal component of $\mathbf{B}$ and the tangential component of $\mathbf{E}$ at the boundary (see Lecture 9). The first is $B_x(\pm L) = B_n$. Tangential electric field vanishes at a conducting boundary. From Ohm's law, $\mathbf{E} = -\mathbf{V} \times \mathbf{B} + \eta\mathbf{J}$, and the velocity boundary condition, we find $\mathbf{E}_{\text{tan}} = \eta\mathbf{J}_{\text{tan}} = 0$. The remaining boundary condition is therefore $dB_z/dx = 0$ at $x = \pm L$.

Fig. 10.2 Flow in a channel formed by two parallel planes with a magnetic field perpendicular to the channel

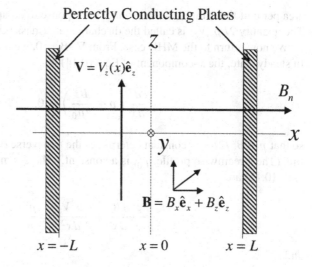

For comparison, we first look at the hydrodynamic case. This occurs when either $\mathbf{B} = 0$ or $\eta \to \infty$; in the latter case, the fluid cannot conduct electric current and the field satisfies $\nabla^2 \mathbf{B} = 0$ (a vacuum field). Then the z-component of Eq. (10.14) is

$$\nu \frac{d^2 V_{z0}}{dx^2} = -\frac{p_0'}{\rho}. \tag{10.16}$$

The subscript "0" denotes that it is the hydrodynamic flow. Recall that p_0' is a constant. The solution of Eq. (10.16) that satisfies the condition $V_{z0}(\pm L) = 0$ is

$$V_{z0}(x) = -\frac{p_0' L^2}{2\nu\rho} \left(1 - \frac{x^2}{L^2}\right). \tag{10.17}$$

Note that the flow profile contains only a single-length scale, the channel half-width L. It is a parabola with its maximum at $x = 0$; a positive flow requires a negative pressure gradient. The mean flow is defined as

$$\langle V_z \rangle = \frac{1}{2L} \int\limits_{-L}^{L} V_z(x) dx. \tag{10.18}$$

In this case we have

$$\langle V_z \rangle_0 = -\frac{1}{3} \frac{L^2}{\nu\rho} p_0'. \tag{10.19}$$

This is called the *Poisieulle formula*. It is an important quantity for flow through channels and pipes. The quantity $\rho \langle V_z \rangle_0$ is the average mass flux (mass per unit area per unit time) through the channel in response to an applied pressure gradient. The quantity $2L\rho \langle V_z \rangle$ is called the discharge rate (mass per unit time).

We now return to the MHD case. From $\nabla \cdot \mathbf{B} = 0$, we have $B_x = \text{constant} = B_n$. In steady state, the x-component of Eq. (10.14) is

$$\frac{d}{dx}\left(p + \frac{B_z^2}{2\mu_0}\right) = 0, \tag{10.20}$$

so that $p + B_z^2/2\mu_0 = \text{constant}$ determines the transverse (to the flow) pressure profile. (The streamwise profile, p_0', is a constant.) The z-components of Eqs. (10.14) and (10.15) are

$$\frac{B_n}{\mu_0 \rho}\frac{dB_z}{dx} + \nu\frac{d^2V_z}{dx^2} = \frac{p_0'}{\rho} \tag{10.21}$$

and

$$B_n\frac{dV_z}{dx} + \frac{\eta}{\mu_0}\frac{d^2B_z}{dx^2} = 0. \tag{10.22}$$

Differentiating Eq. (10.21) with respect to x and using Eq. (10.22), we have

$$\frac{d^3V_z}{dx^3} - \frac{H^2}{L^2}\frac{dV_z}{dx} = 0. \tag{10.23}$$

The non-dimensional parameter

$$H = V_{An}L/\sqrt{\nu D_\eta}, \tag{10.24}$$

where $V_{An} = B_n/\mu_0\rho$ (the Alfvén speed based on the imposed magnetic field) and $D_\eta = \eta/\mu_0$ (the magnetic diffusivity) is called the *Hartmann number*; and the flow we are considering is called *Hartmann flow*. It is the MHD analog of Poissieulle flow in hydrodynamics.[1]

The solution of Eq. (10.23) is of the form $V_z = Ae^{kx} + Be^{-kx} + C$. Substituting this *ansatz* into Eq. (10.23), we find that $k = H/L$. We therefore anticipate that the solutions will exhibit variation on a characteristic length scale $\delta = 1/k = L/H$, which can be small if the Hartmann number is large. Applying the boundary conditions $V_z(\pm L) = 0$, we find

$$V_z(x) = C\left(1 - \frac{\cosh kx}{\cosh kL}\right) = C\left(1 - \frac{\cosh Hx/L}{\cosh H}\right). \tag{10.25}$$

---

[1] See L. D. Landau and E. M. Lifshitz, *Fluid Mechanics*, pp. 55ff, Pergamon Press, London (1959).

The remaining integration constant C is determined by substituting Eq. (10.25) into Eq. (10.21) and requiring $dB_z/dx = 0$ at $x = \pm L$. The result is $C = -p_0'/\rho v k^2$. Then integrating, we find B_z to be

$$B_z(x) = \frac{p_0' L}{V_{An}} B_n \left(\frac{x}{L} - \frac{\sinh Hx/L}{H \cosh H} \right), \tag{10.26}$$

and from Eq. (10.25),

$$V_z(x) = -\frac{p_0' L^2}{v \rho H^2} \left(1 - \frac{\cosh Hx/L}{\cosh H} \right). \tag{10.27}$$

The flow stretches the field lines in the z-direction. This results in a y-directed current density $\mu_0 J_y = -dB_z/dx$, where

$$J_y(x) = -\frac{p_0'}{B_n} \left(1 - \frac{\cosh Hx/L}{\cosh H} \right), \tag{10.28}$$

which has the same profile as the flow. A Lorentz force arises whose z-component, $-B_n J_y$, opposes the flow. The equation for a field line, $\Delta z(x) = z(x) - z(-L)$, is found from integrating the characteristic equation $dz/dx = B_x/B_n$ from $-L$ to x. Using Eq. (10.26), the result is

$$\Delta z(x) = \frac{p_0' L}{V_{An}} \left[\frac{1}{2L} \left(x^2 - L^2 \right) - \frac{L}{H^2 \cosh H} \left(\cosh \frac{Hx}{L} - \cosh H \right) \right]. \tag{10.29}$$

The trajectory of a field line for the case $H = 10$, $L = 1$, and $p_0' L/V_{An} = 1$ is shown in Fig. 10.3.

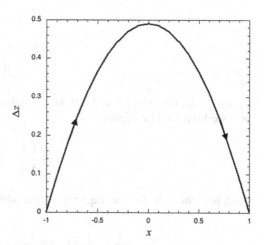

Fig. 10.3 Trajectory of a magnetic field line in Hartmann flow

From Eqs. (10.18) and (10.27), the mean flow in the MHD case is

$$\langle V_z \rangle = -\frac{p_0' L^2}{\nu \rho H^2} \left(1 - \frac{\sinh H}{H \cosh H} \right). \tag{10.30}$$

In the hydrodynamic limit $H \to 0$ (either $B_n \to 0$ or $D_\eta \to \infty$), we recover Eq. (10.19). The ratio of the MHD mean flow to the hydrodynamic Poissieulle formula, Eq. (10.19), is

$$\frac{\langle V_z \rangle}{\langle V_z \rangle_0} = \frac{3}{H^2} \left(1 - \frac{\sinh H}{H \cosh H} \right). \tag{10.31}$$

For $H \ll 1$, $\langle V_z \rangle / \langle V_z \rangle_0 \to 1$, and for $H \gg 1$, $\langle V_z \rangle / \langle V_z \rangle_0 \to 1/H^2$, so that the effect of the magnetic field is to retard the flow. In Fig. 10.4, we plot the ratio $\langle V_z \rangle / \langle V_z \rangle_0$ for a range of Hartmann numbers. This effect can be significant at large Hartmann number (strong field and/or good conductivity). A magnetic field can be used as a non-mechanical valve to throttle the flow of a conducting fluid (such as, perhaps, molten steel) through a channel.

Fig. 10.4 Ratio of the MHD mean flow to the hydrodynamic mean flow as a function of Hartmann number

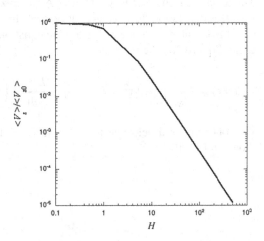

Comparing Eqs. (10.19) and (10.30), we see that the equivalent Poissieulle formula for MHD can be written as

$$\langle V_z \rangle = -\frac{1}{3} \frac{L^2}{\nu_{\text{eff}}} p_0', \tag{10.32}$$

which is of the same form as Eq. (10.19), but with an *effective viscosity* of

$$\nu_{eff} = \frac{\nu H^2}{3 \left[1 - \sinh H / (H \cosh H) \right]}. \tag{10.33}$$

The effect of the magnetic field is to increase the effective viscosity of the fluid flowing in the channel. A plot of v_{eff}/v as a function of the Hartmann number is shown in Fig. 10.5.

Fig. 10.5 Ratio of the effective viscosity to the collisional viscosity as a function of Hartmann number

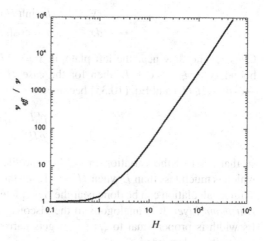

Not only is the discharge rate reduced in the MHD case, but the flow profile itself is flatter. Comparing the slopes of Eqs. (10.17) and (10.27) near the mid-plane $x = 0$, we have

$$\frac{|dV_z/dx|}{|dV_{z0}/dx|} \approx \frac{1}{\cosh H}\left(\frac{H}{L} + \frac{H^2|x|}{6}\right) < 1. \tag{10.34}$$

Plots of the hydrodynamic flow profile and MHD flow profiles for several values of the Hartmann number are shown in Fig. 10.6.

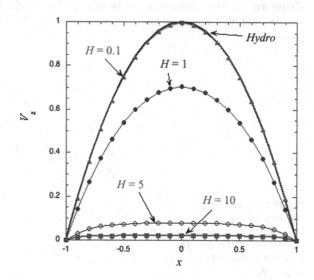

Fig. 10.6 Hydrodynamic flow profile and MHD flow profiles for several values of the Hartmann number. The flows are normalized to $-p_0'L^2/2v\rho$

While the hydrodynamic flow profile is a broad parabola, from Fig. 10.6 we see that, for large H, most of the variation in the MHD flow occurs in a thin region near the plates. From Eq. (10.27), the slope of the MHD profile is

$$\frac{dV_z}{dx} = \frac{p_0'L}{\rho H} \frac{\sinh Hx/L}{\cosh H}. \tag{10.35}$$

Consider the flow near the left plate, at $x = -L$. Defining the distance from the boundary as $\xi = x + L$, then for the case $H \gg 1$, $(\sinh Hx/L)/\cosh H \sim -\exp(-H\xi/L)$ and Eq. (10.35) becomes

$$\frac{dV_z}{dx} \sim -\frac{p_0'L}{\rho H} e^{-H\xi/L}, \tag{10.36}$$

so that most of the variation in the flow profile occurs in a distance $\delta = L/H$, which is much less than L when $H \gg 1$; see the remark preceding Eq. (10.25). (A similar calculation can be done near the right plate at $x = L$.) This layer is called the *Hartmann layer*. It is analogous to the viscous boundary layer of hydrodynamics. Its width is proportional to $\sqrt{\nu D_\eta}$. It gets narrower as either the viscosity or the resistivity is decreased.

Finally, we note from Eq. (10.28) that Hartmann flow generates a net electrical current in the y-direction given by

$$I = \int_{-L}^{L} J_y(x)dx = -\frac{2Lp_0'}{B_n} \left(1 - \frac{\tanh H}{H}\right). \tag{10.37}$$

For $H \gg 1$ we have $I \sim -2Lp_0'/B_n$, while for $H \ll 1$, $I \sim -2Lp_0'H^2/3B_n$. A plot of the normalized current versus Hartmann number is shown in Fig. 10.7. Therefore, a device based on the flow of a conducting fluid across a magnetic field can be used as a non-mechanical current generator. The total current maximizes above about $H = 10$.

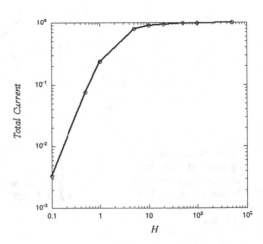

Fig. 10.7 Normalized net current generated by Hartnann flow as a function of the Hartmann number. The current is normalized to $-2p_0'L/B_n$

Lecture 11
Similarity Scaling

> *The universe may have a purpose, but nothing we know*
> *suggests that, if so, it has any similarity to ours.*
> Bertrand Russell

In Lecture 10, we introduced a non-dimensional parameter called the Lundquist number, denoted by S. This is just one of many non-dimensional parameters that can appear in the formulations of both hydrodynamics and MHD. These generally express the ratio of the time scale associated with some dissipative process to the time scale associated with either wave propagation or transport by flow. These are important because they define regions in parameter space that separate flows with different physical characteristics. All flows that have the same non-dimensional parameters behave in the same way. This property is called similarity scaling.

First consider viscous hydrodynamics. The equation of motion is

$$\rho \left(\frac{\partial \mathbf{V}}{\partial t} + \mathbf{V} \cdot \nabla \mathbf{V} \right) = -\nabla p + \rho \nu \nabla^2 \mathbf{V}. \tag{11.1}$$

Here $\nu = \mu/\rho$ is called the *kinematic viscosity*. We now introduce dimensionless variables, signified by a tilde ($\sim$). For example, we choose to measure density in units of ρ_0 (kg/m³), i.e., $\tilde{\rho} = 3$ means $\rho = 3\rho_0$ (kg/m³). We measure lengths in units of L, velocity in units of V_0, pressure in units of p_0, and time in units of t_0. Introducing this ansatz into Eq. (11.1), we have

$$\tilde{\rho} \frac{\partial \tilde{\mathbf{V}}}{\partial \tilde{t}} + \frac{t_0 V_0}{L} \tilde{\rho} \tilde{\mathbf{V}} \cdot \tilde{\nabla} \tilde{\mathbf{V}} = -\frac{t_0}{V_0 \rho_0} \frac{p_0}{L} \tilde{\nabla} \tilde{p} + \frac{t_0 \nu}{L^2} \tilde{\rho} \tilde{\nabla}^2 \tilde{\mathbf{V}}, \tag{11.2}$$

where $\tilde{\nabla} = L\nabla$. We can choose our normalization values so that $t_0 V_0/L = 1$; this sets the characteristic time scale as $t_0 = L/V_0$. Then the coefficient of the first term on the right-hand side becomes $p_0/\rho_0 V_0^2$, which will be unity if we chose to measure the pressure in terms of $p_0 = \rho_0 V_0^2$ (i.e., twice the characteristic kinetic energy). The coefficient of the last term is then $\mu/V_0 L$. It is customary to define the *Reynolds' number* as

$$R_e = \frac{L V_0}{\nu}. \tag{11.3}$$

Schnack, D.D.: *Similarity Scaling*. Lect. Notes Phys. **780**, 65–69 (2009)
DOI 10.1007/978-3-642-00688-3_11   © Springer-Verlag Berlin Heidelberg 2009

so that Eq. (11.2) becomes

$$\frac{\partial \mathbf{V}}{\partial t} + \mathbf{V} \cdot \nabla \mathbf{V} = -\frac{1}{\rho}\nabla p + \frac{1}{R_e}\nabla^2 \mathbf{V}, \tag{11.4}$$

where we have now dropped the tilde notation and all variables are to be considered dimensionless. In this form, it is clear that the solution of Eq. (11.4) depends *only* on the Reynolds' number. That means that *all flows with the same Reynolds' number look the same when scaled to their characteristic variables*. This is called *similarity scaling*. The only thing that matters is the ratio LV_0/μ. Consider a system with characteristic dimension L, flow speed V_0, and kinematic viscosity μ (for example, air with speed V_0 blowing over a ship of length L floating in water), and compare it to another system with the same V_0 (wind speed) and viscosity (air and water), but with length $l \ll L$. The systems will *not* look the same, even qualitatively. This is because the Reynolds' number in the first case is $R_e = LV_0/\nu$, but the Reynolds' number in the second case is $R'_e = lV_0/\nu = (l/L)R_e \ll R_e$; the Reynolds' number is wrong. In order to make them appear the same using the same materials (i.e., air and water), the wind velocity must increase by a factor of L/l. This is why many movie scenes of ships in storms do not look realistic; they were filmed with a model ship in a bathtub, and the Reynolds' number is too small! On the positive side, similarity scaling is the basis for wind tunnel experiments, which have been tremendously important in the development of advanced aircraft.

The kinematic viscosity ν in Eq. (11.1) has the dimensions of a diffusion coefficient, m²/s. Indeed, if we drop the advection and pressure force, the velocity is seen to satisfy a diffusion equation

$$\frac{\partial \mathbf{V}}{\partial t} = \nu \nabla^2 \mathbf{V}.$$

The characteristic time for viscous diffusion is $\tau_\nu = L^2/\nu$, and the ratio of the viscous diffusion time scale to the flow time scale $t_0 = L/V_0$ is $\tau_\nu/t_0 = LV_0/\nu = R_e$. The Reynolds' number is fundamentally a ratio of the characteristic time scales of the system.

Now consider the combination of Faraday's law and Ohm's law (sometimes called the induction equation)

$$\frac{\partial \mathbf{B}}{\partial t} = \nabla \times (\mathbf{V} \times \mathbf{B}) + \frac{\eta}{\mu_0}\nabla^2 \mathbf{B}. \tag{11.5}$$

Introducing non-dimensional variables and applying the same procedure as above, we find

$$\frac{\partial \mathbf{B}}{\partial t} = \nabla \times (\mathbf{V} \times \mathbf{B}) + \frac{1}{R_M}\nabla^2 \mathbf{B}, \tag{11.6}$$

where

$$R_M = \frac{L V_0}{(\eta/\mu_0)} \tag{11.7}$$

is the *magnetic Reynolds' number*. The resistive diffusion time associated with Eq. (11.5) is $\tau_R = L^2/(\eta/\mu_0)$ and so $\tau_R/t_0 = L V_0/(\eta/\mu_0) = R_M$. Again, the magnetic Reynolds' number is the ratio of the resistive diffusion time to the flow time.

If we choose instead $V_0 = V_A$, the (as yet unmotivated) Alfvén velocity, then the magnetic Reynolds' number becomes $S = \tau_R/\tau_A$; this is the *Lundquist number* that was introduced in Lecture 10.

Now consider MHD and, for simplicity of discussion, we let $\rho = \rho_0 =$ constant. We must now retain the Lorentz force $\mathbf{J} \times \mathbf{B}$ in the equation of motion. Measuring the current density in units of $J_0 = B_0/\mu_0 L$ (from $\mu_0 \mathbf{J} = \nabla \times \mathbf{B}$) and transforming to non-dimensional variables, as before, we find that the coefficient of the non-dimensional Lorentz force is

$$\frac{J_0 B_0 t_0}{\rho_0 V_0} = \frac{B_0^2}{\mu_0 \rho_0} \frac{t_0}{L V_0} = \frac{V_A^2}{V_0^2}, \tag{11.8}$$

This strongly suggests measuring the velocity in terms of the Alfvén speed, whose square is $V_A^2 = B_0^2/\mu_0 \rho_0$. Then the pressure is measured in terms of twice the magnetic energy density, $p_0 = \rho_0 V_A^2 = B_0^2/\mu_0$ (which shows that the Alfvén speed is the speed at which the kinetic energy equals the magnetic energy). The Reynolds' number becomes $S_\nu = \tau_\nu/\tau_A = L V_A/\nu$, which we will call the *viscous Lundquist number* (for lack of a better name). With these choices, the (constant density) non-dimensional MHD equations (neglecting energy) become

$$\rho\left(\frac{\partial \mathbf{V}}{\partial t} + \mathbf{V} \cdot \nabla \mathbf{V}\right) = -\nabla p + \mathbf{J} \times \mathbf{B} + \frac{1}{S_\nu} \rho \nabla^2 \mathbf{V} \tag{11.9}$$

and

$$\frac{\partial \mathbf{B}}{\partial t} = \nabla \times (\mathbf{V} \times \mathbf{B}) + \frac{1}{S} \nabla^2 \mathbf{B}. \tag{11.10}$$

Solutions of the coupled MHD system appear the same (i.e., are similar) if both S_ν and S are the same. Situations in which either S_ν or S (or both) are different will behave differently.

There are several other non-dimensional parameters that appear in the literature, which are combinations of S_ν and S. For example, $\mathrm{Pr} = S/S_\nu = \nu/(\eta/\mu_0)$ is called the *magnetic Prandtl number*. It measures the relative effects of viscous and resistive diffusion. Similarly, $H = \sqrt{S S_\nu}$ is called the *Hartmann number* (see Lecture 10). It is important in differentiating regimes in certain MHD flows and also in different operating regimes of some present magnetic fusion experiments.

We have implied that the Reynolds' number (and other non-dimensional parameters) can differentiate regimes in which systems that satisfy the same equations behave quite differently. This can be understood qualitatively as follows. Consider the case of sheared flow. Its effect is to distort the fluid, as shown in Fig. 11.1.

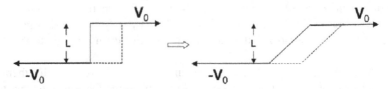

Fig. 11.1 The effect of sheared flow is to stretch, or distort, the fluid

The effect of diffusion is to smooth, or relax, the shear, and hence the distortion, as shown in Fig. 11.2.

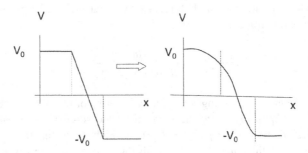

Fig. 11.2 The smoothing effect of diffusion on a distorted flow

Both of these processes are at work simultaneously. The Reynolds' number is the ratio of the time scales associated with the smoothing and distortion processes, $R_e = \tau_v/t_0$. When $R_e \gg 1$ the fluid is distorted faster than it can relax, and when $R_e \ll 1$ the fluid is relaxed faster than it can be distorted. Smoothing and distortion occur on the same time scale when $R_e \sim 1$, or on a length scale $L_0 \sim v/V_0$. Thus, flow with a very large Reynolds' number will tend to look distorted and disorganized, and the velocity field will look "spiky" (also called "turbulent"), while flow with a very low Reynolds' number will be exceedingly smooth, like molasses. Flows with an intermediate Reynolds' number will appear to be smooth, organized, and "laminar." These flow regimes are illustrated in Fig. 11.3. From left to right,

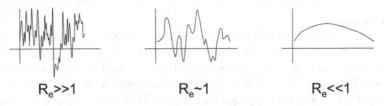

Fig. 11.3 *Left*: "Spiky" structure of flow at large Reynolds' number. *Center*: Laminar structure of flow at a moderate Reynolds' number. *Right*: Smooth structure of flow at a low Reynolds' number

these figures can be thought of either as representing the same scale length with increasing viscosity or representing the same viscosity with decreasing scale length.

Similar remarks apply to the structure of the magnetic field as a function of either the magnetic Reynolds' number R_M or the Lundquist number S. However, in this case the structure in the current density is even sharper than that of the magnetic field, since $J \sim \partial B / \partial x$. The spikes in the structure of the current density are called *current sheets*. These will become of central importance when we discuss reconnection and resistive instabilities.

There are other non-dimensional parameters associated with thermal conduction, rotation, etc., all of which measure the relative importance of various physical effects.

Lecture 12
The Wöltjer Invariants of Ideal MHD, Topological Invariance, Magnetic and Cross-Helicity

The chain of destiny can only be grasped one link at a time.
Winston Churchill

We now return to ideal MHD, so that $\mathbf{E} + \mathbf{V} \times \mathbf{B} = 0$. The magnetic flux through and closed circuit C is

$$\Phi = \int_S \mathbf{B} \cdot \hat{\mathbf{n}} dS, \tag{12.1}$$

where S is any surface bounded by C. Since $\nabla \cdot \mathbf{B} = 0$, we can write $\mathbf{B} = \nabla \times \mathbf{A}$, where $\mathbf{A}$ is the *vector potential*. Then the flux can also be written as

$$\Phi = \int_S \nabla \times \mathbf{A} \cdot \hat{\mathbf{n}} dS = \oint_C \mathbf{A} \cdot d\mathbf{l}. \tag{12.2}$$

Now consider the volume defined by all field lines enclosed by the curve C. This volume V defines a flux tube. The flux Φ within V is constant because $\mathbf{B}$ is everywhere tangent to its boundary. We know that, since $\nabla \cdot \mathbf{B} = 0$, the tube thus defined either closes on itself or fills space ergodically. Any finite volume V_0 contains an infinite number of such flux tubes.

Now consider the following integral:

$$K_l = \int_{V_l} \mathbf{A} \cdot \mathbf{B} dV, \tag{12.3}$$

where V_l is the volume of the lth in V. The flux tube will move about with the fluid velocity $\mathbf{V}$. As it does, Eq. (12.3) changes according to

$$\frac{dK_l}{dt} = \int_{V_l} \left(\frac{\partial \mathbf{A}}{\partial t} \cdot \mathbf{B} dV + \mathbf{A} \cdot \frac{\partial \mathbf{B}}{\partial t} + \mathbf{A} \cdot \mathbf{B} \frac{d}{dt} dV \right). \tag{12.4}$$

Schnack, D.D.: *The Wöltjer Invariants of Ideal MHD, Topological Invariance, Magnetic and Cross-Helicity.* Lect. Notes Phys. **780**, 71–76 (2009)
DOI 10.1007/978-3-642-00688-3_12   © Springer-Verlag Berlin Heidelberg 2009

The last term is evaluated as

$$\frac{d}{dt}dV = \frac{d}{dt}dx_1 dx_2 dx_3 = V_1 dx_2 dx_3 + V_2 dx_2 dx_3 + V_3 dx_1 dx_2$$
$$= \mathbf{V} \cdot \hat{\mathbf{n}} dS. \tag{12.5}$$

Then using Faraday's law, we have

$$\frac{dK_l}{dt} = \int_{V_l} (-\mathbf{E} + \nabla\phi) \cdot \mathbf{B} dV + \int_{V_l} \mathbf{A} \cdot (-\nabla \times \mathbf{E}) \cdot \mathbf{B} dV + \int_{S_l} (\mathbf{A} \cdot \mathbf{B})(\mathbf{V} \cdot \hat{\mathbf{n}}) dS, \tag{12.6}$$

where ϕ is the scalar potential. Now

$$\nabla \cdot (\mathbf{A} \times \mathbf{E}) = \mathbf{E} \cdot \nabla \times \mathbf{A} - \mathbf{A} \cdot \nabla \times \mathbf{E}, \tag{12.7}$$

so that the second integral can be written as

$$\int_{V_l} \mathbf{A} \cdot (\nabla \times \mathbf{E}) \cdot \mathbf{B} dV = \int_{V_l} \mathbf{E} \cdot \mathbf{B} dV - \int_{V_l} \nabla \cdot (\mathbf{A} \times \mathbf{E}) dV$$
$$= \int_{V_l} \mathbf{E} \cdot \mathbf{B} dV - \int_{S_l} (\mathbf{A} \times \mathbf{E}) \cdot \hat{\mathbf{n}} dS. \tag{12.8}$$

Similarly, the first integral can be rewritten as

$$\int_{V_l} \nabla\phi \cdot \mathbf{B} dV = \int_{V_l} \nabla \cdot (\phi\mathbf{B}) dV$$
$$= \int_{S_l} \phi\mathbf{B} \cdot \hat{\mathbf{n}} dS = 0, \tag{12.9}$$

because $\nabla \cdot \mathbf{B} = 0$ and $\mathbf{B} \cdot \hat{\mathbf{n}} = 0$ on S_l by definition since V_l is a flux tube. Therefore

$$\frac{dK_l}{dt} = -2 \int_{V_l} \mathbf{E} \cdot \mathbf{B} dV + \int_{S_l} (\mathbf{A} \times \mathbf{E}) \cdot \hat{\mathbf{n}} dS + \int_{S_l} (\mathbf{A} \cdot \mathbf{B})(\mathbf{V} \cdot \hat{\mathbf{n}}) dS. \tag{12.10}$$

Now invoking ideal MHD, $\mathbf{E} = -\mathbf{V} \times \mathbf{B}$, Eq. (12.10) becomes

$$\frac{dK_l}{dt} = - \int_{S_l} \left[(\mathbf{A} \cdot \mathbf{B})(\mathbf{V} \cdot \hat{\mathbf{n}}) - (\mathbf{A} \cdot \mathbf{V})(\mathbf{B} \cdot \hat{\mathbf{n}}) \right] dS = 0, \tag{12.11}$$

since both $\mathbf{B} \cdot \hat{\mathbf{n}}$ and $\mathbf{V} \cdot \hat{\mathbf{n}}$ vanish on S_l. Therefore $K_l =$ constant *for each and every flux tube in the system*. The K_l are called the *Wöltjer invariants*.[1] They depend on $\mathbf{E} = -\mathbf{V} \times \mathbf{B}$ (ideal MHD) and $\mathbf{B} \cdot \hat{\mathbf{n}} = \mathbf{V} \cdot \hat{\mathbf{n}} = 0$ on S_l. Of the latter two equalities, the first is a property of the flux tube and the second is also a consequence of ideal MHD (the flux tube moves with the fluid).

Fig. 12.1 The topological linking of two flux tubes. This cannot be altered in ideal MHD

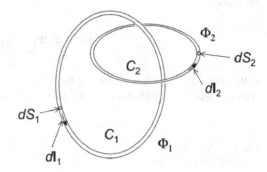

It is possible to give a physical interpretation of the Wöltjer invariants.[2] Consider the linked flux tubes shown in Fig. 12.1. Flux tube C_1 contains flux Φ_1. Flux tube C_2 contains flux Φ_2. The Wöltjer invariant for tube C_1 is

$$K_1 = \int_{V_1} \mathbf{A} \cdot \mathbf{B} dV. \tag{12.12}$$

For this flux tube, we have

$$\begin{aligned}
\mathbf{B} dV &= (B_1\hat{\mathbf{e}}_1 + B_2\hat{\mathbf{e}}_2 + B_3\hat{\mathbf{e}}_3)dx_1dx_2dx_3 \\
&= \hat{\mathbf{e}}_1 dx_1 (B_1 dx_2 dx_3) + \hat{\mathbf{e}}_2 dx_2 (B_2 dx_1 dx_3) + \hat{\mathbf{e}}_3 dx_3 (B_3 dx_1 dx_2) \\
&= (\mathbf{B} \cdot \hat{\mathbf{n}} dS) d\mathbf{l},
\end{aligned} \tag{12.13}$$

so that Eq. (12.12) becomes

$$K_1 = \int_{S_1} \mathbf{B} \cdot \hat{\mathbf{n}} dS \oint_{C_1} \mathbf{A} \cdot d\mathbf{l}. \tag{12.14}$$

The first integral is just Φ_1, the flux contained *within* tube C_1. From Eq. (12.1), the second integral is the flux *enclosed, or linked, by the curve* C_1, which is Φ_2 if the

[1] L. Wöltjer, Proc. Nat. Acad. Sciences **44**, 489 (1958).

[2] See H. K. Moffatt, *Magnetic Field Generation in Electrically Conducting Fluids*, Cambridge University Press, Cambridge, UK (1978).

tubes have "right-hand" linkage, $-\Phi_2$ if the tubes have "left-hand" linkage, and 0 if the tubes are not linked. For now we write

$$K_1 = \Phi_1 \Phi_2. \tag{12.15}$$

Similarly,

$$K_2 = \Phi_2 \Phi_1 = K_1. \tag{12.16}$$

If the tubes are linked N times, we have $K_1 = K_2 = \pm N \Phi_2 \Phi_1$. The same results are obtained for a single-knotted flux tube, as shown in Fig. 12.2.

Fig. 12.2 A very rough sketch of a single-knotted flux tube

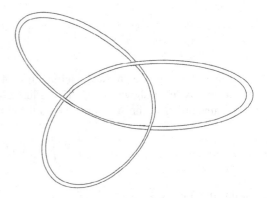

The Wöltjer invariants are thus a direct measure of the linkage, or *topology*, of the flux tubes. Since the K_l are constant in ideal MHD, it means that *the topology of the flux tubes cannot change and is preserved for all time*. This property is called *topological invariance*. (It is really just another way of saying that the magnetic field is co-moving with the fluid.) It is a result of the ideal MHD Ohm's law, $\mathbf{E} + \mathbf{V} \times \mathbf{B} = 0$, and places a very strong constraint on the allowable motions of the fluid.

Now consider a fixed volume V of fluid (no longer a flux tube). The volume integral

$$K_{\mathrm{M}} = \int_V \mathbf{A} \cdot \mathbf{B} dV \tag{12.17}$$

is called the *magnetic helicity* associated with the volume V. We remark that the integrand $\mathbf{A} \cdot \mathbf{B}$ contains the vector potential, and hence depends on the choice of gauge. Letting $\mathbf{A}' = \mathbf{A} + \nabla \chi$, we have

$$K'_M = \int_V \mathbf{A}' \cdot \mathbf{B} dV$$

$$= \int_V (\mathbf{A} + \nabla\chi) \cdot \mathbf{B} dV$$

$$= \int_V \mathbf{A} \cdot \mathbf{B} dV + \int_V \nabla\chi \cdot \mathbf{B} dV$$

$$= K_M + \int_V \nabla \cdot (\chi\mathbf{B}) dV$$

$$= K_M + \oint_S \chi\mathbf{B} \cdot \hat{\mathbf{n}} dS. \tag{12.18}$$

Therefore, $K'_M = K_M$ only if the surface integral vanishes. There are many practical cases where this is true. Examples are periodic boundary conditions or perfectly conducting boundaries. (However, if the geometry is not simply connected, as in a torus, the flux within the fluid may link some external flux, and this must be taken into account. We will discuss more on this when we take on MHD relaxation.)

Nonetheless, in future topics we will find it useful to have a definition of magnetic helicity that is manifestly gauge invariant. This can be obtained by defining

$$K_{M0} = \int_V (\mathbf{A} + \mathbf{A}_0) \cdot (\mathbf{B} - \mathbf{B}_0) dV, \tag{12.19}$$

where $\mathbf{B}_0 = \nabla \times \mathbf{A}_0$ is a reference field, to be defined. Letting $\mathbf{A}' = \mathbf{A} + \nabla\chi$, it is easy to show that

$$K'_{M0} = K_{M0} + \oint_S \chi (\mathbf{B} \cdot \hat{\mathbf{n}} - \mathbf{B}_0 \cdot \hat{\mathbf{n}}) dS. \tag{12.20}$$

If we then choose the reference field such that $\mathbf{B}_0 \cdot \hat{\mathbf{n}} = \mathbf{B} \cdot \hat{\mathbf{n}}$ on S, K_{M0} will be gauge invariant. (This holds true if we also introduce $\mathbf{A}'_0 = \mathbf{A}_0 + \nabla\phi$.) A straightforward calculation then shows that

$$\frac{dK_{m0}}{dt} = -2 \int_V (\mathbf{E} \cdot \mathbf{B} - \mathbf{E}_0 \cdot \mathbf{B}_0) dV, \tag{12.21}$$

where $\mathbf{E}_0 = -\partial\mathbf{A}_0/\partial t$. Then if $\mathbf{E} = -\mathbf{V} \times \mathbf{B}$ and $\mathbf{E}_0 = -\mathbf{V} \times \mathbf{B}_0$, K_{M0} remains constant for all time. This is called the *generalized magnetic helicity*. It is conserved in ideal MHD.

Since generalized helicity is conserved, it is tempting to interpret the integrand $\mathbf{A} \cdot \mathbf{B}$ as a helicity density. This can be misleading. From the discussion of this lecture, it is clear that helicity only has physical meaning as a volume integral. Attempts to assign some physical meaning to the local quantity $\mathbf{A} \cdot \mathbf{B}$ have not led to significant insights.

We close this lecture with the derivation of another integral invariant that explicitly involves both the flow velocity and the magnetic field. It is called the *cross helicity* and appears in theories of MHD turbulence. It is defined as

$$H_C = \int \mathbf{V} \cdot \mathbf{B} dV. \tag{12.22}$$

Assuming ideal MHD and no dissipation, its time derivative is

$$\frac{dH_C}{dt} = \int \left(\mathbf{B} \cdot \frac{\partial \mathbf{V}}{\partial t} + \mathbf{V} \cdot \frac{\partial \mathbf{B}}{\partial t} \right) dV$$

$$= \int \left[\mathbf{B} \cdot \left(-\mathbf{V} \cdot \nabla \mathbf{V} - \frac{1}{\rho} \nabla p + \frac{1}{\rho} \mathbf{J} \times \mathbf{B} \right) + \mathbf{V} \cdot \nabla \times (\mathbf{V} \times \mathbf{B}) \right] dV. \tag{12.23}$$

The third term in the integrand, involving the Lorentz force, vanishes identically. If we assume the adiabatic law $p \sim \rho^{\Gamma}$, the second term is

$$\frac{1}{\rho} \mathbf{B} \cdot \nabla p = \nabla \cdot \left(\frac{\Gamma}{\Gamma - 1} \frac{p}{\rho} \mathbf{B} \right), \tag{12.24}$$

since $\nabla \cdot \mathbf{B} = 0$. Using the vector identities $\mathbf{V} \cdot \nabla \mathbf{V} = \nabla(V^2/2) - \mathbf{V} \times \nabla \times \mathbf{V}$ and $\mathbf{V} \cdot \nabla \times (\mathbf{V} \times \mathbf{B}) = \nabla \cdot [\mathbf{V} \times (\mathbf{V} \times \mathbf{B})] + (\mathbf{V} \times \mathbf{B}) \cdot \nabla \times \mathbf{V}$, the first and fourth terms combine to form a divergence:

$$-\mathbf{B} \cdot (\mathbf{V} \cdot \nabla \mathbf{V}) + \mathbf{V} \cdot \nabla \times (\mathbf{V} \times \mathbf{B}) = -\nabla \cdot \left[\frac{1}{2} V^2 \mathbf{B} - \mathbf{V} \times (\mathbf{V} \times \mathbf{B}) \right]. \tag{12.25}$$

Using the divergence theorem, the rate of change of H_C is

$$\frac{dH_C}{dt} = -\oint_S \hat{\mathbf{n}} \cdot \left[\left(\frac{1}{2} V^2 + \frac{\Gamma}{\Gamma - 1} \frac{p}{\rho} \right) \mathbf{B} - \mathbf{V} \times (\mathbf{V} \times \mathbf{B}) \right] dS, \tag{12.26}$$

which vanishes when $\hat{\mathbf{n}} \cdot \mathbf{B} = \hat{\mathbf{n}} \cdot \mathbf{V} = 0$ on the boundary S. Under these circumstances, the cross-helicity is an invariant in ideal MHD.

Lecture 13
Reduced MHD[1]

> *Our life is frittered away by detail. Simplify, simplify.*
> Henry David Thoreau

One often encounters situations in which the magnetic field is strong and *almost* uni-directional. Since a constant field does not produce a current density, these fields are sometimes said to be *almost potential*. Examples are the magnetic fields in loops in the solar corona and tokamaks. What current density exists arises from small variations of the field from uniformity. This situation is sketched in Fig. 13.1.

Fig. 13.1 An "almost potential" magnetic field

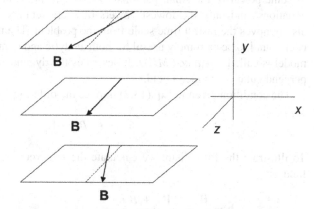

Historically, the development of the reduced MHD model was motivated by some properties of the full MHD equations that we have not yet discussed. These involve issues of force balance and time and space scales associated with various manifestations of plasma dynamics. It is therefore not clear when it is appropriate to introduce reduced MHD into this course of study. I have chosen to present it here, at the end of the development of various MHD models, rather than waiting until the details of the important waves and instabilities have been worked out. This will require us to look ahead a little and anticipate some of the important results. The derivation given here will therefore be more heuristic than formal. I hope this does not lead to too much confusion.

---

[1] Much of this presentation follows that of Dieter Biskamp, *Nonlinear Magnetohydrodynamics*, Cambridge University Press, Cambridge, UK (1993).

Schnack, D.D.: *Reduced MHD*. Lect. Notes Phys. **780**, 77–83 (2009)
DOI 10.1007/978-3-642-00688-3_13   © Springer-Verlag Berlin Heidelberg 2009

When we say that the field is strong, we mean that both the kinetic and internal energy densities are much smaller than the magnetic energy density, i.e.,

$$\rho V^2 \sim p \ll \frac{B^2}{2\mu_0}. \tag{13.1}$$

Since the magnetic field is almost uniform and uni-directional, the field has one almost uniform component (B_z, say, taken to be positive) that is much larger than the other components. It is customary to denote these other components collectively by the notation $\mathbf{B}_\perp$, meaning the components of $\mathbf{B}$ perpendicular to the strong, nearly uniform component. [In this choice of coordinate system, these are the (x, y) components.] The magnetic field is thus characterized by the condition

$$\frac{B_\perp}{B_z} \sim \varepsilon \ll 1. \tag{13.2}$$

We will seek a simplified MHD model that describes the dynamics under these conditions. Formally, the variables appearing in the MHD equations are ordered as some power of the small parameter ε. This *ansatz* is introduced into the MHD equations, and only the lowest powers of ε are retained. The formal procedure also removes the fastest time scale from the problem. The resulting equations have been found to be extremely useful for both analytic and numerical calculations. The model is called a *reduced MHD*. It describes the dynamics of system in the plane perpendicular to the mean field.

The condition given in Eq. (13.2) implies the ordering

$$B_\perp \sim \varepsilon, \quad B_z \sim 1. \tag{13.3}$$

To illustrate the formalism, we calculate the unit vector parallel to the magnetic field as:

$$\begin{aligned}
\hat{\mathbf{b}} = \frac{\mathbf{B}}{B} &= \frac{\mathbf{B}_\perp + B_z \hat{\mathbf{e}}_z}{\left(B_z^2 + B_\perp^2\right)^{1/2}} \\
&= \frac{\varepsilon \mathbf{B}_\perp + B_z \hat{\mathbf{e}}_z}{B_z \left(1 + \varepsilon^2 \frac{B_\perp^2}{B_z^2}\right)^{1/2}} \approx \frac{\varepsilon \mathbf{B}_\perp + B_z \hat{\mathbf{e}}_z}{B_z} \left(1 - \frac{1}{2}\varepsilon^2 \frac{B_\perp^2}{B_z^2}\right) \\
&= \hat{\mathbf{e}}_z + \varepsilon \frac{\mathbf{B}_\perp}{B_z} + O(\varepsilon^2) \approx \hat{\mathbf{e}}_z,
\end{aligned} \tag{13.4}$$

to lowest order in ε.

We now assume that the dynamics of the fluid and the field in the plane perpendicular to the mean field lead to approximate *energy equipartition*, i.e.,

$$\rho V_\perp^2 \sim p \sim \frac{B_\perp^2}{2\mu_0}. \tag{13.5}$$

Since $B_\perp \sim \varepsilon$, we require for consistency

$$V_\perp \sim \varepsilon, \ p \sim \varepsilon^2. \tag{13.6}$$

We will see that the most important motions in systems of this sort have little spatial variation along the mean magnetic field. Most of their spatial structure is in the plane perpendicular to the mean field [the (x, y) plane]. Using this hindsight, we introduce the ordering

$$\frac{\lambda_\parallel}{\lambda_\perp} = \frac{k_\perp}{k_\parallel} \sim \varepsilon, \tag{13.7}$$

which implies that

$$\nabla_\perp \sim 1, \ \frac{\partial}{\partial z} \sim \varepsilon. \tag{13.8}$$

Again with hindsight, we assume that the dynamics parallel to the mean magnetic field occur on a much shorter time scale than the dynamics in the perpendicular plane. (For example, sound waves will propagate rapidly along the field and smooth out significant variations in that direction.) In this case, we expect approximate force balance to be maintained in the parallel direction on the time scale of the perpendicular dynamics, i.e.,

$$\hat{\mathbf{b}} \cdot \nabla \left(p + \frac{B^2}{2\mu_0} \right) \approx 0, \tag{13.9}$$

so that $dV_z/dt \approx 0$ or $V_z \approx$ constant; we choose $V_z = 0$. Using Eq. (13.3), this implies

$$\frac{\partial p}{\partial z} + \frac{1}{\mu_0} B_z \frac{\partial B_z}{\partial z} = 0. \tag{13.10}$$

Since B_z is almost uniform, we can write $B_z = B_{z0} + \tilde{B}_z(x, y, z)$, where $B_{z0} =$ constant. Then

$$\frac{\partial p}{\partial z} + \frac{1}{\mu_0} B_{z0} \frac{\partial \tilde{B}_z}{\partial z} = 0, \tag{13.11}$$

or $p \sim (B_{z0}/\mu_0) \tilde{B}_z$, which, in light of Eq. (13.6), yields the ordering

$$\tilde{B}_z \sim \varepsilon^2. \tag{13.12}$$

Finally, we are interested in situations in which the resistivity is small, so we order

$$\eta \sim \varepsilon. \tag{13.13}$$

The *reduced MHD ordering* is then summarized as

$$V_z = 0, \ \nabla_\perp \sim 1, \frac{\partial}{\partial z} \sim \varepsilon, \eta \sim \varepsilon, \tag{13.14}$$
$$V_\perp \sim \varepsilon, \ p \sim \varepsilon^2, \ \tilde{B}_z \sim \varepsilon^2.$$

It is also customary to take $\rho = \rho_0 = \text{constant} \sim 1$.

We now proceed with the derivation. The current density is

$$\mu_0 \mathbf{J} = \nabla \times \mathbf{B}$$
$$= \left(\nabla_\perp + \varepsilon \hat{\mathbf{e}}_z \frac{\partial}{\partial z} \right) \times (\varepsilon \mathbf{B}_\perp + B_{z0} \hat{\mathbf{e}}_z)$$
$$= \varepsilon \nabla_\perp \times \mathbf{B}_\perp + O(\varepsilon^2),$$

so that

$$\mu_0 J_z \sim \varepsilon, \quad \mu_0 \mathbf{J}_\perp \sim \varepsilon^2. \tag{13.15}$$

The magnetic field is written as

$$\mathbf{B} = \hat{\mathbf{e}}_z \times \nabla \psi + B_{z0} \hat{\mathbf{e}}_z, \tag{13.16}$$

and the condition $\nabla \cdot \mathbf{B} = 0$ is satisfied to the lowest order in ε, i.e.,

$$\nabla \cdot \mathbf{B} = \nabla \cdot (\mathbf{B}_\perp + B_{z0} \hat{\mathbf{e}}_z)$$
$$= \nabla \cdot \mathbf{B}_\perp$$
$$= \nabla \cdot (\hat{\mathbf{e}}_z \times \nabla \psi)$$
$$= \hat{\mathbf{e}}_z \cdot \nabla \times \nabla \psi + \nabla \psi \cdot \nabla \times \hat{\mathbf{e}}_z$$
$$= 0.$$

In this representation, the current density is

$$\mu_0 \mathbf{J} = \nabla \times (\hat{\mathbf{e}}_z \times \nabla \psi) = \hat{\mathbf{e}}_z \nabla^2 \psi + O(\varepsilon),$$

so that

$$\mu_0 J_z = \nabla^2 \psi. \tag{13.17}$$

The flux function ψ is related to the vector potential by

$$\begin{aligned}
\mathbf{B} &= \varepsilon \mathbf{B}_\perp + B_{z0}\hat{\mathbf{e}}_z = \nabla \times \mathbf{A} \\
&= \left(\nabla_\perp + \varepsilon \hat{\mathbf{e}}_z \frac{\partial}{\partial z} \right) \times (\mathbf{A}_\perp + A_z \hat{\mathbf{e}}_z) \\
&= \nabla_\perp \times \mathbf{A}_\perp - \hat{\mathbf{e}}_z \times \nabla A_z + \varepsilon \frac{\partial}{\partial z}(\hat{\mathbf{e}}_z \times \mathbf{A}_\perp) \\
&\approx \underbrace{\hat{\mathbf{e}}_z \times \nabla(-A_z)}_{\perp \text{ direction}} + \underbrace{\nabla_\perp \times \mathbf{A}_\perp}_{z \text{ direction}},
\end{aligned}$$

so that $\psi = -A_z$. Further, since $B_z = B_{z0} + \varepsilon^2 \tilde{B}_z$, we have $\nabla_\perp \times \mathbf{A}_\perp \approx 0$, and we *choose* $\mathbf{A}_\perp = 0$.

Now look at the induction equation (Faraday's law and Ohm's law). Since $\partial \mathbf{B}/\partial t = -\nabla \times \mathbf{E}$ and $\mathbf{E} = -\mathbf{V} \times \mathbf{B} + \eta \mathbf{J}$, using the choices of the previous paragraph we have

$$\hat{\mathbf{e}}_z \frac{\partial A_z}{\partial t} = \mathbf{V} \times \mathbf{B} - \eta \mathbf{J} - \nabla \chi, \tag{13.18}$$

where χ is the scalar potential. The parallel and perpendicular components of Eq. (13.18) are

$$-\frac{\partial \psi}{\partial t} = \hat{\mathbf{e}}_z \cdot \mathbf{V} \times \mathbf{B} - \eta J_z - \frac{\partial \chi}{\partial z} \tag{13.19}$$

and

$$0 = (\mathbf{V} \times \mathbf{B})_\perp - \nabla_\perp \chi. \tag{13.20}$$

In Eq. (13.19), we have used the fact that $\eta \mathbf{J}_\perp \sim \varepsilon^3$. Now, since $V_z = 0$, $\mathbf{V} \times \mathbf{B} = \mathbf{V}_\perp \times \mathbf{B}_\perp + B_{z0}\mathbf{V}_\perp \times \hat{\mathbf{e}}_z$, and therefore

$$\hat{\mathbf{e}}_z \cdot \mathbf{V} \times \mathbf{B} = \hat{\mathbf{e}}_z \cdot \mathbf{V}_\perp \times \mathbf{B}_\perp \tag{13.21}$$

and

$$(\mathbf{V} \times \mathbf{B})_\perp = B_{z0}\mathbf{V}_\perp \times \hat{\mathbf{e}}_z. \tag{13.22}$$

Then using Eqs. (13.20) and (13.22), $\mathbf{V}_\perp \times \hat{\mathbf{e}}_z = \nabla_\perp \chi$, and so

$$\mathbf{V}_\perp = \hat{\mathbf{e}}_z \times \nabla \phi, \tag{13.23}$$

where $\phi = \chi/B_{z0}$ is the *stream function* for the velocity. With this representation we see that

$$\nabla_\perp \cdot \mathbf{V}_\perp = \nabla_\perp \cdot (\hat{\mathbf{e}}_z \times \nabla\phi) = 0, \tag{13.24}$$

so that the flow is incompressible in the perpendicular plane.

Using Eqs. (13.16), (13.17), and (13.21), we find that Eq. (13.19), the parallel component of the induction equation, becomes

$$\frac{\partial\psi}{\partial t} = -\mathbf{V}_\perp \cdot \nabla\psi + \frac{\eta}{\mu_0}\nabla^2\psi - B_{z0}\frac{\partial\phi}{\partial z}. \tag{13.25}$$

Equation (13.25) describes the evolution of the magnetic field. It contains the unknown potential functions ψ and ϕ. It remains to find an equation for the evolution of the flow. For this type of problem (almost two-dimensional) it is convenient to use the z-component of the curl of the equation of motion, i.e.,

$$\hat{\mathbf{e}}_z \cdot \rho_0\left[\frac{\partial}{\partial t}\nabla \times \mathbf{V} + \nabla \times (\mathbf{V} \cdot \nabla\mathbf{V})\right] = \hat{\mathbf{e}}_z \cdot \nabla \times (\mathbf{J} \times \mathbf{B}). \tag{13.26}$$

We use the useful identity $\mathbf{V} \cdot \nabla\mathbf{V} = \nabla\left(V^2/2\right) - \mathbf{V} \times \nabla \times \mathbf{V}$ and define the *vorticity* as $\boldsymbol{\omega} = \nabla \times \mathbf{V}$. Then

$$\begin{aligned}\nabla \times (\mathbf{V} \cdot \nabla\mathbf{V}) &= -\nabla \times (\mathbf{V} \times \boldsymbol{\omega})\\ &= -\boldsymbol{\omega} \cdot \nabla\mathbf{V} + \mathbf{V} \cdot \nabla\boldsymbol{\omega} + \boldsymbol{\omega}\nabla \cdot \mathbf{V}.\end{aligned}$$

Since $V_z = \nabla \cdot \mathbf{V} = 0$, the z-component of this expression is $\mathbf{V} \cdot \nabla\omega$, where $\omega = \hat{\mathbf{e}}_z \cdot \nabla \times \mathbf{V}$ is the z-component of the vorticity or, in light of Eq. (13.23), $\omega = \nabla_\perp^2\phi$. Then Eq. (13.26) can be written as

$$\rho_0\left(\frac{\partial\omega}{\partial t} + \mathbf{V} \cdot \nabla\omega\right) = \hat{\mathbf{e}}_z \cdot \nabla \times (\mathbf{J} \times \mathbf{B}). \tag{13.27}$$

The right-hand side simplifies according to

$$\nabla \times (\mathbf{J} \times \mathbf{B}) = \mathbf{B} \cdot \nabla(J_z\hat{\mathbf{e}}_z),$$

since $\nabla \cdot \mathbf{B} = \nabla \cdot \mathbf{J} = 0$, and $J_z\partial B_z/\partial z \approx 0$ to the lowest order in ε. The resulting equation is

$$\rho_0\left(\frac{\partial\omega}{\partial t} + \mathbf{V} \cdot \nabla\omega\right) = \mathbf{B} \cdot \nabla(J_z). \tag{13.28}$$

The reduced MHD model therefore consists of Eqs. (13.25) and (13.28):

$$\frac{\partial \psi}{\partial t} = -\mathbf{V}_\perp \cdot \nabla \psi + \frac{\eta}{\mu_0} \nabla^2 \psi - B_{z0} \frac{\partial \phi}{\partial z} \qquad (13.29)$$

and

$$\rho_0 \left(\frac{\partial \omega}{\partial t} + \mathbf{V} \cdot \nabla \omega \right) = \mathbf{B} \cdot \nabla (J_z). \qquad (13.30)$$

These equations are closed by the subsidiary relations

$$\omega = \nabla_\perp^2 \phi, \qquad (13.31)$$

$$\mu_0 J_z = \nabla^2 \psi, \qquad (13.32)$$

and

$$\mathbf{V}_\perp = \hat{\mathbf{e}}_z \times \nabla \phi. \qquad (13.33)$$

We note the following:

1. Reduced MHD consists of six equations in the six unknowns $\psi, \omega, \phi, J_z,$ and $\mathbf{V}_\perp$.
2. We need to only deal with scalar functions.
3. There are no parallel dynamics; these fast time scales have been ordered out of the problem.
4. Two of the Eqs. (13.31) and (13.32) are of the Poisson type.
5. The pressure does not enter the equations; it has been ignored since $\beta = 2\mu_0 p / B_z^2 \sim \varepsilon^2$. However, the model can be extended to include the ordering $\beta \sim \varepsilon$ (called the "finite-β" equations).
6. Reduced MHD forms the basis of much of modern tokamak theory.

Lecture 14
Equilibrium: General Considerations—The Virial Theorem

Be still my beating heart.

Sting

In MHD we are often interested in situations of equilibrium, or force balance. This is because, in magnetic fusion we seek to confine a hot plasma for a very long time. Clearly, a state of equilibrium is a minimum condition for this type of fusion to occur.

We distinguish between the cases stationary and non-stationary equilibrium. In stationary equilibrium, the flow velocity vanishes and the condition for force balance is $d\mathbf{V}/dt = \partial\mathbf{V}/\partial t = \mathbf{F}/\rho = 0$. In non-stationary equilibrium there is a finite flow velocity, so the condition is $\partial\mathbf{V}/\partial t = 0 = \mathbf{F}/\rho - \mathbf{V} \cdot \nabla\mathbf{V}$. From now on, when we talk about equilibrium we mean *stationary* equilibrium (unless otherwise noted).

In hydrodynamics, stationary equilibrium is relatively simple. Consider the case of a fluid in a gravitational field. The condition for stationary equilibrium is $-\nabla p + \rho\mathbf{g} = 0$. In one spatial dimension, with $\mathbf{g} = -g\hat{\mathbf{e}}_x$,

$$\frac{dp}{dx} = -\rho g \tag{14.1}$$

or

$$p(x) = p_0 - g \int_{x_0}^{x} \rho(x')dx'. \tag{14.2}$$

We can find the pressure if the density profile is specified or if there is a relationship $p = p(\rho)$. In the latter case $dp/dx = C_s^2 d\rho/dx$, where $C_s^2 \equiv dp/d\rho$ is the square of the sound speed. If p and ρ are related linearly, then $C_s^2 = $ constant, and the solution is

$$\rho(x) = \rho_0 e^{-gx/C_s^2}. \tag{14.3}$$

For the case of a non-stationary equilibrium in hydrodynamics, we have $\rho\mathbf{V} \cdot \nabla\mathbf{V} = -\nabla p$, so that finite flow can lead to non-uniform pressure. In one dimension, this is just

Schnack, D.D.: *Equilibrium: General Considerations—The Virial Theorem*. Lect. Notes Phys. **780**, 85–89 (2009)
DOI 10.1007/978-3-642-00688-3_14　　　　　© Springer-Verlag Berlin Heidelberg 2009

$$\frac{d}{dx}\left(p + \frac{1}{2}\rho V_x^2\right) = 0 \tag{14.4}$$

or $p + \rho V_x^2/2 = $ constant. This is a special case of *Bernoulli's theorem*.

In ideal MHD, the pressure gradient can be balanced by the Lorentz force. This is the basis for "magnetic confinement." This situation is of great importance and must be studied in more detail. The condition for stationary equilibrium in ideal MHD is

$$\nabla p = \mathbf{J} \times \mathbf{B}. \tag{14.5}$$

We note immediately that $\mathbf{B} \cdot \nabla p = 0$ and $\mathbf{J} \cdot \nabla p = 0$, so that the pressure gradient must be perpendicular to both the magnetic field $\mathbf{B}$ and the current density $\mathbf{J}$. The first condition means that *the pressure must be constant along the direction of the magnetic field*. This means that magnetic field lines lie everywhere within regions of constant pressure and implies the possibility that these regions could be two-dimensional surfaces. The second condition means that the existence of a current that is not parallel to the magnetic field requires a pressure gradient and vice versa.

We will immediately be more general.[1] In MHD, the momentum evolves according to

$$\frac{\partial}{\partial t}\rho \mathbf{V} = -\nabla \cdot \mathbf{T}, \tag{14.6}$$

so that the condition for equilibrium can be expressed as

$$\nabla \cdot \mathbf{T} = 0 \tag{14.7}$$

or

$$\frac{\partial T_{ki}}{\partial x_k} = 0, \tag{14.8}$$

where, with $\mathbf{V} = 0$,

$$T_{ik} = T_{ki} = \left(p + \frac{B^2}{2\mu_0}\right)\delta_{ik} - \frac{1}{\mu_0}B_i B_k \tag{14.9}$$

is the total stress tensor. It is convenient to rewrite T_{ik} as

$$T_{ik} = \left(p + \frac{B^2}{2\mu_0}\right)\delta_{ik} - \frac{B^2}{\mu_0}\frac{B_i B_k}{B^2}. \tag{14.10}$$

---

[1] The discussion of the Virial Theorem follows that of V. D. Shafranov, "Plasma Equilibrium in a Magnetic Field", in *Reviews of Plasma Physics*, M. A. Leontovich (ed.), Consultants Bureau, New York (1965).

Then we can define the effective *perpendicular* and *parallel pressures* as

$$p_\perp = p + \frac{B^2}{2\mu_0} \tag{14.11}$$

and

$$p_\parallel = p - \frac{B^2}{2\mu_0}, \tag{14.12}$$

so that $p_\parallel - p_\perp = -B^2/\mu_0$. The stress tensor can then be written as

$$T_{ik} = p_\perp \delta_{ik} + \left(p_\parallel - p_\perp\right) \frac{B_i B_k}{B^2}. \tag{14.13}$$

Therefore, in MHD the stress tensor acts as though the pressure were *anisotropic*. For the special case $\mathbf{B} = B\hat{\mathbf{e}}_z$, the components of the stress tensor are

$$\begin{array}{lll} T_{xx} = p_\perp & T_{xy} = 0 & T_{xz} = 0 \\ T_{yx} = 0 & T_{yy} = p_\perp & T_{yz} = 0 \\ T_{zx} = 0 & T_{zy} = 0 & T_{zz} = p_\parallel \end{array}$$

Since $p_\perp > p_\parallel$, the fluid feels a greater pressure perpendicular to the magnetic field than it does parallel to the field. This behaviur is completely different from our usual experience with an unmagnetized fluid.

We now consider the expression

$$\begin{aligned} \frac{\partial}{\partial x_k}(x_i T_{ik}) &= T_{ik}\frac{\partial x_i}{\partial x_k} + x_i \frac{\partial T_{ik}}{\partial x_k} \\ &= T_{ik}\delta_{ik} \\ &= T_{ii}, \end{aligned} \tag{14.14}$$

where we have used the equilibrium condition, Eq. (14.8). (Note that T_{ii} is the *trace* of the tensor $\mathbf{T}$, i.e., the sum of the diagonal elements.)

Now introduce an *isolated* fluid permeated by a magnetic field. In equilibrium, Eq. (14.14) must hold. Integrating over a volume and applying Gauss' theorem, we have

$$\int_V T_{ii} dV = \oint_S x_i T_{ik} dS_k, \tag{14.15}$$

where V is the volume of integration and S is the surface area of that volume (not necessarily the same as the volume and surface of the fluid). Evaluating the terms in Eq. (14.15), we find

$$\int_V 3\left(p + \frac{B^2}{2\mu_0}\right) dV = \oint_S \mathbf{x} \cdot \left[\left(p + \frac{B^2}{2\mu_0}\right)\mathbf{I} - \frac{\mathbf{BB}}{\mu_0}\right] \cdot d\mathbf{S}. \tag{14.16}$$

The left-hand side of this equation is positive definite. We now seek to estimate the sign and magnitude the right-hand side.

We consider the case where all the current density $\mathbf{J}$ and the pressure p are completely *within* the fluid, and the surface S is completely *outside* the fluid. The magnetic field arises from currents within the plasma only. This is shown in Fig. 14.1. We now take $S \rightarrow \infty$, so that only terms involving $\mathbf{B}$ contribute to the right-hand side of Eq. (14.16). The vector potential due to currents within the fluid is

$$\mathbf{A}(\mathbf{r}) = \frac{\mu_0}{4\pi} \int_V \frac{\mathbf{J}(\mathbf{r}')}{|\mathbf{r} - \mathbf{r}'|} dV'. \tag{14.17}$$

If $\mathbf{r}$ is far from the source $\mathbf{r}'$, we can expand $1/|\mathbf{r} - \mathbf{r}'|$ as

$$\frac{1}{|\mathbf{r} - \mathbf{r}'|} = \frac{1}{|\mathbf{r}|} + \frac{\mathbf{r} \cdot \mathbf{r}'}{|\mathbf{r} - \mathbf{r}'|^3} + \cdots \cdots \tag{14.18}$$

Inserting this expansion into Eq. (14.17), and using the fact that, since $\nabla \cdot \mathbf{J} = 0$,

$$\int_V \mathbf{J}(\mathbf{r}') dV' = 0, \tag{14.19}$$

we find that $A \sim 1/r^2$, $B \sim 1/r^3$ (since $\mathbf{B} = \nabla \times \mathbf{A}$), and $B^2 \sim 1/r^6$. The other term is $\mathbf{x} \cdot d\mathbf{S}$, which scales like r^3, so the right-hand side of Eq. (14.16) scales like $1/r^3$, which vanishes as $r \rightarrow \infty$.

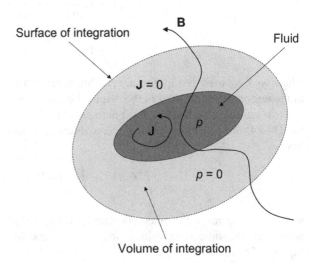

Fig. 14.1 An electrically conducting fluid permeated by a magnetic field. The surface of integration lies completely outside the fluid

Then, as $S \rightarrow \infty$, the left-hand side of Eq. (14.16) remains positive, while the right-hand side goes to zero. This contradiction means that, under the stated conditions, Eq. (14.8) cannot be satisfied and *equilibrium is impossible*! This important and general result is called the *Virial Theorem*. It says that *a magnetized fluid cannot be in MHD equilibrium under forces generated by its own internal currents*. It implies that any MHD equilibrium must be supported by external currents.

Of course, the Virial Theorem is satisfied in all laboratory experiments that contain external coils to produce magnetic fields. It may not be satisfied under astrophysical conditions. However, we already know that the universe is a dynamic place!

Lecture 15
Simple MHD Equilibria

*The path of precept is long, that of example is short and
effectual.*

Seneca

In this lecture we will examine some simple examples of MHD equilibrium configurations. These will all be in cylindrical geometry. They form the basis for more
complicated equilibrium states in toroidal geometry.

Many MHD equilibrium configurations (including tokamaks, spheromaks, and
RFPs) are based on the *pinch effect*, which results from the attractive nature of
parallel currents (exceptions are mirrors and stellarators). Consider two elements
carrying currents $\mathbf{J}_1$ and $\mathbf{J}_2$ in the z-direction, as shown in Fig. 15.1.

Fig. 15.1 Illustration of the
attractive nature of the
Lorentz force between two
elements carrying parallel
currents

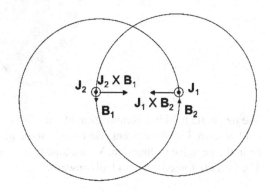

The magnetic field $\mathbf{B}_1$ produced by $\mathbf{J}_1$ encircles element 1 according to the right-
hand rule, and similarly for $\mathbf{B}_2$ and $\mathbf{J}_2$. The Lorentz force $\mathbf{J}_1 \times \mathbf{B}_2$ acting in element 1
is directed toward element 2. Similarly, the Lorentz force $\mathbf{J}_2 \times \mathbf{B}_1$ acting on element
2 is directed toward element 1.

If the current $\mathbf{J}$ is distributed continuously in space, the net effect of the Lorentz
force will be to pull the fluid together, compressing it and thereby increasing the
pressure. This process will cease when the increase in the pressure force tending
to expand the fluid just balances the Lorentz force tending to compress the fluid or
$\nabla p = \mathbf{J} \times \mathbf{B}$. If the current flows in a column, the column will tend to contract
or *pinch* in a direction perpendicular to its axis until the equilibrium condition is
reached. This is called the pinch effect, which is shown in Fig. 15.2.

Schnack, D.D.: *Simple MHD Equilibria*. Lect. Notes Phys. **780**, 91–98 (2009)
DOI 10.1007/978-3-642-00688-3_15   © Springer-Verlag Berlin Heidelberg 2009

Fig. 15.2 Illustration of the
pinch effect, which tends to
make the current channel
contract

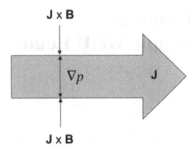

Equilibrium configurations based on the pinch effect are named after the direction
of the current, not the magnetic field. We will now examine several of these in
cylindrical geometry.

In the *theta-pinch* (or θ-pinch), the current flows only in the azimuthal, or θ,
direction. This produces a magnetic field in the z-direction, as shown in the Fig. 15.3.

Fig. 15.3 A theta-pinch
equilibrium

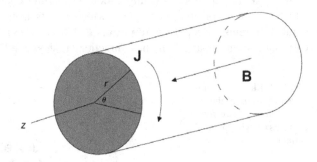

The magnetic field is a combination of a uniform, externally generated field and the
field produced by the current. We take B_z to be in the positive z-direction and J_θ to
be in the negative θ-direction. We assume that all quantities are functions of r only.
The Lorentz force is then radially inward, i.e.,

$$\mathbf{J} \times \mathbf{B} = -J_\theta \hat{\mathbf{e}}_\theta \times B_z \hat{\mathbf{e}}_z = -J_\theta B_z \hat{\mathbf{e}}_r. \tag{15.1}$$

We will generally use three equations to analyze an equilibrium configuration.[1]
These are the following:

1. $\nabla \cdot \mathbf{B} = 0$.
2. Ampére's law, $\mu_0 \mathbf{J} = \nabla \times \mathbf{B}$.
3. Force balance, $\nabla p = \mathbf{J} \times \mathbf{B}$.

---

[1] This discussion follows that of Jeffrey P. Freidberg, *Ideal Magnetohydrodynamics*, Plenum Press,
New York (1987).

These are now considered for the case of the theta-pinch as follows:

1. $\nabla \cdot \mathbf{B} = 0$. In cylindrical geometry, this is

$$\frac{1}{r}\frac{\partial}{\partial r}(r B_r) + \frac{1}{r}\frac{\partial B_\theta}{\partial \theta} + \frac{\partial B_z}{\partial z} = 0. \tag{15.2}$$

Since the configuration depends only on r, and $B_r = 0$, we require

$$\frac{\partial B_z}{\partial z} = 0, \tag{15.3}$$

which is satisfied automatically.

2. *Ampére's law*, $\mu_0 \mathbf{J} = \nabla \times \mathbf{B}$. Under these conditions, this becomes

$$\mu_0 J_\theta = -\frac{d B_z}{dr}. \tag{15.4}$$

3. *Force balance*, $\nabla p = \mathbf{J} \times \mathbf{B}$. This becomes

$$\frac{dp}{dr} = J_\theta B_z. \tag{15.5}$$

Substituting Eq. (15.4) into Eq. (15.5), we have

$$\frac{dp}{dr} = B_z \left(-\frac{1}{\mu_0}\frac{d B_z}{dr} \right) = -\frac{d}{dr}\left(\frac{B_z^2}{2\mu_0} \right)$$

or

$$\frac{d}{dr}\left(p + \frac{B_z^2}{2\mu_0} \right) = 0. \tag{15.6}$$

The second term in parentheses is the magnetic pressure. This equation can be integrated to yield

$$p + \frac{B_z^2}{2\mu_0} = \frac{B_0^2}{2\mu_0}, \tag{15.7}$$

so that $B_z = B_0$ when $p = 0$, i.e., outside the fluid. The constant B_0 is thus the externally generated component of the axial magnetic field. Note that Eq. (15.7) is a single equation containing two unknowns, B_z and p. We are free to specify one and then determine the other. This will be a general property of MHD equilibria. An example of a solution of Eq. (15.7) is

$$p(r) = p_0 e^{-r^2/a^2} \tag{15.8}$$

and

$$B_z(r) = B_0 \left(1 - \beta_0 e^{-r^2/a^2}\right)^{1/2}, \tag{15.9}$$

where $\beta_0 = 2\mu_0 p_0/B_0^2$ and $r = a$ is the radius of the outer boundary. These solutions are sketched (very roughly!) in Fig. 15.4.

Fig. 15.4 Rough sketch of pressure and magnetic field profiles in a theta-pinch

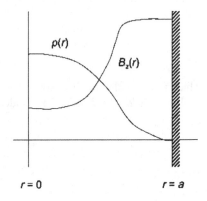

We now consider the *linear z-pinch*. The current now flows in the z-direction and the magnetic field is in the θ-direction, as shown in Fig. 15.5.

Fig. 15.5 A z-pinch equilibrium

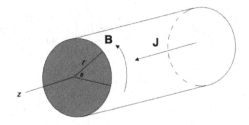

We again assume that there is only r-dependence and proceed as with the θ-pinch.

1. $\nabla \cdot \mathbf{B} = 0$.

$$\frac{1}{r}\frac{\partial}{\partial r}(r B_r) + \frac{1}{r}\frac{\partial B_\theta}{\partial \theta} + \frac{\partial B_z}{\partial z} = 0. \tag{15.10}$$

We have $B_r = B_z = 0$, so we require

$$\frac{\partial B_\theta}{\partial \theta} = 0, \tag{15.11}$$

which is automatically satisfied if $B_\theta = B_\theta(r)$.

2. *Ampére's law,* $\mu_0 \mathbf{J} = \nabla \times \mathbf{B}$.

$$\mu_0 J_z = \frac{1}{r}\frac{d}{dr}(r B_\theta) . \tag{15.12}$$

3. *Force balance,* $\nabla p = \mathbf{J} \times \mathbf{B}$.

$$\frac{dp}{dr} = -J_z B_\theta. \tag{15.13}$$

Using Eq. (15.12),

$$\frac{dp}{dr} = -\frac{B_\theta}{\mu_0 r}\frac{d}{dr}(r B_\theta)$$

$$= -\frac{B_\theta}{\mu_0}\frac{d B_\theta}{dr} - \frac{B_\theta^2}{\mu_0 r}$$

or

$$\frac{d}{dr}\left(p + \frac{B_\theta^2}{2\mu_0}\right) = -\frac{B_\theta^2}{\mu_0 r}. \tag{15.14}$$

This looks like the result for the θ-pinch, Eq. (15.16), with the addition of a term on the right-hand side. This term is called the hoop stress and arises from the curvature of the magnetic field lines. (In the θ-pinch, the field lines are straight.)

Consider the curve shown in Fig. 15.6.

Fig. 15.6 Illustration of the change in the tangent vector along a magnetic field line

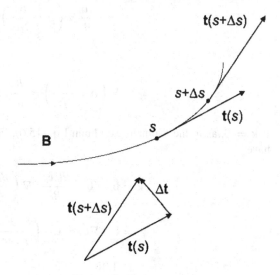

Let s be the distance along the curve and define $\mathbf{t}$ as a unit vector tangent to the curve. The *curvature vector* is then defined as

$$\kappa = \frac{d\mathbf{t}}{ds}, \tag{15.15}$$

the rate of change of the tangent vector as we move along the curve. The *radius of curvature* at a point s is defined as

$$R_c = \frac{1}{|\kappa|}. \tag{15.16}$$

If the curve is a magnetic field line, the unit tangent vector is $\hat{\mathbf{b}} = \mathbf{B}/B$, and $d/ds = \hat{\mathbf{b}} \cdot \nabla$, so the *curvature of a magnetic field line* is

$$\kappa = \hat{\mathbf{b}} \cdot \nabla \hat{\mathbf{b}}. \tag{15.17}$$

A straightforward calculation yields

$$\hat{\mathbf{b}} \times \left(\nabla \times \hat{\mathbf{b}} \right) = -\hat{\mathbf{b}} \cdot \nabla \hat{\mathbf{b}} = -\kappa. \tag{15.18}$$

Then another straightforward calculation using Ampére's law and force balance leads to

$$\kappa = -\hat{\mathbf{b}} \times \left(\nabla \times \hat{\mathbf{b}} \right)$$

$$= \frac{\mu_0}{B^2} \nabla p + \frac{1}{B} \nabla_\perp B$$

$$= \frac{\mu_0}{B^2} \nabla \left(p + \frac{B^2}{2\mu_0} \right)$$

or

$$\nabla \left(p + \frac{B^2}{2\mu_0} \right) = \frac{B^2}{\mu_0} \kappa. \tag{15.19}$$

If $\kappa = 0$, as in the θ-pinch, we obtain Eq. (15.6). For the case of the z-pinch, we have

$$\kappa = \hat{\mathbf{b}} \cdot \nabla \hat{\mathbf{b}} = \frac{B_\theta \hat{\mathbf{e}}_\theta}{B_\theta} \cdot \nabla \left(\frac{B_\theta \hat{\mathbf{e}}_\theta}{B_\theta} \right)$$

$$= \hat{\mathbf{e}}_\theta \cdot \nabla \hat{\mathbf{e}}_\theta = \hat{\mathbf{e}}_\theta \cdot \left(\frac{\hat{\mathbf{e}}_\theta}{r} \frac{\partial \hat{\mathbf{e}}_\theta}{\partial r} \right)$$

$$= \frac{1}{r} \frac{\partial \hat{\mathbf{e}}_\theta}{\partial r} = -\frac{\hat{\mathbf{e}}_r}{r}. \tag{15.20}$$

Then Eq. (15.19) becomes

$$\nabla \left(p + \frac{B^2}{2\mu_0} \right) = -\frac{B^2}{\mu_0 r} \hat{\mathbf{e}}_r$$

or, for our one-dimensional configuration,

$$\frac{d}{dr} \left(p + \frac{B_\theta^2}{2\mu_0} \right) = -\frac{B_\theta^2}{\mu_0 r}, \tag{15.21}$$

which agrees with Eq. (15.14). The hoop stress, or tension force, balances the gradient of the total pressure. Again, this is one equation in two unknowns. One of the unknowns can be specified arbitrarily.

The case that contains both $B_\theta(r)$ and $B_z(r)$ (and, consequently, both J_θ and J_z) is called the *general screw pinch*, because the field lines wrap around the cylinder in a helical fashion, like the threads on a screw. For this configuration:

1. $\nabla \cdot \mathbf{B} = 0$.

$$\frac{1}{r} \frac{\partial B_\theta}{\partial \theta} + \frac{\partial B_z}{\partial z} = 0, \tag{15.22}$$

which is trivially satisfied if the fields are functions of r only.

2. *Ampére's law,* $\mu_0 \mathbf{J} = \nabla \times \mathbf{B}$.

$$\mu_0 J_\theta = -\frac{d B_z}{dr} \tag{15.23}$$

and

$$\mu_0 J_z = \frac{1}{r} \frac{d}{dr} (r B_\theta). \tag{15.24}$$

3. *Force balance,* $\nabla p = \mathbf{J} \times \mathbf{B}$.

$$\frac{dp}{dr} = J_\theta B_z - J_z B_\theta$$

$$= -\frac{d}{dr} \left(\frac{B_z^2}{2\mu_0} \right) - \frac{d}{dr} \left(\frac{B_\theta^2}{2\mu_0} \right) - \frac{B_\theta^2}{\mu_0 r}$$

or

$$\frac{d}{dr} \left(p + \frac{B_\theta^2 + B_z^2}{2\mu_0} \right) = -\frac{B_\theta^2}{\mu_0 r}. \tag{15.25}$$

We now have one equation in three unknowns, so that two of the functions can be specified.

We will follow the same procedure for analyzing the more complicated situation of toroidal equilibrium.

Finally, we remark that in each of the examples considered in this lecture, the cylinder is infinitely long in the z-direction, i.e., each example is purely two-dimensional. Recall that the Virial Theorem proven in Lecture 14 assumed that a surface of integration could be taken completely outside the fluid. This is clearly impossible if the fluid extends to infinity in some direction. We thus do not expect the Virial Theorem to apply to these simple examples.

Lecture 16
Poloidal Beta, Paramagnetism, and Diamagnetism

If you want high pressure, you must choke off waste.
Joseph Farrell

In the lecture we present some fundamental characteristics of MHD equilibria. For simplicity, we illustrate these concepts in cylindrical geometry, although they are generally applicable to other configurations as well.

Consider the equilibrium for the general screw pinch, presented in Lecture 15:

$$\frac{d}{dr}\left(\mu_0 p + \frac{B_\theta^2 + B_z^2}{2}\right) + \frac{B_\theta^2}{r} = 0. \tag{16.1}$$

We envision a plasma with radius r_0 surrounded by a conducting wall with radius a. Integrating Eq. (16.1) from $r = a$ to $r = r_0$, we have

$$2\mu_0 [p(r_0) - p(a)] + B_\theta^2(r_0) - B_\theta^2(a) + B_z^2(r_0) - B_z^2(a) = -2\int_a^{r_0} \frac{B_\theta^2}{r} dr. \tag{16.2}$$

We define the volume average of a function as

$$\langle f \rangle = \frac{2\pi \int_0^a f(r)r\,dr}{2\pi \int_0^a r\,dr} = \frac{2}{a^2}\int_0^a f(r)r\,dr. \tag{16.3}$$

Taking the volume average of Eq. (16.2), using r_0 (the plasma radius) as the independent variable,

$$2\mu_0 [\langle p \rangle - p(a)] + \langle B_\theta^2 \rangle - B_\theta^2(a) + \langle B_z^2 \rangle - B_z^2(a) =$$

$$-2\left(\frac{2}{a^2}\right)\int_0^a r_0\,dr_0 \int_a^{r_0} \frac{B_\theta^2}{r} dr. \tag{16.4}$$

Schnack, D.D.: *Poloidal Beta, Paramagnetism, and Diamagnetism.* Lect. Notes Phys. **780**, 99–101 (2009)
DOI 10.1007/978-3-642-00688-3_16   © Springer-Verlag Berlin Heidelberg 2009

The term on the right-hand side can be integrated by parts according to

$$u = \int_a^{r_0} \frac{B_\theta^2}{r} dr, \qquad du = \frac{du}{dr_0} dr_0 = \frac{B_\theta^2(r_0)}{r_0} dr_0,$$

$$dv = r_0 dr_0, \qquad v = \frac{1}{2} r_0^2.$$

We then have

$$\int_0^a r_0 dr_0 \int_a^{r_0} \frac{B_\theta^2}{r} dr = \left(\frac{1}{2} r_0^2\right) \left(\int_a^{r_0} \frac{B_\theta^2}{r} dr\right)\Bigg|_{r_0=0}^{r_0=a} - \int_0^a \frac{1}{2} r_0^2 \frac{B_\theta^2(r_0)}{r_0} dr_0.$$

In the first term on the right-hand side, the first factor vanishes when $r_0 = 0$ and the second factor vanishes when $r_0 = a$. Therefore

$$\int_0^a r_0 dr_0 \int_a^{r_0} \frac{B_\theta^2}{r} dr = -\frac{1}{2} \int_0^a B_\theta^2(r_0) r_0 dr_0 = -\frac{1}{2} \left(\frac{a^2}{2}\right) \langle B_\theta^2 \rangle$$

$$= -\frac{a^2}{4} \langle B_\theta^2 \rangle. \tag{16.5}$$

Equation (16.4) is then

$$2\mu_0 [\langle p \rangle - p(a)] = B_\theta^2(a) - \langle B_z^2 \rangle + B_z^2(a)$$

or, with $p(a) = 0$,

$$2\mu_0 \langle p \rangle = B_\theta^2(a) + \left[B_z^2(a) - \langle B_z^2 \rangle \right]. \tag{16.6}$$

Defining $\beta_p = 2\mu_0 \langle p \rangle / B_\theta^2(a)$, we have

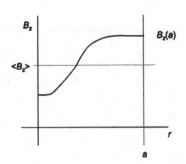

Fig. 16.1 Exclusion of the magnetic field by the plasma when $\beta_p > 1$

$$\beta_p = 1 + \frac{B_z^2(a) - \langle B_z^2 \rangle}{B_\theta^2(a)}. \tag{16.7}$$

The parameter β_p is called *poloidal beta*. It is an important parameter in magnetic fusion energy concepts. It measures the ratio of the internal energy in the fluid, $\langle p \rangle$, to the energy in the poloidal field, $B_\theta^2(a)$, as measured at the outer boundary. As seen from Eq. (16.1), the poloidal field provides the confinement by balancing the outward force of the total pressure gradient.

If $\beta_p > 1$, then $B_z^2(a) > \langle B_z^2 \rangle$, and the presence of the fluid tends to *exclude* the axial field, as shown in Fig. 16.1.

In this case the fluid is said to be *diamagnetic*.

If $\beta_p < 1$, then $B_z^2(a) < \langle B_z^2 \rangle$, and the field excludes the plasma, as shown in Fig. 16.2.

Fig. 16.2 Penetration of the magnetic field into the plasma when $\beta_p > 1$

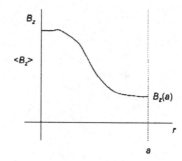

In this case the fluid is said to be *paramagnetic*.

These concepts can be generalized to toroidal geometry.

Lecture 17
"Force-Free" Fields

> *Persuasion is often more effectual than force.*
>
> Aesop

An equilibrium of some interest is the case of force-free fields. Of course, by definition all equilibrium situations are "force-free," but in MHD that description is usually reserved for the special case where the Lorentz force vanishes, i.e.,

$$\mathbf{J} \times \mathbf{B} = 0. \tag{17.1}$$

The pressure is constant ($\nabla p = 0$) and the current is everywhere parallel to the magnetic field, i.e.,

$$\mathbf{J} = \alpha(\mathbf{x})\mathbf{B} \tag{17.2}$$

or

$$\nabla \times \mathbf{B} = \frac{\alpha}{\mu_0}\mathbf{B}. \tag{17.3}$$

(Vector fields with the property that they are everywhere parallel to their curl are called *Beltrami fields*.) Taking the divergence of Eq. (17.3) and using $\nabla \cdot \mathbf{B} = 0$ yields

$$\mathbf{B} \cdot \nabla\alpha = 0, \tag{17.4}$$

while taking the curl along with $\nabla \cdot \mathbf{B} = 0$ gives

$$\nabla^2\mathbf{B} - \left(\frac{\alpha}{\mu_0}\right)^2 \mathbf{B} = \frac{1}{\mu_0}\nabla\alpha \times \mathbf{B}. \tag{17.5}$$

Equation (17.4) says that $\alpha(\mathbf{x})$ is constant along a magnetic field line. Together, Eqs. (17.4) and (17.5) are two equations that, in principle, can be solved for the unknowns $\mathbf{B}$ and α. (In practice, however, this presents some very subtle mathematical issues.[1])

---

[1] See J. J. Aly, Ap. J. **283**, 349 (1984).

Schnack, D.D.: *"Force-Free" Fields*. Lect. Notes Phys. **780**, 103–105 (2009)
DOI 10.1007/978-3-642-00688-3_17

A case of particular interest occurs when $\alpha = $ constant and the geometry is cylindrical. We also introduce the notation $\lambda \equiv \alpha/\mu_0$, which has units of L^{-1}. Then Eq. (17.4) is satisfied automatically and, with $B_r = 0$, the θ and z components of Eq. (17.5) are

$$r^2 \frac{d^2 B_\theta}{dr^2} + r \frac{d B_\theta}{dr} - \left(1 - \lambda^2 r^2\right) B_\theta = 0 \qquad (17.6)$$

and

$$r^2 \frac{d^2 B_z}{dr^2} + r \frac{d B_z}{dr} + \lambda^2 r^2 B_z = 0. \qquad (17.7)$$

These are both forms of *Bessel's equation*. The solution of Eq. (17.6) is

$$B_\theta = B_0 J_1 \left(\lambda r\right) \qquad (17.8)$$

and the solution of Eq. (17.7) is

$$B_z = B_0 J_0 \left(\lambda r\right), \qquad (17.9)$$

where J_0 and J_1 are called Bessel functions of the zeroth and first order, respectively. We remark that $B_z(r)$ changes sign when $\lambda r = j_{0,1} = 2.4048$, where $j_{n,k}$ denotes the kth zero of the nth order Bessel function. (These are extensively tabulated.) These solutions are shown in Fig. 17.1 (where B_z is labeled as B_ϕ).

Fig. 17.1 The Bessel function model for force-free magnetic fields. The crosses indicate some experimental measurements

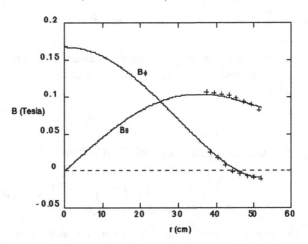

Let a conducting wall be located at $r = a$. A large variety of screw-pinch equilibria are represented by varying the parameter λa. From the preceding remarks, we know that the axial field changes sign ("reverses") inside the wall if $\lambda a > 2.4048$. This parameter regime is called the *reversed-field pinch* (RFP). If $\lambda a \ll 1$, then since $J_\nu(z) \sim (z/2)^\nu/\Gamma(\nu + 1)$ for $|z| \ll 1$, we have $B_\theta \sim \lambda r/2$ and $B_z \sim 1$; the poloidal (θ) component of the magnetic field increases linearly (implying that the axial current density J_z is constant), and the axial (z) component of the magnetic field is constant. This parameter regime is called the *tokamak*. The intermediate range of λa (sometimes called the *paramagnetic pinch*, since the axial field is larger in the center than at the edge) has proven to be less interesting experimentally.

Of course, there is no a priori reason to expect that λ will be constant. In general, it must be determined from the solution of Eqs. (17.4) and (17.5), along with appropriate boundary conditions. (The allowable form the boundary conditions for these equations is a subtle mathematical problem.) We will return to this point when we discuss MHD relaxation later in this course.

Lecture 18
Toroidal Equilibrium; The Grad–Shafranov Equation[1]

> *Donuts. Is there anything they can't do?*
> Matt Groening, *The Simpsons*

We have just considered three examples of MHD equilibria in cylindrical geometry. These were the θ-pinch, the z-pinch, and the screw pinch. In all of these cases, the cylinder was considered to be infinitely long. In practice, or course, a cylinder must have finite length. The achievement of MHD equilibrium was possible because, in ideal MHD, the fluid cannot flow freely across the magnetic field. However, the fluid (or plasma) can flow freely along the field, and this allows the fluid to exit the apparatus through the ends of the cylinder. These inherent end losses have proven to be detrimental to achieving fluid confinement in finite cylindrical geometry. (A possible exception of the field-reversed configuration, or FRC, but many of its interesting properties arise from non-MHD effects, and we will not discuss it further in this course.)

An ingenious solution to the end-loss problem is to connect the ends of the cylinder to each other, transforming the cylinder into a torus (shaped like a donut). This is shown in Fig. 18.1.

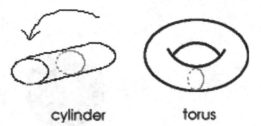

Fig. 18.1 Wrapping a periodic cylinder into a torus

cylinder   torus

The end losses are thus completely eliminated; all the magnetic field lines remain within the boundaries of the system.

We are thus motivated to study equilibrium in a toroidal configuration. With a torus, it is usual to work in a cylindrical coordinate system (R, ϕ, Z) in which the

---

[1] We are again motivated by Jeffrey P. Freidberg, *Ideal Magnetohydrodynamics*, Plenum Press, New York (1987).

Schnack, D.D.: *Toroidal Equilibrium; The Grad–Shafranov Equation.* Lect. Notes Phys. **780**, 107–120 (2009)
DOI 10.1007/978-3-642-00688-3_18

cross-sectional area of the fluid lies in the (R, Z) plane (called the *poloidal plane*), and ϕ is the angle of rotation about the Z-axis. The important situation in which all quantities are independent of ϕ is called *axisymmetric*. This is the equivalent of z-independence (or translational symmetry) in the straight cylinder. The radius of the center of the poloidal cross section is called the major radius. The radius of the outer boundary with respect to the major radius is called the minor radius. These are shown in Fig. 18.2.

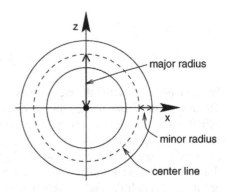

Fig. 18.2 Top view illustrating the major and minor radii of a torus

Unfortunately, when a cylindrical MHD equilibrium is bent into a torus it is no longer an equilibrium. Instead, it tends to expand outward in the major radial (R) direction. There are two reasons for this. First, a straight cylinder is symmetric about its central axis. The pressure force are therefore distributed equally on all parts of the outer boundary. However, in a torus the outer part of the surface has a larger surface area $(S_2 \sim R_2)$ than the inner surface $(S_1 \sim R_1)$, as shown in Fig. 18.3.

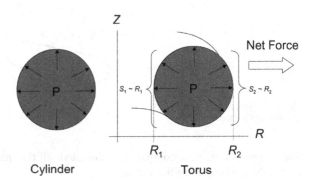

Fig. 18.3 Illustration of the net force in the direction of the major radius exerted by pressure forces in a torus

Second, just as parallel currents attract each other by means of the Lorentz force, antiparallel currents repel each other. Each current element at angular location ϕ repels (and is repelled by) the current element at angular location $\phi + \pi$. This results in a net outward force in the radial (R) direction, as shown in Fig. 18.4.

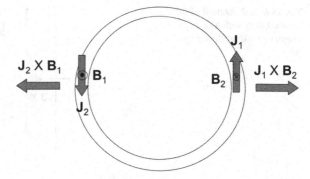

Fig. 18.4 Outward expansion of a toroidal ring of current due to the Lorentz force

Each of these forces makes the torus expand in major radius. Some externally supplied currents and fields are necessary for equilibrium to be maintained. The system is now finite in extent and the Virial Theorem applies.

One way to provide these required external fields is to enclose the minor cross section of the torus in an electrically conducting shell. If the shell is a perfect conductor, then as the toroidal fluid tries to expand outward the field lines enclosing the fluid will not be able to penetrate the shell, and they will be compressed between the fluid and the shell along the outer (in major radius) part of the torus (called the *outboard side*). This will appear as an increase in magnetic pressure on the outboard side, thus opposing the expansion. A new state of equilibrium will be reached in which the fluid is shifted outward with respect to the geometric center line; the magnetic axis (∼the center of concentric poloidal field) no longer coincides with the geometric axis. This is called the *Shafranov shift*, and its magnitude us usually denoted by Δ. This is shown in Fig. 18.5.

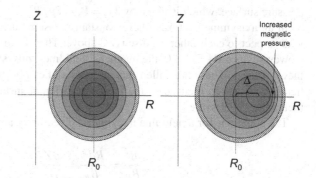

Fig. 18.5 Stabilization of the outward expansion with a perfectly conducting shell. The outward shift of the magnetic axis is called the Shafranov shift

Another way to provide the external field necessary for toroidal equilibrium is with current carrying Helmholtz coils that induce a field in the Z-direction. If properly oriented, this field can interact with the toroidal current (J_ϕ) in the fluid to provide an inward Lorentz force that balances the outward expanding tendency of the torus, as shown in Fig. 18.6.

Fig. 18.6 Stabilization of the
outward shift with a vertical
magnetic field produced by
coils

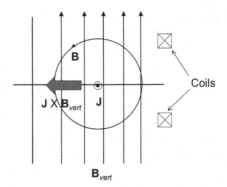

Note that the effect of the vertical field is to amplify the field due to the plasma current on the outboard side, and decrease the field on the inboard side. It thus provides the same mechanism as the conducting shell. In the former case, the vertical field is produced by image currents that flow in the shell.

In order to proceed beyond these simple cartoons, we will have to develop some more general ideas about toroidal equilibria. From now on, we will assume that the configurations are axisymmetric, i.e., all quantities are independent of the toroidal angle ϕ.

We have seen that in a straight (infinitely long) cylinder, the pressure is constant on concentric cylindrical surfaces, i.e., $p = p(r)$. Since $\nabla p = \mathbf{J} \times \mathbf{B}$, we have $\mathbf{B} \cdot \nabla p = 0$, so that the pressure is constant along the direction of $\mathbf{B}$. Conversely, the field lines of $\mathbf{B}$ must lie in constant pressure surfaces, i.e., they must wrap around a cylindrical surface. Since $\mathbf{J} \cdot \nabla p = 0$, the current must also lie in these surfaces. However, it need not be aligned with $\mathbf{B}$; if there is a pressure gradient across these surfaces, there will be a component perpendicular to $\mathbf{B}$, also within the constant pressure surface, which is given by $\mathbf{J}_\perp = \mathbf{B} \times \nabla p / B^2$.

In an axisymmetric torus, these constant pressure surfaces are shifted outward with respect to each other, as discussed above. They form nested toroidal surfaces. However, since $\mathbf{B} \cdot \nabla p = 0$, the magnetic field lines must still lie completely within these surfaces. These are called flux surfaces and can be "labeled" by any variable that is constant on them, e.g., the pressure; different surfaces can be identified by their value of pressure.

The equations for a field line in cylindrical geometry are

$$\frac{dR}{B_R} = \frac{R d\phi}{B_\phi} = \frac{dZ}{B_Z}.$$

(18.1)

Consider a field line that begins at coordinates (R_0, ϕ_0, Z_0). This point will make an angle θ_0 with respect to an (R, ϕ) plane through the center of the concentric surfaces of constant pressure. Now integrate this field line once around the torus, i.e., follow its trajectory until $\phi_1 = \phi_0 + 2\pi$. In general this will intersect the poloidal plane at R_1 and Z_1, which are different from R_0 and Z_0, and which make a different angle

$\theta_0 + \Delta\theta$ with respect to the axis. This field line can be said to *map* the point (R_0, Z_0) into the point (R_1, Z_1). This is shown in Fig. 18.7.

Fig. 18.7 Locus of the points of intersection of a field line with the poloidal plane, defined by ϕ = constant

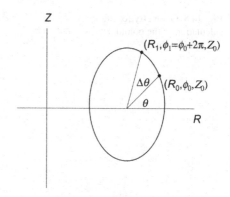

We know that the pressure at (R_1, Z_1) must be the same as the pressure at point (R_0, Z_0). There are two possible types of trajectory (or mapping) for a given field line. One is that it fills the entire volume ergodically. In that case, the pressure must be constant throughout the volume. This is not consistent with confinement. While this can occur dynamically during the evolution of the magnetoplasma system, we will not consider it as part of our discussion of equilibrium. The second case is that the field line maps out a two-dimensional surface, which corresponds to a constant pressure surface. Then there are two further possibilities. The first is that the field line, while remaining on the surface, nonetheless never returns to its original position. These field lines fill the two-dimensional surface ergodically, but they do not close upon themselves. Surfaces on which the field lines are ergodic are said to be *irrational* (for reasons that will be seen below). The second possibility is that the field line returns *exactly* to its initial coordinates (closes upon itself) after N turns around the torus. These surfaces are said to be *rational*.

These concepts can be quantified by introducing the *rotational transform*

$$\iota \equiv \lim_{N \to \infty} \frac{1}{N} \sum_{n=1}^{N} \Delta\theta_n, \tag{18.2}$$

where $\Delta\theta_n$ is the change in the angle θ during the nth toroidal circuit. *If $\iota/2\pi$ is a rational number* (i.e., the ratio of two integers), *then the field line is closed and the surface is rational.* Otherwise, the field line is not closed and the surface is irrational. If ι is a rational number, it is the number of times the field line must transit the torus in the toroidal (ϕ) direction (sometimes called the *long way around*) for it to make one complete transit about the surface in the poloidal (R, Z) plane (the *short way around*). The quantity $q \equiv 2\pi/\iota$ is called the *safety factor*. It is important in the theories of equilibrium and stability of confined plasmas. We will see it again.

It is possible to define fluxes based on the poloidal field (i.e., B_R and B_Z) and toroidal field (B_ϕ) components. We define $d\mathbf{S}_t$ and $d\mathbf{S}_p$ as surface elements extending between constant pressure surfaces oriented in the toroidal and poloidal directions, respectively, as shown in Fig. 18.8.

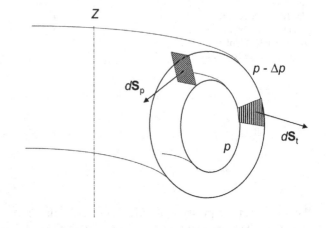

Fig. 18.8 Geometry for the calculation of the poloidal and toroidal flux

The *poloidal flux* is defined as

$$\psi_P(p) = \int \mathbf{B} \cdot d\mathbf{S}_p. \tag{18.3}$$

Since ψ_P is a function of the pressure, we can (and usually will) adopt ψ_P as a surface label. (Any function $f(p)$ that is constant on a flux surface is called a *surface function*, and can equally well be adopted as a surface label.) We define the *toroidal flux* as

$$\psi_t(p) = \int \mathbf{B} \cdot d\mathbf{S}_t. \tag{18.4}$$

It is useful to also define the *toroidal current*

$$I_t(p) = \int \mathbf{J} \cdot d\mathbf{S}_t \tag{18.5}$$

and the *poloidal current*

$$I_p(p) = \int \mathbf{J} \cdot d\mathbf{S}_p. \tag{18.6}$$

Since these are all functions of the pressure, they are all surface functions and could serve as surface labels. Finally, we can define the *volume contained within a constant pressure surface*. It is often useful to use the coordinate system shown in Fig. 18.9.

Fig. 18.9 An alternative coordinate system for calculations in toroidal geometry

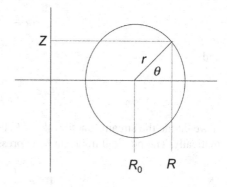

The coordinates of a point can be equally well written as (R, Z) or (r, θ), where

$$R = R_0 + r\cos\theta \tag{18.7}$$

and

$$Z = r\sin\theta. \tag{18.8}$$

Then $\psi_p(r, \theta) = $ constant defines a flux surface. We assume that an inverse transformation exists (although it may be difficult to compute), i.e., the radius of a flux surface is given by $r = \hat{r}(\theta, \psi_p)$. Then the volume contained within the surface with label ψ_p is given by

$$V(\psi) = \int_0^{2\pi} d\phi \int_0^{2\pi} d\theta \int_0^{\hat{r}(\theta, \psi)} (R_0 + r\cos\theta)\, r\, dr. \tag{18.9}$$

We now proceed to derive the equations that describe axially symmetric force balance in a torus. We proceed as in Lecture 15, i.e.,

1. $\nabla \cdot \mathbf{B} = 0$.
2. Ampére's law, $\mu_0 \mathbf{J} = \nabla \times \mathbf{B}$.
3. Force balance, $\nabla p = \mathbf{J} \times \mathbf{B}$.

1. $\nabla \cdot \mathbf{B} = 0$. The total magnetic field is $\mathbf{B} = \mathbf{B_P} + B_\phi \hat{\mathbf{e}}_\phi$, where $\mathbf{B_P}$ is the poloidal field containing the R and Z components. Since the system is independent of ϕ, we have

$$\frac{1}{R}\frac{\partial}{\partial R}(RB_R) + \frac{\partial B_Z}{\partial Z} = 0. \tag{18.10}$$

Since $\mathbf{B} = \nabla \times \mathbf{A}$, we have

$$B_R = -\frac{\partial A_\phi}{\partial Z} \qquad (18.11)$$

and

$$B_Z = \frac{1}{R}\frac{\partial}{\partial R}\left(RA_\phi\right). \qquad (18.12)$$

If we define the stream function $\psi = RA_\phi$, then Eq. (18.10) will be satisfied automatically. The poloidal field can be expressed as

$$\mathbf{B}_P = \frac{1}{R}\nabla\psi \times \hat{\mathbf{e}}_\phi. \qquad (18.13)$$

The stream function can be related to the poloidal flux by noting that the latter is a measure of the flux of B_Z passing through the mid-plane of torus ($Z = 0$) between the shifted center of the surfaces, R_a, and another radius $R_b > R_a$, as shown in Fig. 18.10.

Fig. 18.10 Magnetic field lines passing through the strip $R_0 \le R \le R_a$ in the plane $Z = 0$

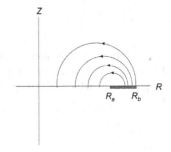

Then

$$\psi_P = \int\limits_0^{2\pi} d\phi \int\limits_{R_a}^{R_b} RdRB_z(R,0)$$

$$= 2\pi \int RdR\frac{1}{R}\frac{\partial\psi}{\partial R}\bigg|_{Z=0}$$

$$= 2\pi\psi\left(R_b,0\right), \qquad (18.14)$$

where we have set $\psi(R_a, 0) = 0$. Therefore, we can, and will from now on, label the flux surfaces with ψ.

2 *Ampére's law,* $\mu_0\mathbf{J} = \nabla \times \mathbf{B}$. Using the identities $\nabla \cdot \hat{\mathbf{e}}_\phi = 0$, $\nabla \times \hat{\mathbf{e}}_\phi = \hat{\mathbf{e}}_Z/R$, and $\nabla\hat{\mathbf{e}}_\phi = -\hat{\mathbf{e}}_\phi\hat{\mathbf{e}}_R/R$, we have

$$\mu_0 \mathbf{J} = \nabla \times \left(\frac{1}{R} \nabla \psi \times \hat{\mathbf{e}}_\phi + B_\phi \hat{\mathbf{e}}_\phi \right)$$

$$= \mu_0 J_\phi \hat{\mathbf{e}}_\phi + \frac{1}{R} \nabla \left(R B_\phi \right) \times \hat{\mathbf{e}}_\phi, \tag{18.15}$$

where the toroidal current density is

$$\mu_0 J_\phi = -\nabla \cdot \left(\frac{1}{R} \nabla \psi \right) - \frac{1}{R^2} \frac{\partial \psi}{\partial R}. \tag{18.16}$$

It is customary to define the operator $\Delta^* \psi$ as

$$\Delta^* \psi \equiv R \nabla \cdot \left(\frac{1}{R} \nabla \psi \right) - \frac{1}{R} \frac{\partial \psi}{\partial R} = R \frac{\partial}{\partial R} \left(\frac{1}{R} \frac{\partial \psi}{\partial R} \right) + \frac{\partial^2 \psi}{\partial Z^2}, \tag{18.17}$$

so that

$$\mu_0 J_\phi = -\frac{1}{R} \Delta^* \psi. \tag{18.18}$$

3 *Force balance*, $\nabla p = \mathbf{J} \times \mathbf{B}$. Since there is no ϕ dependence, we have $\mathbf{B}_P \cdot \nabla p = 0$. Using Eq. (18.13), $(\nabla \psi \times \hat{\mathbf{e}}_\phi) \cdot \nabla p = 0$ or

$$(\nabla \psi \times \nabla p) \cdot \hat{\mathbf{e}}_\phi = 0. \tag{18.19}$$

This expression vanishes identically if $p = p(\psi)$, as it must since, by construction, the pressure is constant on flux surfaces.

Similarly, since $\mathbf{J} \cdot \nabla p = 0$, it follows from Eq. (18.15) that

$$\left[\nabla \left(R B_\phi \right) \times \nabla p \right] \cdot \hat{\mathbf{e}}_\phi = 0, \tag{18.20}$$

so that $R B_\phi = F(\psi)$, which we could not anticipate. The function $F(\psi)$ is related to the *total poloidal current* (plasma plus coil) flowing between the major axis of the torus, $R = 0$, and any radius R_b:

$$I_P = \int \mathbf{J}_P \cdot d\mathbf{S}$$

$$= \int_0^{2\pi} d\phi \int_0^{R_b} R \, dR \, J_Z(R, 0)$$

$$= 2\pi \int_0^{R_b} R \, dR \, \frac{1}{R} \frac{\partial}{\partial R} \left(R B_\phi \right) \Big|_{Z=0}$$

$$= 2\pi R B_\phi(R_b, 0)$$

$$= 2\pi F(\psi). \tag{18.21}$$

The expression for force balance, $\nabla p = \mathbf{J} \times \mathbf{B}$, is then

$$p' \nabla \psi = \left(J_\phi \hat{\mathbf{e}}_\phi + \frac{1}{R\mu_0} F' \nabla \psi \times \hat{\mathbf{e}}_\phi \right) \times \left(\frac{1}{R} \nabla \psi \times \hat{\mathbf{e}}_\phi + B_\phi \hat{\mathbf{e}}_\phi \right),$$

where $(..)'$ denotes differentiation with respect to ψ. After some vector algebra, this becomes

$$p' = -\frac{1}{\mu_0 R^2} \left(\Delta^* \psi + F F' \right)$$

or

$$\Delta^* \psi = -\mu_0 R^2 p' - F F'. \tag{18.22}$$

Equation (18.22) is called the *Grad–Shafranov equation*. It is one of the most famous equations arising from MHD. It is a second-order partial differential equation that, given the functions $p(\psi)$ and $F(\psi)$, describes *equilibrium in an axisymmetric torus*. The functions $p(\psi)$ and $F(\psi)$ are *completely arbitrary* and must be determined from considerations other than theoretical force balance. (For example, they could be determined experimentally, or from a transport calculation, or simply fabricated from whole cloth.) We have seen this sort of a situation before in Lecture 14 when we discussed the equilibrium in the general cylindrical screw pinch.

At least in principle, given the functions $p(\psi)$ and $F(\psi)$, along with the appropriate boundary conditions (generally that ψ is specified on some boundary), Eq. (18.22) can be solved for $\psi(R, Z)$. This gives the equilibrium flux distribution. However, it is important to note that the functions p and F can be (and generally are) *nonlinear*, so that *these solutions are not guaranteed to either exist or to be unique*; there may be no solution or many solutions, satisfying both (18.22) and the boundary conditions.

As an example of the character of the solutions of the Grad–Shafranov equation, we consider the linear case $F = \text{constant}$ ($F' = 0$) and

$$p' = \text{constant} = \frac{8\psi_0}{\mu_0 R_0^2} \left(1 + \alpha^2 \right), \tag{18.23}$$

where $\psi_0 = \psi(R_0, 0)$ and α is a constant. Then it can be verified that the solution of Eq. (18.22) is

$$\psi(R, Z) = \frac{\psi_0 R^2}{R_0^4} \left(2R_0^2 - R^2 - 4\alpha^2 Z^2\right). \tag{18.24}$$

Surfaces of constant ψ for $\alpha = 1$ are sketched in Fig. 18.11.

Fig. 18.11 Sketch of the contours of constant flux (flux surfaces) Shafranov toroidal equilibrium, Eq. (18.24)

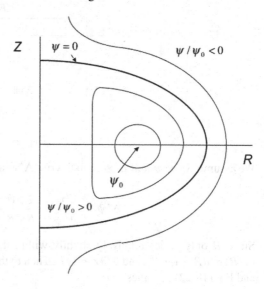

The flux surfaces are closed and nearly circular near the magnetic axis at R_0. They remain closed but become non-circular ("shaped") when $\psi/\psi_0 > 0$, and become open when $\psi/\psi_0 < 0$. The surface $\psi/\psi_0 = 0$ is called a *separatrix*; it separates the regions of closed an open flux surfaces. The constant α determines the shape of the closed flux surfaces. As α increases from 1, they become more elongated and vice versa. The boundary of the plasma is defined as $p = 0$. The function $p(\psi)$ can be adjusted (by adding a constant) so that any surface $\psi/\psi_0 = $ constant can be the boundary. Finally, since $F = $ constant, $B_\phi \sim 1/R$.

Since the Grad–Shafranov equation is nonlinear, there are no general existence or uniqueness proofs available. There may be one solution, no solutions, or multiple solutions depending on the specific forms of $p(\psi)$ and $F(\psi)$. Points in parameter space where solutions coalesce or disappear are called *bifurcation points*. We will now present a specific example of this behavior.

Consider the case of a tall, thin toroidal plasma with large aspect ratio.[2] The plasma is surrounded by a conducting wall, as shown in Fig. 18.12.

---

[2] This example is due to K. D. Marx (private communication).

Fig. 18.12 Tall, thin toroidal
equilibrium configuration

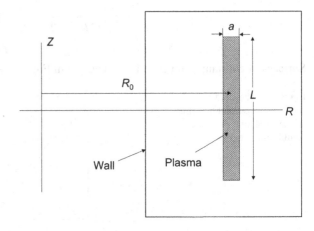

We assume $R_0 \gg a$ and $L \gg a$, and write $\Delta^* \psi$ as

$$\Delta^* \psi = \frac{\partial^2 \psi}{\partial R^2} - \frac{1}{R} \frac{\partial \psi}{\partial R} + \frac{\partial^2 \psi}{\partial Z^2}. \tag{18.25}$$

Since R only varies relatively slightly within the plasma, we have $\partial/\partial R \sim 1/a$, $(1/R)\,\partial/\partial R \sim 1/aR_0$, and $\partial/\partial Z \sim 1/L$. Then to the lowest order, $\Delta^* \psi \sim \partial^2 \psi/\partial R^2$, and Eq. (18.22) becomes

$$\frac{d^2 \psi}{dR^2} = -\mu_0 R^2 p' - FF'$$
$$= -S', \tag{18.26}$$

where $S(\psi) = \mu_0 R^2 p(\psi) + F^2(\psi)/2$ is a nonlinear function of ψ, here chosen to be

$$S(\psi) = -C\,(\psi - \psi_P) \text{ for } \psi < \psi_P$$
$$= 0 \text{ otherwise} \tag{18.27}$$

The flux at the plasma boundary is ψ_P and the flux at the wall is zero; ψ is negative everywhere; C is a constant. This is sketched in Fig. 18.13.

Fig. 18.13 Nonlinear source
function for the
Grad–Shafranov equation for
the tall, thin equilibrium

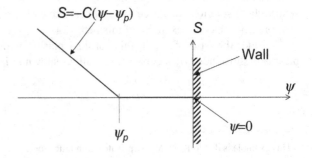

The function S' is sketched in Fig. 18.14.

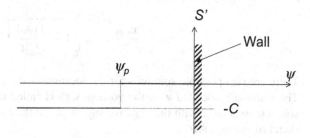

Fig. 18.14 The derivative of the Grad–Shafranov source function

Now let $x = R - R_0$. Then Eq. (18.26) is

$$\frac{d^2\psi}{dx^2} = 0, \psi > \psi_P \tag{18.28a}$$

$$= C, \psi < \psi_P. \tag{18.28b}$$

In the vacuum region, $\psi > \psi_P$, the solution is

$$\psi_> = \alpha x + \beta, \tag{18.29}$$

and in the plasma, $\psi < \psi_P$, the solution is

$$\psi_< = \frac{1}{2}Cx^2 + \gamma x + \delta. \tag{18.30}$$

Let the wall be located at $x_{\text{wall}} = R_{\text{wall}} - R_0$. The solution must satisfy the boundary condition $\psi_> = 0$ at $x = x_{\text{wall}}$. Since Eq. (18.28) is symmetric in x, the solution must be symmetric about $x = 0$. The solution must be continuous at $x = x_P$. Further, $B_z = d\psi/dx$ must be continuous across the boundary of the plasma at $x = x_P$. The character of the solution is sketched in Fig. 18.15.

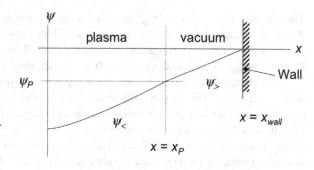

Fig. 18.15 Sketch of the solution of the Grad–Shafranov equation for the tall, thin toroidal equilibrium

Combining these conditions yields a quadratic equation for the location of the plasma boundary, x_P, whose solution is

$$x_P = \frac{x_{\text{wall}}}{2} \left[1 \pm \sqrt{1 - \frac{4\,|\psi_P|}{x_{\text{wall}}^2 C}} \right]. \tag{18.31}$$

There are two possible solutions for the location of the plasma/vacuum boundary. The solution associated with the positive sign is called the *deep solution*, and the solution associated with the negative sign is called the *shallow solution*. These are sketched in Fig. 18.16.

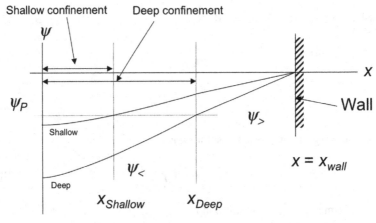

Fig. 18.16 Sketch of the deep and shallow solutions of the nonlinear Grad–Shafranov equation

In the shallow solution, the plasma is confined in the region $0 \le x \le x_{\text{Shallow}}$. For the deep solution, the confinement region is $0 \le x \le x_{Deep}$.

From Eq. (18.31), we see that when $4\,|\psi_P|/x_{\text{wall}}^2 C > 1$ there are no real solutions and *equilibrium is impossible*. This occurs when

$$C < \frac{4\,|\psi_P|}{x_{\text{wall}}^2}. \tag{18.32}$$

The quantity $C_0 = 4\,|\psi_P|/x_{\text{wall}}^2$ is called a *bifurcation point* for the equilibrium. Above this value of C there are two solutions. These solutions merge at the bifurcation. For values of C below the bifurcation point there are no solutions. When $C \to \infty$, $x_{Deep} \to x_{\text{wall}}$ and $x_{\text{Shallow}} \to 0$, so that the deep solution survives.

It is reasonable to ask which of the two possible equilibrium solutions nature will decide upon. This is generally determined by the reason of stability and not equilibrium. Often one of the solutions has more energy than the other. Since nature likes to seek low-energy states, one might guess that the solution with the lowest energy is the one that will be observed. However, this must be tested by performing stability studies on the candidate equilibria. That is one of the topics that we will now address.

Lecture 19
Behavior of Small Displacements in Ideal MHD

> *Every big problem was at one time a wee disturbance.*
> Anonymous

We have just studied MHD equilibrium, states in which the forces are exactly in balance. We now begin studies of states that are slightly displaced from equilibrium.

We assume that the fluid and magnetic field are in an equilibrium state, $\mathbf{F} = 0$. Now let a fluid element be displaced by a small amount $\boldsymbol{\xi}$ from its equilibrium position. That is, a fluid element with equilibrium position $\mathbf{r}$ is displaced slightly to a new position $\mathbf{r}' = \mathbf{r} + \boldsymbol{\xi}$, as shown in Fig. 19.1.

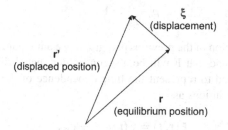

Fig. 19.1 Geometry for displacement from equilibrium

After this displacement, the system is no longer in equilibrium, i.e., $\mathbf{F} \neq 0$. Since $\mathbf{F} = 0$ when $\boldsymbol{\xi} = 0$, we can write

$$\mathbf{F} = \mathbf{F}\{\boldsymbol{\xi}\}. \tag{19.1}$$

(The curly bracket notation is standard here; it will be explained in Lecture 25.) Further, for small displacements we expect $\mathbf{F}$ to be a *linear* function of $\boldsymbol{\xi}$.

Now suppose that $\boldsymbol{\xi} \cdot \mathbf{F} < 0$. Then the displacement and the force are in opposite directions. The force tends to restore the system to its original equilibrium position; it opposes the displacement. In this case, we might expect the system to oscillate about its equilibrium position, so that the system can be said to be *stable* (of course, this must be proven). On the other hand, when $\boldsymbol{\xi} \cdot \mathbf{F} > 0$, then the force is in the same direction as the displacement. The force tends to amplify the displacement and drive the system further from its equilibrium position. In this case, we might expect the original displacement to grow in time, so that the system can be said to be *unstable* (again, this must be proven). A third possibility is that $\boldsymbol{\xi} \cdot \mathbf{F} = 0$, so that the force

Schnack, D.D.: *Behavior of Small Displacements in Ideal MHD*. Lect. Notes Phys. **780**, 121–128 (2009)
DOI 10.1007/978-3-642-00688-3_19

and the displacement are orthogonal. The system is then said to be *neutrally stable*. In ideal MHD, these are the only possibilities. (For example, when $\boldsymbol{\xi} \cdot \mathbf{F} < 0$ the restoring force could be so large as to amplify the oscillations, leading to growing oscillations. This case is called *overstable*. We will see that this is not allowed in ideal MHD.)

We can be more specific about these concepts. Let the subscript $(..)_0$ denote equilibrium quantities. If the equilibrium is stationary ($\mathbf{V}_0 = 0$) and the displacements are small, we can ignore the quadratic term $\mathbf{V} \cdot \nabla \mathbf{V}$, and the equation of motion is

$$\rho_0 \frac{\partial \mathbf{V}}{\partial t} = \mathbf{F}. \tag{19.2}$$

We introduce the displacement vector $\boldsymbol{\xi}$, defined by

$$\mathbf{V} \equiv \frac{\partial \boldsymbol{\xi}}{\partial t}. \tag{19.3}$$

Then the equation of motion is

$$\rho_0 \frac{\partial^2 \boldsymbol{\xi}}{\partial t^2} = \mathbf{F} \{\boldsymbol{\xi}\}. \tag{19.4}$$

In light of the discussion of the previous paragraph, we anticipate that Eq. (19.4) will be a *wave equation* and that $\mathbf{F}$ will be a second-order spatial differential operator. We are thus motivated to represent the time dependence of the displacement $\boldsymbol{\xi}$ in terms of wave-like solutions as

$$\boldsymbol{\xi}(\mathbf{r}, t) = \boldsymbol{\xi}(\mathbf{r}) e^{i\omega t} + c.c., \tag{19.5}$$

where $\boldsymbol{\xi}(\mathbf{r})$ is now a complex function and *c.c.* denotes the complex conjugate. (This is required to make the physical displacement real. We will often leave it out of the ensuing formulas, but it is always implied. Alternatively, one can interpret the behavior of the physical displacement as being represented by the real part of the formulas.) Introducing the *ansatz* (19.5) into (19.4), we have

$$-\rho_0 \omega^2 \boldsymbol{\xi} = \mathbf{F} \{\boldsymbol{\xi}\}. \tag{19.6}$$

Since $\mathbf{F}$ is linear in $\boldsymbol{\xi}$, we can write symbolically $\mathbf{F} \{\boldsymbol{\xi}\} \equiv \mathbf{F} \cdot \boldsymbol{\xi}$, so that $\mathbf{F}$ is now represented as a tensor (or matrix). Then

$$-\rho_0 \omega^2 \boldsymbol{\xi} = \mathbf{F} \cdot \boldsymbol{\xi} \tag{19.7}$$

is the *linear equation of motion for small displacements from stationary equilibrium*. Since $\boldsymbol{\xi}$ is a as yet undetermined function of space and time, $\mathbf{F} \{\boldsymbol{\xi}\}$ is called a *functional* of $\boldsymbol{\xi}$ (see Lecture 25). It is called the *ideal MHD force operator*. (Heuristically, we can see that if $\boldsymbol{\xi} \cdot \mathbf{F} \{\boldsymbol{\xi}\} < 0$, this suggests that $\omega^2 > 0$ and the motion is oscillatory; and if $\boldsymbol{\xi} \cdot \mathbf{F}\{\boldsymbol{\xi}\} > 0$, this suggests that $\omega^2 < 0$, $\omega = \pm i\gamma$, and the motion will have exponentially growing behavior. However, this still needs to be proven.) Again, we emphasize that formulas such as Eq. (19.7) only have physical meaning when combined with their complex conjugate or their real part is taken.

Equation (19.7) is a linear system of the form $\mathbf{A} \cdot \mathbf{x} = \lambda \mathbf{x}$, where $\mathbf{A}(\to \mathbf{F})$ is a linear operator (matrix, tensor, or differential), $\mathbf{x}(\to \boldsymbol{\xi})$ is a vector, and $\lambda(\to -\omega^2)$ is a constant. The problem is to find non-trivial solutions for $\mathbf{x}$ (i.e., $\mathbf{x} \neq 0$). This special important problem is called an *eigenvalue problem*. Non-trivial solutions are possible only for certain special values of λ that are roots of the equation $\det[(\mathbf{A} - \lambda\mathbf{I}) \cdot \mathbf{x}] = 0$. These special values of λ are called the *eigenvalues* of $\mathbf{A}$, and the corresponding non-trivial solutions $\mathbf{x}$ are called the *eigenvectors* of $\mathbf{A}$.

Equation (19.7) can be written as the homogenous system

$$\left(\mathbf{F} + \rho_0\omega^2\mathbf{I}\right) \cdot \boldsymbol{\xi} = 0. \tag{19.8}$$

This suggests that Eq. (19.8) has non-trivial solutions for $\boldsymbol{\xi}$ only for special values of the (negative of the square of the) frequency $-\omega^2$ that satisfy

$$\det\left[\mathbf{F} + \rho_0\omega^2\mathbf{I}\right] = 0. \tag{19.9}$$

The frequencies $-\omega^2$ are the eigenvalues of $\mathbf{F}$, and the corresponding displacements are the eigenvectors.

A Diversion on Homogenous Systems and Eigenvalue Problems

Equation (19.8) stands for three simultaneous homogenous equations in the three unknowns ξ_x, ξ_y, and ξ_z. It is of the form

$$\left(\mathbf{A} - \omega^2\mathbf{I}\right) \cdot \boldsymbol{\xi} = 0. \tag{19.10}$$

The obvious solution of Eq. (19.10) is trivial: $\boldsymbol{\xi} = 0$. We enquire as to what is needed to find non-trivial (i.e., $\boldsymbol{\xi} \neq 0$) solutions of this equation.

First consider the simple prototype equation $ax = b$, where all variables are scalars. The solution is $x = a/b$. If $b = 0$ the equation is homogeneous, and it is clear that then the only possible way to have $x \neq 0$ is to have $a = 0$. Then x can be anything; in particular, it can always be written as a multiple of the solution $x = 1$.

Now consider the 2×2 system

$$a_{11}x_1 + a_{12}x_2 = b_1, \tag{19.11}$$

$$a_{21}x_1 + a_{22}x_2 = b_2. \tag{19.12}$$

The solution is

$$x_1 = \frac{\begin{vmatrix} b_1 & a_{12} \\ b_2 & a_{22} \end{vmatrix}}{\begin{vmatrix} a_{11} & a_{12} \\ a_{21} & a_{22} \end{vmatrix}}, \qquad (19.13)$$

$$x_2 = \frac{\begin{vmatrix} a_{11} & b_1 \\ a_{21} & b_2 \end{vmatrix}}{\begin{vmatrix} a_{11} & a_{12} \\ a_{21} & a_{22} \end{vmatrix}}. \qquad (19.14)$$

In analogy with the case $ax = b$, if $b_1 = b_2 = 0$, the solution is $x_1 = x_2 = 0$ unless

$$\begin{vmatrix} a_{11} & a_{12} \\ a_{21} & a_{22} \end{vmatrix} = a_{11}a_{22} - a_{21}a_{12} \equiv \det \mathbf{A} = 0. \qquad (19.15)$$

Equation (19.15) means that *equations* (19.11) *and* (19.12) *are no longer independent equations*; one is a linear combination of the other. In that case, we can determine the *ratio* x_1/x_2 from *either* Eq. (19.11) *or* Eq. (19.12); they must give the same result. From Eq. (19.11) we have $x_1/x_2 = -a_{12}/a_{11}$, and from Eq. (19.12) $x_1/x_2 = -a_{22}/a_{21}$. For these to be the same we require $a_{21}/a_{11} = a_{22}/a_{21}$, which is identical to Eq. (19.15). If the system were $N \times N$, the vanishing of the determinant would ensure that *only $N-1$ of the equations are independent.*

Now consider the problem

$$(\mathbf{A} - \lambda \mathbf{I}) \cdot \mathbf{x} = 0, \qquad (19.16)$$

[see, for example, Eq. (19.10)]. By the above argument, the only solution of Eq. (19.16) is $\mathbf{x} = 0$, unless

$$\det(\mathbf{A} - \lambda \mathbf{I}) = 0. \qquad (19.17)$$

This is called the *characteristic* (or *eigenvalue*) *equation*. (Apparently the German word *eigen* best translates to English as something like *peculiar*, meaning "characteristic of an individual." The mathematical usage is often attributed to David Hilbert, and who are we to argue?) It can be solved for λ; this determines *special values of λ for which the system of equations* (19.16) *are no longer independent*. These special values of λ are called *eigenvalues* or characteristic (or, perhaps, *peculiar*) values. Again consider the 2×2 system

$$\begin{pmatrix} a_{11} - \lambda & a_{12} \\ a_{21} & a_{22} - \lambda \end{pmatrix} \begin{pmatrix} x_1 \\ x_2 \end{pmatrix} = 0. \qquad (19.18)$$

The characteristic equation is

$$(a_{11} - \lambda)(a_{22} - \lambda) - a_{12}a_{21} = 0. \qquad (19.19)$$

It has two roots, λ_1 and λ_2. When $\lambda = \lambda_1$ or $\lambda = \lambda_2$, the two equations represented by (19.18) are no longer independent. For example, when $\lambda = \lambda_1$, *either* of the equations

$$(a_{11} - \lambda_1)x_1 + a_{12}x_2 = 0 \qquad (19.20)$$

or

$$a_{21}x_1 + (a_{22} - \lambda_1)x_2 = 0 \qquad (19.21)$$

can be solved for the ratio x_1/x_2. For these to yield the same result requires

$$(a_{11} - \lambda_1)(a_{22} - \lambda_1) - a_{12}a_{21} = 0, \qquad (19.22)$$

which is just Eq. (19.19). The same holds for $\lambda = \lambda_2$, but the ratio x_1/x_2 will be different.

There are two vectors $\mathbf{x}^{(1)}$ and $\mathbf{x}^{(2)}$, corresponding to the two numbers λ_1 and λ_2, whose components x_1 and x_2 have the ratios x_1/x_2 determined by the procedure just given. These are called *eigenvectors*. Only the ratio of their components can be determined.

We anticipate that, in differential form, $\mathbf{F}$ will contain the vector operator ∇. For the important special case of an infinite (or periodic) system, we can let $\nabla \rightarrow i\mathbf{k}$, where $\mathbf{k}$ is the wave vector in the direction of wave propagation with amplitude $k = 2\pi/\lambda$. (Here λ is the wavelength, *not* to be confused with the eigenvalue of the previous paragraph.) Equation (19.9) then becomes a set of linear algebraic equations whose roots can be written as

$$\omega^2 = \omega^2(\mathbf{k}). \qquad (19.23)$$

This is called the *dispersion relation* for the system under investigation. The roots (19.23) are also called the "characteristic oscillations," or the "normal modes."

If the system is not spatially periodic, or the substitition $\nabla \rightarrow i\mathbf{k}$ cannot be made for any other reason, then Eq. (19.8) is a differential equation that must be solved subject to the proper boundary conditions.

The procedure for studying the behavior of small displacements from stationary equilibrium in ideal MHD is therefore:

1. Find the functional form $\mathbf{F}\{\boldsymbol{\xi}\}$.
2. Determine the eigenvalues of $(\mathbf{F} + \rho_0\omega^2\mathbf{I}) \cdot \boldsymbol{\xi} = 0$. This may require solving a differential equation or the algebraic equation $\det[\mathbf{F} + \rho_0\omega^2\mathbf{I}] = 0$.
3. Examine the behavior of the eigenvalues of this equation with respect to their implications for oscillatory of exponentially growing behavior.

In Lecture 2 we introduced the concept of the *adjoint*, $\mathbf{F}^\dagger$, of an operator $\mathbf{F}$. The adjoint has the property that if $\mathbf{u} = \mathbf{F} \cdot \mathbf{x}$, then $\mathbf{u}^* = \mathbf{F}^\dagger \cdot \mathbf{x}^*$, where $(..)^*$ denotes the complex conjugate. If $\mathbf{F}$ is a matrix, then $F_{ij}^\dagger = F_{ji}^*$. If $\mathbf{F} = \mathbf{F}^\dagger$, then $\mathbf{F}$ is said to be *self-adjoint*. An important property of a self-adjoint operator is that it satisfies

$$\int dV \mathbf{u}^* \cdot \mathbf{F} \cdot \mathbf{v} = \int dV \mathbf{v}^* \cdot \mathbf{F} \cdot \mathbf{u}. \tag{19.24}$$

(More generally, $(\mathbf{u}, \mathbf{F} \cdot \mathbf{v}) = (\mathbf{v}, \mathbf{F} \cdot \mathbf{u})$, where $(\mathbf{x}, \mathbf{y})$ denotes an *inner product* in function space.)

We will soon prove that *the ideal MHD force operator is self-adjoint*. This has important consequences for the behavior of small oscillations in ideal MHD. First, let $\boldsymbol{\xi}_i$ and $\boldsymbol{\xi}_j$ be two eigenvectors of $\mathbf{F}$ corresponding to the eigenvalues $-\omega_i^2$ and $-\omega_j^2$, respectively, i.e.,

$$\mathbf{F} \cdot \boldsymbol{\xi}_i = -\rho_0\omega_i^2\boldsymbol{\xi}_i \tag{19.25}$$

and

$$\mathbf{F} \cdot \boldsymbol{\xi}_j = -\rho_0\omega_j^2\boldsymbol{\xi}_j. \tag{19.26}$$

The complex conjugate of Eq. (19.26) is

$$(\mathbf{F} \cdot \boldsymbol{\xi}_j)^* = \mathbf{F} \cdot \boldsymbol{\xi}_j^* = -\rho_0 (\omega_j^2)^* \boldsymbol{\xi}_j^*, \tag{19.27}$$

since $\mathbf{F}$ is assumed to be self-adjoint. Now dot $\boldsymbol{\xi}_j^*$ with Eq. (19.25), and $\boldsymbol{\xi}_i$ with Eq. (19.26), subtract, and integrate over all space:

$$\int \left(\boldsymbol{\xi}_j^* \cdot \mathbf{F} \cdot \boldsymbol{\xi}_i - \boldsymbol{\xi}_i \cdot \mathbf{F} \cdot \boldsymbol{\xi}_j^*\right) dV = -\rho_0 \left[\omega_i^2 - (\omega_j^2)^*\right] \int \boldsymbol{\xi}_i \cdot \boldsymbol{\xi}_j^* dV. \tag{19.28}$$

The left-hand side vanishes by Eq. (19.24), since $\mathbf{F}$ is assumed to be self-adjoint. Therefore,

$$\rho_0 \left[\omega_i^2 - (\omega_j^2)^*\right] \int \boldsymbol{\xi}_i \cdot \boldsymbol{\xi}_j^* dV = 0. \tag{19.29}$$

There are two non-trivial possibilities for satisfying Eq. (19.29). First, let $i = j$. Then, since $|\boldsymbol{\xi}_i|^2 = \boldsymbol{\xi}_i^* \cdot \boldsymbol{\xi}_i \neq 0$ because $\boldsymbol{\xi}_i$ is a *non-trivial* solution of Eq. (19.8), we require

$$\omega_i^2 = \left(\omega_i^2\right)^*, \tag{19.30}$$

i.e., *the eigenvalues of* $\mathbf{F}$ *are real*. Therefore, *in ideal MHD the normal modes are either purely oscillating or purely growing (or damped)*. Overstable modes are impossible. If $\omega_i^2 > 0$, then $\omega_i = \pm\Omega_i$, and the displacement evolves according to $\boldsymbol{\xi} \sim e^{\pm i\Omega_i t}$, so the normal modes are pure oscillations. If $\omega_i^2 < 0$, then $\omega_i = \pm i\gamma_i$, the displacement evolves according to $\boldsymbol{\xi} \sim e^{\pm\gamma t}$, and one of the normal modes exhibits pure exponential growth.

The second possibility for satisfying Eq. (19.29) is that $i \neq j$ and $\omega_i^2 \neq \omega_j^2$. Then

$$\int \boldsymbol{\xi}_i \cdot \boldsymbol{\xi}_j^* dV = 0, \tag{19.31}$$

so that *the eigenvectors of* $\mathbf{F}$ *are orthogonal*. Further, they can be normalized so that

$$\int \boldsymbol{\xi}_i \cdot \boldsymbol{\xi}_j^* dV = \delta_{ij}, \tag{19.32}$$

in which case they are said to be *orthonormal*.

Finally, it can be shown that the eigenvectors $\boldsymbol{\xi}_i$ form a *complete set*. By this we mean that any "piecewise continuous" function $\boldsymbol{\xi}(\mathbf{r})$ can be approximated "in the mean" arbitrarily closely by a linear combination of the eigenfunctions $\boldsymbol{\xi}_i$. That is, for any reasonably behaved displacement $\boldsymbol{\xi}(\mathbf{r})$, we can write

$$\boldsymbol{\xi}(\mathbf{r}, t) = \sum_{k=1}^{\infty} a_k \boldsymbol{\xi}_k(\mathbf{r}, t), \tag{19.33}$$

where a_k are complex numbers called *expansions coefficients*. Dotting Eq. (19.33) with $\boldsymbol{\xi}_j^*$ and integrating, we have

$$\int \boldsymbol{\xi}_j^* \cdot \boldsymbol{\xi} dV = \sum_{k=1}^{\infty} a_k \int \boldsymbol{\xi}_j^* \cdot \boldsymbol{\xi} dV,$$

so that, in light of Eq. (19.32),

$$a_j = \int \boldsymbol{\xi}_j^* \cdot \boldsymbol{\xi} dV. \tag{19.34}$$

Therefore, the behavior of any arbitrary displacement can be obtained by knowing the behavior of the eigenvectors $\boldsymbol{\xi}_k$. If we find the eigenvectors and eigenvalues, we will know the behavior of the system.

Recall that the time dependence of the eigenvectors is given by $\boldsymbol{\xi}_k(\mathbf{r}, t) = \boldsymbol{\xi}_k(\mathbf{r})e^{i\omega_k t}$, so that

$$\boldsymbol{\xi}(\mathbf{r}, t) = \sum_{k=1}^{\infty} a_k e^{i\omega_k t} \boldsymbol{\xi}_k(\mathbf{r}). \tag{19.35}$$

Therefore, if *all* of the quantities ω_k are *real* ($\omega_k^2 > 0$), then the system exhibits oscillatory behavior about its equilibrium position; it is *stable*. However, if *one* of the ω_k is *imaginary* ($\omega_k^2 < 0$), then the system exhibits exponential deviation from its equilibrium position. The existence of a single unstable eigenvector renders the entire system *unstable*.

Now we just need to find the functional form of $\mathbf{F}\{\boldsymbol{\xi}\}$ and prove that it is self-adjoint!

Lecture 20
Linearized Equations and the Ideal MHD Force Operator

Be wise. Linearize.

Ed Greitzer

The ideal MHD equations are

$$\frac{\partial \rho}{\partial t} = -\nabla \cdot \rho \mathbf{V}, \tag{20.1}$$

$$\rho \left(\frac{\partial \mathbf{V}}{\partial t} + \mathbf{V} \cdot \nabla \mathbf{V} \right) = -\nabla p + \frac{1}{\mu_0} (\nabla \times \mathbf{B}) \times \mathbf{B}, \tag{20.2}$$

$$\frac{\partial p}{\partial t} + \mathbf{V} \cdot \nabla p = -\Gamma p \nabla \cdot \mathbf{V}, \tag{20.3}$$

and

$$\frac{\partial \mathbf{B}}{\partial t} = \nabla \times (\mathbf{V} \times \mathbf{B}). \tag{20.4}$$

Since we are interested in the behavior of a system when it is perturbed only slightly from its equilibrium state, we write all dependent variables in the form

$$f(\mathbf{r}, t) = f_0(\mathbf{r}) + f_1(\mathbf{r}, t), \tag{20.5}$$

where f_0 is the value in the equilibrium state (i.e., the solution of Eqs. (20.1, 20.2, 20.3, 20.4) when $\partial/\partial t = 0$) and f_1 is a small perturbation, i.e., $|f_1/f_0| \ll 1$. When we substitute the *ansatz* (20.5) into Eqs. (20.1, 20.2, 20.3, 20.4), the nonlinear terms (like $\mathbf{V} \cdot \nabla \mathbf{V}$ and $\mathbf{V} \times \mathbf{B}$, for example) will behave as

$$
\begin{aligned}
uv &= (u_0 + u_1)(v_0 + v_1) \\
&= u_0 v_0 \left[1 + \frac{u_1}{u_0} + \frac{v_1}{v_0} + \left(\frac{u_1}{u_0} \right) \left(\frac{v_1}{v_0} \right) \right] \\
&\approx u_0 v_0 + v_0 u_1 + u_0 v_1,
\end{aligned} \tag{20.6}
$$

Schnack, D.D.: *Linearized Equations and the Ideal MHD Force Operator.* Lect. Notes Phys. **780**, 129–131 (2009)
DOI 10.1007/978-3-642-00688-3_20

since the product $(u_1/u_0)(v_1/v_0)$ is much, much less than either u_1/u_0 or v_1/v_0. The resulting equations will be *linear* in the perturbed quantities f_1; not surprisingly, the formal process is called *linearization*. Assuming a stationary ($\mathbf{V}_0 = 0$) equilibrium state ($\nabla p_0 = \mathbf{J}_0 \times \mathbf{B}_0$), the linearized ideal MHD equations are

$$\frac{\partial \rho_1}{\partial t} = -\nabla \cdot \rho_0 \mathbf{V}_1, \tag{20.7}$$

$$\rho_0 \frac{\partial \mathbf{V}_1}{\partial t} = -\nabla p_1 + \mathbf{J}_0 \times \mathbf{B}_1 + \mathbf{J}_1 \times \mathbf{B}_0, \tag{20.8}$$

$$\frac{\partial p_1}{\partial t} = -\mathbf{V}_1 \cdot \nabla p_0 - \Gamma p_0 \nabla \cdot \mathbf{V}_1, \tag{20.9}$$

and

$$\frac{\partial \mathbf{B}_1}{\partial t} = \nabla \times (\mathbf{V}_1 \times \mathbf{B}_0), \tag{20.10}$$

where $\mu_0 \mathbf{J}_0 = \nabla \times \mathbf{B}_0$ and $\mu_0 \mathbf{J}_1 = \nabla \times \mathbf{B}_1$. Given the equilibrium state $\rho_0(\mathbf{r})$, $p_0(\mathbf{r})$, and $\mathbf{B}_0(\mathbf{r})$, these are eight equations that can be solved for the eight unknowns ρ_1, $\mathbf{V}_1$, p_1, and $\mathbf{B}_1$.

Following Lecture 19, our goal is to write Eqs. (20.7), (20.8), (20.9), (20.10) in the form of a wave equation

$$\rho_0 \frac{\partial^2 \boldsymbol{\xi}}{\partial t^2} = \mathbf{F}\{\boldsymbol{\xi}\}, \tag{20.11}$$

where the displacement $\boldsymbol{\xi}$ is defined by

$$\mathbf{V}_1 = \frac{\partial \boldsymbol{\xi}}{\partial t}. \tag{20.12}$$

Substituting Eq. (20.12) into the continuity equation, Eq. (20.7), we have

$$\begin{aligned}
\frac{\partial \rho_1}{\partial t} &= -\nabla \cdot \rho_0 \mathbf{V}_1 \\
&= -\mathbf{V}_1 \cdot \nabla \rho_0 - \rho_0 \nabla \cdot \mathbf{V}_1 \\
&= -\frac{\partial \boldsymbol{\xi}}{\partial t} \cdot \nabla \rho_0 - \rho_0 \nabla \cdot \frac{\partial \boldsymbol{\xi}}{\partial t}
\end{aligned}$$

or, integrating with respect to time,

$$\rho_1 = -\boldsymbol{\xi} \cdot \nabla \rho_0 - \rho_0 \nabla \cdot \boldsymbol{\xi}. \tag{20.13}$$

Similarly, the energy equation, Eq. (20.9), becomes

$$p_1 = -\boldsymbol{\xi} \cdot \nabla p_0 - \Gamma p_0 \nabla \cdot \boldsymbol{\xi}, \tag{20.14}$$

and Eq. (20.10) is

$$\mathbf{Q} \equiv \mathbf{B}_1 = \nabla \times (\boldsymbol{\xi} \times \mathbf{B}_0). \tag{20.15}$$

(This is the standard notation.) Substituting Eqs. (20.12), (20.13), (20.14), (20.15) into the perturbed equation of motion, Eq. (20.8), we finally find

$$\rho_0 \frac{\partial^2 \boldsymbol{\xi}}{\partial t^2} = \mathbf{J}_0 \times [\nabla \times (\boldsymbol{\xi} \times \mathbf{B}_0)] + \frac{1}{\mu_0} \{\nabla \times [\nabla \times (\boldsymbol{\xi} \times \mathbf{B}_0)]\} \times \mathbf{B}_0$$

$$+ \nabla (\boldsymbol{\xi} \cdot \nabla p_0) + \Gamma \nabla (p_0 \nabla \cdot \boldsymbol{\xi}) \tag{20.16}$$

$$\equiv \mathbf{F} \{\boldsymbol{\xi}\}. \tag{20.17}$$

Equation (20.16) is the *ideal MHD wave equation*, and Eq. (20.17) defines the *ideal MHD force operator*. It will soon be important to remember that $\mathbf{F}$ depends on $\boldsymbol{\xi}$, but *not* on $\dot{\boldsymbol{\xi}}$.

Lecture 21
Boundary Conditions for Linearized Ideal MHD

BOUNDARY, n. An imaginary line between two nations, separating the imaginary rights of one from the imaginary rights of the other.

Ambrose Bierce, *The Devil's Dictionary*

Before we prove that $\mathbf{F}\{\boldsymbol{\xi}\}$ is self-adjoint, we must derive the boundary conditions to be imposed on the solutions of the linearized ideal MHD wave equation

$$\rho_0\ddot{\boldsymbol{\xi}} = \mathbf{J}_0 \times \mathbf{Q} + \frac{1}{\mu_0}(\nabla \times \mathbf{Q}) \times \mathbf{B}_0 + \nabla(\boldsymbol{\xi} \cdot \nabla p_0) + \Gamma\nabla(p_0\nabla \cdot \boldsymbol{\xi}), \qquad (21.1)$$

where $\mathbf{Q} = \nabla \times (\boldsymbol{\xi} \times \mathbf{B}_0)$. We consider the system to consist of a conducting fluid (a plasma) surrounded by a vacuum, all enclosed within a perfectly conducting boundary, as shown in Fig. 21.1.

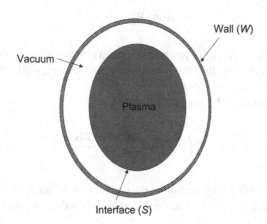

Fig. 21.1 A plasma in equilibrium surrounded by a vacuum and a perfectly conducting boundary

The conducting boundary (i.e., wall) is denoted as W, and the surface separating the plasma and the vacuum is denoted as S. The magnetic field is everywhere tangent to this surface, by construction.

Schnack, D.D.: *Boundary Conditions for Linearized Ideal MHD*. Lect. Notes Phys. **780**, 133–136 (2009)
DOI 10.1007/978-3-642-00688-3_21

Let $\hat{\mathbf{B}} = \nabla \times \hat{\mathbf{A}}$ be the magnetic field in the vacuum. Since there can be no electric current in the vacuum,

$$\nabla \times \nabla \times \hat{\mathbf{A}} = 0, \tag{21.2}$$

where $\hat{\mathbf{A}}$ is the vector potential in the vacuum region. At the wall W, the tangential component of the electric field must vanish: $\hat{\mathbf{n}} \times \hat{\mathbf{E}} = 0$, where $\hat{\mathbf{n}}$ is the outward drawn normal. Since $\hat{\mathbf{E}} = -\partial \hat{\mathbf{A}} / \partial t$, we require that

$$\hat{\mathbf{n}} \times \hat{\mathbf{A}}\Big|_{W} = 0. \tag{21.3}$$

This is the boundary condition to be applied to Eq. (21.2) at W.

Now consider the plasma/vacuum interface S. Here we define $\hat{\mathbf{n}}$ to point from the plasma into the vacuum. From Maxwell's equations, we know that, at the interface S, the magnetic field must satisfy the conditions

$$[\![\hat{\mathbf{n}} \cdot \mathbf{B}]\!] = 0, \tag{21.4}$$

$$[\![\hat{\mathbf{n}} \times \mathbf{B}]\!] = \mu_0 \mathbf{K}, \tag{21.5}$$

and

$$\hat{\mathbf{n}} \times [\![\mathbf{E}]\!] = 0, \tag{21.6}$$

where $[\![f]\!] \equiv f_V - f_P$ is the jump in a quantity across the interface and $\mathbf{K}$ is a surface current flowing within the interface S. If $\mathbf{K} = 0$, then all components of $\mathbf{B}$ are continuous across S.

The equilibrium force balance condition is

$$\nabla \left(p + \frac{B^2}{2\mu_0} \right) = \frac{\mathbf{B} \cdot \nabla \mathbf{B}}{\mu_0}. \tag{21.7}$$

Now define a variable n that measures the distance in the local direction of $\hat{\mathbf{n}}$, i.e., across the interface at $n = 0$. Then $\hat{\mathbf{n}} \cdot \nabla \equiv \partial / \partial n$, and the normal component of Eq. (21.7) is

$$\frac{\partial}{\partial n} \left(p + \frac{B^2}{2\mu_0} \right) = \frac{1}{\mu_0} \hat{\mathbf{n}} \cdot (\mathbf{B} \cdot \nabla \mathbf{B}). \tag{21.8}$$

Integrating this expression across the interface S, we have

$$\int_{-\varepsilon}^{\varepsilon} \frac{\partial}{\partial n} \left(p + \frac{B^2}{2\mu_0} \right) dn = \frac{1}{\mu_0} \int_{-\varepsilon}^{\varepsilon} \hat{\mathbf{n}} \cdot (\mathbf{B} \cdot \nabla \mathbf{B}) \, dn. \tag{21.9}$$

When $\varepsilon \to 0$, this becomes

$$\left[\!\!\left[p + \frac{B^2}{2\mu_0} \right]\!\!\right] = \frac{1}{\mu_0} \lim_{\varepsilon \to 0} \int_{-\varepsilon}^{\varepsilon} \hat{\mathbf{n}} \cdot (\mathbf{B} \cdot \nabla \mathbf{B}) \, dn = 0, \tag{21.10}$$

since the integrand on the right-hand side is continuous across S.

The condition (21.10) is called the *pressure balance condition*. Let $\mathbf{r}_0$ be the equilibrium position of the plasma/vacuum interface S_0. Then, from Eq. (21.10), the equilibrium variables must satisfy

$$p_0(\mathbf{r}_0) + \frac{B_0^2(\mathbf{r}_0)}{2\mu_0} = \frac{\hat{B}_0^2(\mathbf{r}_0)}{2\mu_0}, \tag{21.11}$$

since the pressure in the vacuum must vanish. During the displacement of the plasma, the plasma/vacuum interface will also be displaced to a new location $\mathbf{r}' = \mathbf{r}_0 + \boldsymbol{\xi}$. At this perturbed boundary S', the pressure balance condition is

$$p_0(\mathbf{r}_0 + \boldsymbol{\xi}) + p_1(\mathbf{r}_0 + \boldsymbol{\xi}) + \frac{1}{2\mu_0} [\mathbf{B}_0(\mathbf{r}_0 + \boldsymbol{\xi}) + \mathbf{B}_1(\mathbf{r}_0 + \boldsymbol{\xi})]^2 =$$
$$\frac{1}{2\mu_0} \left[\hat{\mathbf{B}}_0(\mathbf{r}_0 + \boldsymbol{\xi}) + \hat{\mathbf{B}}_1(\mathbf{r}_0 + \boldsymbol{\xi}) \right]^2. \tag{21.12}$$

To the lowest order in small quantities, this becomes

$$p_0(\mathbf{r}') + p_1(\mathbf{r}') + \frac{1}{2\mu_0} \left[B_0^2(\mathbf{r}') + 2\mathbf{B}_0(\mathbf{r}') \cdot \mathbf{B}_1(\mathbf{r}') \right] =$$
$$\frac{1}{2\mu_0} \left[\hat{B}_0^2(\mathbf{r}') + 2\hat{\mathbf{B}}_0(\mathbf{r}') \cdot \hat{\mathbf{B}}_1(\mathbf{r}') \right]. \tag{21.13}$$

All quantities at S' can be related to quantities at S_0 by

$$f(\mathbf{r}') = f(\mathbf{r}_0) + \boldsymbol{\xi} \cdot \nabla f(\mathbf{r}_0), \tag{21.14}$$

and as a consequence, to the lowest order $\mathbf{B}_0(\mathbf{r}') \cdot \mathbf{B}_1(\mathbf{r}') = \mathbf{B}_0(\mathbf{r}_0) \cdot \mathbf{B}_1(\mathbf{r}_0)$. The perturbed pressure is determined from the linearized energy equation

$$p_1 = -\boldsymbol{\xi} \cdot \nabla p_0 - \Gamma p_0 \nabla \cdot \boldsymbol{\xi}. \tag{21.15}$$

Finally, substituting all this into Eq. (21.13) and using the equilibrium pressure balance condition, Eq. (21.11), we find that the condition

$$-\Gamma p_0 \nabla \cdot \boldsymbol{\xi} + \frac{1}{\mu_0} (\boldsymbol{\xi} \cdot \nabla \mathbf{B}_0 + \mathbf{Q}) \cdot \mathbf{B}_0 = \frac{1}{\mu_0} \left(\boldsymbol{\xi} \cdot \nabla \hat{\mathbf{B}}_0 + \nabla \times \hat{\mathbf{A}} \right) \cdot \mathbf{B}_0 \qquad (21.16)$$

must be satisfied at the equilibrium interface S_0.

A second condition to be satisfied at S_0 is found from Eq. (21.6). Let the boundary be moving with velocity $\mathbf{V}_1 = \partial \boldsymbol{\xi} / \partial t$, $\mathbf{E}$ be the electric field measured in the stationary frame of reference, and $\mathbf{E}^* = \mathbf{E} + \mathbf{V}_1 \times \mathbf{B}_0$ be the electric field seen in the frame moving with the boundary. In the plasma, $\mathbf{E} = -\mathbf{V}_1 \times \mathbf{B}_0$, by Ohm's law. Therefore $\mathbf{E}^* = 0$; the electric field in the frame moving with the boundary must vanish on the plasma side of the interface. Then Eq. (21.6) requires that $\hat{\mathbf{n}} \times \hat{\mathbf{E}}^* = 0$, i.e., the tangential electric field in the frame moving with the boundary must vanish in the vacuum. Since $\hat{\mathbf{E}}^* = \hat{\mathbf{E}} + \mathbf{V}_1 \times \hat{\mathbf{B}}_0$ by the (non-relativistic) law of transformation of the electromagnetic fields, we have

$$\hat{\mathbf{n}} \times \hat{\mathbf{E}} = \hat{\mathbf{B}}_0 (\hat{\mathbf{n}} \cdot \mathbf{V}_1) - \mathbf{V}_1 \left(\hat{\mathbf{n}} \cdot \hat{\mathbf{B}}_0 \right). \qquad (21.17)$$

But on S_0, $\hat{\mathbf{n}} \cdot \mathbf{B}_0 = \hat{\mathbf{n}} \cdot \hat{\mathbf{B}}_0 = 0$, by construction. Then using $\hat{\mathbf{E}} = -\partial \hat{\mathbf{A}} / \partial t$ and $\mathbf{V}_1 = \partial \boldsymbol{\xi} / \partial t$, we require that

$$\hat{\mathbf{n}} \times \hat{\mathbf{A}} = -(\hat{\mathbf{n}} \cdot \boldsymbol{\xi}) \hat{\mathbf{B}}_0 \qquad (21.18)$$

be satisfied on S_0.

Then the *boundary value problem* for determining the normal modes of the system can be stated as follows. In the fluid, solve

$$\rho_0 \ddot{\boldsymbol{\xi}} = \mathbf{J}_0 \times \mathbf{Q} + \frac{1}{\mu_0} (\nabla \times \mathbf{Q}) \times \mathbf{B}_0 + \nabla (\boldsymbol{\xi} \cdot \nabla p_0) + \Gamma \nabla (p_0 \nabla \cdot \boldsymbol{\xi}), \qquad (21.19)$$

where $\mathbf{Q} = \nabla \times (\boldsymbol{\xi} \times \mathbf{B}_0)$, and in the vacuum, solve

$$\nabla \times \nabla \times \hat{\mathbf{A}} = 0, \qquad (21.20)$$

subject to the *matching conditions*

$$-\Gamma p_0 \nabla \cdot \boldsymbol{\xi} + \frac{1}{\mu_0} (\boldsymbol{\xi} \cdot \nabla \mathbf{B}_0 + \mathbf{Q}) \cdot \mathbf{B}_0 = \frac{1}{\mu_0} \left(\boldsymbol{\xi} \cdot \nabla \hat{\mathbf{B}}_0 + \nabla \times \hat{\mathbf{A}} \right) \cdot \mathbf{B}_0 \qquad (21.21)$$

and

$$\hat{\mathbf{n}} \times \hat{\mathbf{A}} = -(\hat{\mathbf{n}} \cdot \boldsymbol{\xi}) \hat{\mathbf{B}}_0 \qquad (21.22)$$

at the vacuum/fluid interface, and the *boundary condition*

$$\hat{\mathbf{n}} \times \hat{\mathbf{A}} = 0 \qquad (21.23)$$

at the perfectly conducting wall.

Lecture 22
Proof that the Ideal MHD Force Operator is Self-Adjoint

> *The mind's first step to self awareness must be through the body.*
>
> George Sheehan

The ideal MHD equation of motion is

$$\rho_0 \ddot{\boldsymbol{\xi}} = \mathbf{F}\{\boldsymbol{\xi}\}. \tag{22.1}$$

We will now prove that the ideal MHD force operator $\mathbf{F}\{\boldsymbol{\xi}\}$ is self-adjoint.[1] That is, we will demonstrate that *for any two vector fields $\boldsymbol{\xi}$ and $\boldsymbol{\eta}$ satisfying the same boundary conditions,*

$$\int dV \boldsymbol{\eta} \cdot \mathbf{F}\{\boldsymbol{\xi}\} = \int dV \boldsymbol{\xi} \cdot \mathbf{F}\{\boldsymbol{\eta}\}. \tag{22.2}$$

Proof The total energy of the system is

$$U = \int \left(\frac{1}{2}\rho V^2 + \frac{B^2}{2\mu_0} + \frac{p}{\Gamma - 1} \right) dV. \tag{22.3}$$

We have seen that this quantity is conserved in ideal MHD, so that

$$U = K + \delta W = \text{constant}. \tag{22.4}$$

Here, K is the kinetic energy and δW is the change in the potential energy of the system, as a result of the displacement. Therefore

$$\dot{U} = \dot{K} + \delta\dot{W} = 0. \tag{22.5}$$

---

[1] This proof is due to I. B. Bernstein, E. A. Frieman, M. D. Kruskal, and R. M. Kulsrud, *Proc. Roy. Soc.* (London), **A244**, 17 (1958).

Schnack, D.D.: *Proof that the Ideal MHD Force Operator is Self-Adjoint.* Lect. Notes Phys. **780**, 137–140 (2009)
DOI 10.1007/978-3-642-00688-3_22

Now the kinetic energy is

$$K = \frac{1}{2} \int \rho_0 V_1^2 dV = \frac{1}{2} \int \rho_0 \dot{\xi}^2 dV, \tag{22.6}$$

or

$$K\{\dot{\xi}, \dot{\xi}\} = \frac{1}{2} \int \rho_0 \dot{\xi} \cdot \dot{\xi} dV, \tag{22.7}$$

so that K is a *symmetric functional of* $\dot{\xi}$. Then the rate of change of kinetic energy is

$$\dot{K} = \int \rho_0 \dot{\xi} \cdot \ddot{\xi} dV = 2K\{\dot{\xi}, \ddot{\xi}\}$$
$$= 2K\{\ddot{\xi}, \dot{\xi}\} \tag{22.8}$$

because of the symmetry of the arguments. In light of Eq. (22.1), this can be written as

$$\dot{K} = 2K\left\{\frac{1}{\rho_0} \mathbf{F}\{\xi\}, \dot{\xi}\right\} \tag{22.9}$$

or, from Eq. (22.6),

$$\dot{K} = -\delta \dot{W}$$
$$= -\delta W\{\xi, \dot{\xi}\} - \delta W\{\dot{\xi}, \dot{\xi}\}. \tag{22.10}$$

Let $\mathbf{r}_0$ stands for the equilibrium position of the system and let $W(\mathbf{r}_0)$ be the equilibrium potential energy of the system. The potential energy after a small displacement ξ will change according to

$$W(\mathbf{r}_0 + \xi) = W(\mathbf{r}_0) + \sum_i \left(\frac{\partial W}{\partial r_i}\right)_0 \xi_i + \frac{1}{2} \sum_i \sum_j \left(\frac{\partial^2 W}{\partial r_i \partial r_j}\right)_0 \xi_i \xi_j + \cdots \tag{22.11}$$

But equilibrium is an extremum of the potential energy, so that $(\partial W/\partial r_i)_0 = 0$. Therefore

$$\delta W = W(\mathbf{r}_0 + \xi) - W(\mathbf{r}_0)$$
$$= \frac{1}{2} \sum_i \sum_j \left(\frac{\partial^2 W}{\partial r_i \partial r_j}\right)_0 \xi_i \xi_j$$
$$= \delta W\{\xi, \xi\}, \tag{22.12}$$

which is a *symmetric quadratic form*, so that δW is also a symmetric function of its arguments. For any two vector functions ξ and η satisfying the same boundary conditions,

$$\delta W \{\xi, \eta\} = \delta W \{\eta, \xi\}. \tag{22.13}$$

Applying this to Eq. (22.10), we have

$$\dot{K} = -2\delta W \{\dot{\xi}, \xi\} \tag{22.14}$$

or, using Eq. (22.9),

$$K \left\{ \frac{1}{\rho_0} F \{\xi\}, \dot{\xi} \right\} = -\delta W \{\dot{\xi}, \xi\}. \tag{22.15}$$

Now, it is important to recognize that ξ and $\dot{\xi}$ are *independent functions*. To avoid confusion on this point, we replace $\dot{\xi}$ by η, so that Eq. (22.15) becomes

$$K \left\{ \frac{1}{\rho_0} F \{\xi\}, \eta \right\} = -\delta W \{\eta, \xi\}. \tag{22.16}$$

Equation (22.16) must hold for arbitrary ξ and η. In particular, it must hold if we *interchange* ξ and η, i.e.,

$$K \left\{ \frac{1}{\rho_0} F \{\eta\}, \xi \right\} = -\delta W \{\xi, \eta\}. \tag{22.17}$$

Using Eqs. (22.13) and (22.16),

$$\delta W \{\xi, \eta\} = \delta W \{\eta, \xi\}$$
$$= -K \left\{ \frac{1}{\rho_0} F \{\xi\}, \eta \right\}. \tag{22.18}$$

Therefore, from Eq. (22.9),

$$K \left\{ \frac{1}{\rho_0} F \{\eta\}, \xi \right\} = -\delta W \{\xi, \eta\}$$
$$= K \left\{ \frac{1}{\rho_0} F \{\xi\}, \eta \right\}. \tag{22.19}$$

But, by definition,

$$K \{\xi, \eta\} = \frac{1}{2} \int dV \rho_0 \xi \cdot \eta,$$

so that

$$\int dV \eta \cdot F \{\xi\} = \int dV \xi \cdot F \{\eta\}, \tag{22.20}$$

which is identical to Eq. (22.2). Therefore, $F \{\xi\}$ is self-adjoint. *Q.E.D.*

This ingenious proof relies only on the conservation of energy, on $\mathbf{F}$ being independent of $\dot{\boldsymbol{\xi}}$, and on K and δW being symmetric functions of their arguments. It does not depend on any specific form of $\mathbf{F}$. In ideal MHD, $\mathbf{F}$ depends on $\mathbf{V} = \dot{\boldsymbol{\xi}}$ only through $\mathbf{B}_1$, defined by Faraday's law and Ohm's law; the explicit dependence on $\dot{\boldsymbol{\xi}}$ integrates out through the definition $\mathbf{Q} = \nabla \times (\boldsymbol{\xi} \times \mathbf{B}_0)$. Note that, if the equilibrium is not stationary, so that $\mathbf{V}_0 \neq 0$, then the explicit dependence on $\dot{\boldsymbol{\xi}}$ will remain, and $\mathbf{F}$ is no longer self-adjoint. Similarly, the so-called two-fluid extensions of Ohm's law negate the convenient time integration that occurs in ideal MHD, and self-adjointness is also lost in this case.

From Eq. (22.19), we have (with $\boldsymbol{\xi} = \boldsymbol{\eta}$)

$$\delta W\{\boldsymbol{\xi}, \boldsymbol{\xi}\} = -K\left\{\frac{1}{\rho_0}\mathbf{F}\{\boldsymbol{\xi}\}, \boldsymbol{\xi}\right\}$$

$$= -\frac{1}{2}\int dV \boldsymbol{\xi} \cdot \mathbf{F}\{\boldsymbol{\xi}\}. \tag{22.21}$$

Therefore, if $\boldsymbol{\xi} \cdot \mathbf{F}\{\boldsymbol{\xi}\} > 0$ the displacement decreases the potential energy of the system and causes instability. This is consistent with the previous discussions.

In Lecture 19, we discussed some consequences of the self-adjointness of $\mathbf{F}$. These are

1. The eigenfunctions $-\omega_k^2$ are *real*.
2. The eigenfunctions $\boldsymbol{\xi}_k$ are *orthogonal*.
3. The eigenfunctions $\boldsymbol{\xi}_k$ form a *complete set*.

A further important consequence is that an *energy principle* exists. Substitute the eigenfunction expansion $\boldsymbol{\xi} = \sum_j a_j \boldsymbol{\xi}_j$ into the expression for the change in the potential energy, Eq. (22.21), and use Eq. (22.1) and the fact that $\mathbf{F}$ is linear:

$$\delta W = \frac{1}{2}\sum_j\sum_k a_j a_k \omega_k^2 \int \rho_0 \boldsymbol{\xi}_j \cdot \boldsymbol{\xi}_k dV$$

$$= \frac{1}{2}\sum_j\sum_k a_j a_k \omega_k^2 \delta_{jk}$$

$$= \frac{1}{2}\sum_j a_j^2 \omega_j^2. \tag{22.22}$$

Therefore, $\delta W < 0$ if and only if there exists at least one unstable eigenmode (with $\omega_j^2 < 0$), i.e., $\delta W < 0$ *is a necessary and sufficient condition for instability*. Mathematically, this means that we can use a *variational principle* to test whether the system is unstable without having to solve the underlying boundary value problem. This is a very powerful tool for theoretical and computational analysis. However, it requires a discussion of the calculus of variations. We will defer this discussion until later in this course.

Lecture 23
Waves in a Uniform Medium: Special Cases

I'm pickin' up good vibrations.

The Beach Boys

We consider the special case of an infinite, uniform medium with $\mathbf{B}_0 = B_0 \hat{\mathbf{e}}_z$ and $\mathbf{J}_0 = 0$. In that case we can expand an arbitrary displacement in plane wave solutions as

$$\boldsymbol{\xi}(\mathbf{r}, t) = \sum_{\mathbf{k}} \boldsymbol{\xi}_{\mathbf{k}} e^{i(\mathbf{k}\cdot\mathbf{r}+\omega_k t)}, \tag{23.1}$$

where $\mathbf{k}$ is the wave vector and the addition of the complex conjugate is implied. When we substitute this into the ideal MHD wave equation, we find that $\nabla \rightarrow i\mathbf{k}$ and $\partial/\partial t \rightarrow i\omega$, so that the problem is reduced to algebra. The result of the substitution is

$$\omega^2 \boldsymbol{\xi} = V_A^2 \left\{ \mathbf{k} \times \left[\mathbf{k} \times (\boldsymbol{\xi} \times \hat{\mathbf{b}}) \right] \right\} \times \hat{\mathbf{b}} + C_S^2 (\mathbf{k}\cdot\boldsymbol{\xi}) \boldsymbol{\xi}, \tag{23.2}$$

where $V_A^2 = B_0^2/2\mu_0\rho_0$ is the square of the Alfvén speed, $C_S^2 = \Gamma p_0/\rho_0$ is the square of the sound speed, and $\hat{\mathbf{b}}(= \hat{\mathbf{e}}_z)$ is a unit vector in the direction of the equilibrium magnetic field. This represents three coupled homogenous equations of the form $(\mathbf{A} - \omega^2 \mathbf{I}) \cdot \boldsymbol{\xi} = 0$ in the three unknowns ξ_x, ξ_y, and ξ_z. It has non-trivial solutions when

$$\det(\mathbf{A} - \omega^2 \mathbf{I}) = 0. \tag{23.3}$$

Equation (23.3) is the *dispersion relation for waves in a uniform medium*. The solutions $\omega^2(\mathbf{k})$ are the characteristic vibrational frequencies of the system

Expanding the vector identities, Eq. (23.2) can be rewritten as

$$\left[\omega^2 - V_A^2 (\mathbf{k}\cdot\hat{\mathbf{b}})^2 \right] \boldsymbol{\xi} = \left[(C_S^2 + V_A^2)(\mathbf{k}\cdot\boldsymbol{\xi}) - V_A^2 (\mathbf{k}\cdot\hat{\mathbf{b}})(\boldsymbol{\xi}\cdot\hat{\mathbf{b}}) \right] \mathbf{k}$$
$$- V_A^2 (\mathbf{k}\cdot\hat{\mathbf{b}})(\mathbf{k}\cdot\boldsymbol{\xi}) \hat{\mathbf{b}}. \tag{23.4}$$

Schnack, D.D.: *Waves in a Uniform Medium: Special Cases.* Lect. Notes Phys. **780**, 141–145 (2009)
DOI 10.1007/978-3-642-00688-3_23 © Springer-Verlag Berlin Heidelberg 2009

This can be put in another form by defining a unit vector in the direction of $\mathbf{k}$ as $\mathbf{k} = k\hat{\mathbf{k}}$, and defining θ as the angle between the direction of propagation and the magnetic field, $\hat{\mathbf{k}} \cdot \hat{\mathbf{b}} = \cos\theta$. The result is

$$\left[\frac{\omega^2}{k^2} - V_A^2\cos^2\theta\right]\boldsymbol{\xi} = \left[\left(C_S^2 + V_A^2\right)\left(\hat{\mathbf{k}} \cdot \boldsymbol{\xi}\right) - \left(\boldsymbol{\xi} \cdot \hat{\mathbf{b}}\right) V_A^2\cos\theta\right]\hat{\mathbf{k}}$$
$$- \left(\hat{\mathbf{k}} \cdot \boldsymbol{\xi}\right) V_A^2\cos\theta\,\hat{\mathbf{b}}. \qquad (23.5)$$

Equation (23.5) contains three *independent* vectors: the displacement $\boldsymbol{\xi}$, the direction of propagation $\hat{\mathbf{k}}$, and the direction of the magnetic field $\hat{\mathbf{b}}$. The properties of the waves depend on their relative orientation. In this lecture, we will consider several special cases.

Case I. Transverse Waves:
Here the displacement is perpendicular to the direction of propagation or $\hat{\mathbf{k}} \cdot \boldsymbol{\xi} = 0$, as shown in Fig. 23.1.

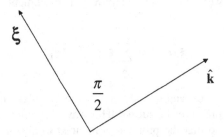

Fig. 23.1 Displacement ($\boldsymbol{\xi}$) and propagation vector ($\hat{\mathbf{k}}$) for transverse waves

Then Eq. (23.5) becomes

$$\left[\frac{\omega^2}{k^2} - V_A^2\cos^2\theta\right]\boldsymbol{\xi} + \left(\boldsymbol{\xi} \cdot \hat{\mathbf{b}}\right) V_A^2\cos\theta\,\hat{\mathbf{k}} = 0. \qquad (23.6)$$

Since $\boldsymbol{\xi}$ and $\hat{\mathbf{k}}$ are linearly independent, their coefficients must vanish individually.
Case IA: We first examine the vanishing of the coefficient of $\hat{\mathbf{k}}$,

$$\left(\boldsymbol{\xi} \cdot \hat{\mathbf{b}}\right) V_A^2\cos\theta = 0. \qquad (23.7)$$

There are two possibilities, which are as follows:.

1. $\boldsymbol{\xi} \cdot \hat{\mathbf{b}} = 0$. If $\cos\theta = \hat{\mathbf{k}} \cdot \hat{\mathbf{b}} \neq 0$, then $\hat{\mathbf{b}}$ must be perpendicular to $\boldsymbol{\xi}$, so $\boldsymbol{\xi}$ is perpendicular to both $\hat{\mathbf{k}}$ and $\hat{\mathbf{b}}$, as shown in Fig. 23.2.
2. $\cos\theta = \hat{\mathbf{k}} \cdot \hat{\mathbf{b}} = 0$ or $\theta = \pm\pi/2$. Then $\hat{\mathbf{k}}, \hat{\mathbf{b}}$, and $\boldsymbol{\xi}$ are mutually perpendicular. This is just a special case of possibility (1), above. Together, (1) and (2) determine the *polarization* of the wave.

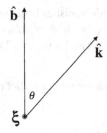

Fig. 23.2 The displacement (ξ) is perpendicular to both the direction of propagation ($\hat{k}$) and the direction of the magnetic field ($\hat{b}$)

Case IB: The vanishing of the coefficient of ξ in Eq. (23.6) leads to

$$\frac{\omega}{k} = \pm V_A \cos\theta \qquad (23.8)$$

or

$$\omega = \pm k_\parallel V_A, \qquad (23.9)$$

where $k_\parallel = k\cos\theta$ is the component of **k** parallel to the magnetic field. The polarization of this wave is given by IA1 and IA2, above. It is a *transverse wave* that propagates *along* the magnetic field. It does *not* propagate *across* the magnetic field. This is another manifestation of the anisotropy introduced by the presence of the magnetic field. Its phase velocity is $\omega/k_\parallel = V_A$. Note that if there is no magnetic field, $V_A = 0$ and the wave does not exist. It only occurs in magnetized fluids.[1] This is the famous *Alfvén wave* for which Hannes Alfvén won the Nobel Prize in 1970.

The behavior of the perturbed magnetic field in the Alfvén wave is found from Faraday's law and Ohm's law. In this case, the former is

$$\omega \mathbf{B}_1 = -\mathbf{k} \times \mathbf{E}_1 \qquad (23.10)$$

and the latter is

$$\mathbf{E}_1 = -i\omega B_0 \left(\xi \times \hat{b}\right). \qquad (23.11)$$

These can be combined as

$$\mathbf{B}_1 = ikB_0 \left[\xi \left(\hat{k} \cdot \hat{b}\right) - \hat{b} \left(\hat{k} \cdot \xi\right)\right]. \qquad (23.12)$$

---

[1] We have remarked that the Alfvén wave is a new type of wave, i.e., a shear wave that can propagate in a fluid. It is actually the remnant of a light wave in the very low-frequency limit. The inertia of the medium plays the role of the displacement current. Viewed this way, its polarization is not surprising.

The last term on the right-hand side vanishes because the wave is transverse in this case, i.e., $\hat{\mathbf{k}} \cdot \boldsymbol{\xi} = 0$. Then $\mathbf{B}_1 = ik B_0 \left(\hat{\mathbf{k}} \cdot \hat{\mathbf{b}} \right) \boldsymbol{\xi} = ik B_0 \cos \theta \boldsymbol{\xi}$. The perturbed magnetic field $\mathbf{B}_1$ is thus in the same direction as $\boldsymbol{\xi}$, but is $\pi/2$ out of phase (due to the i).

Case II. Longitudinal Waves:

For a longitudinal wave, the displacement is parallel to the direction of propagation, or $\boldsymbol{\xi} = \xi \hat{\mathbf{k}}$, as shown in Fig. 23.3.

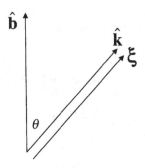

Fig. 23.3 In a longitudinal wave, the displacement ($\boldsymbol{\xi}$) is parallel to the direction of propagation ($\hat{\mathbf{k}}$) but makes an angle with the direction of the magnetic field ($\hat{\mathbf{b}}$)

After some algebra, Eq. (23.5) becomes

$$\left[\frac{\omega^2}{k^2} - \left(C_S^2 + V_A^2 \right) \right] \hat{\mathbf{k}} + V_A^2 \cos \theta \hat{\mathbf{b}} = 0. \tag{23.13}$$

We consider the following special cases.

Case IIA: Here $\cos \theta = \hat{\mathbf{k}} \cdot \hat{\mathbf{b}} = 0$, so that $\hat{\mathbf{k}}$ is perpendicular to $\hat{\mathbf{b}}$ and $\theta = \pm \pi/2$. Then Eq. (23.13) reduces to

$$\omega = \pm \left(C_S^2 + V_A^2 \right)^{1/2} k. \tag{23.14}$$

This is a longitudinal wave that propagates across the magnetic field. The square of its phase velocity is the sum of the squares of the sound speed and the Aflvén speed. It is called the *magneto-acoustic (MA)* wave.

The perturbed magnetic field is found from Eq. (23.12) with $\hat{\mathbf{k}} \cdot \hat{\mathbf{b}} = 0$ and $\hat{\mathbf{k}} \cdot \boldsymbol{\xi} = \xi$:

$$\mathbf{B}_1 = -ik B_0 \xi \hat{\mathbf{b}}. \tag{23.15}$$

The perturbed field is parallel to the mean field and $-\pi/2$ out of phase with the displacement. The perturbed field thus reinforces the mean field during a part of the cycle and weakens it during another part. This causes a perturbed magnetic pressure that acts in the same manner as the perturbed fluid pressure. It can support a longitudinal wave.

Case IIB: In this special case, $\hat{\mathbf{b}}$, $\hat{\mathbf{k}}$, and $\boldsymbol{\xi}$ are all parallel. The Eq. (23.13) becomes

$$\omega = \pm k C_{\mathrm{s}}. \tag{23.16}$$

The perturbed magnetic field vanishes:

$$\mathbf{B}_1 = ikB_0 \left[\boldsymbol{\xi} \left(\hat{\mathbf{k}} \cdot \hat{\mathbf{b}} \right) - \hat{\mathbf{b}} \left(\hat{\mathbf{k}} \cdot \boldsymbol{\xi} \right) \right] = ikB_0\xi \left(\cos\theta - \cos\theta \right) = 0. \tag{23.17}$$

This is a *sound wave* propagating parallel to the magnetic field.

Lecture 24
Waves in a Uniform Medium: Arbitrary Angle of Propagation

> *Crude classifications and false generalizations are the curse of organized life.*
>
> George Bernard Shaw

In Lecture 23, we showed that in an infinite, uniform medium, the solutions of the ideal MHD wave equation could be decomposed into plane wave solutions $\xi_{\mathbf{k}} e^{i(\mathbf{k}\cdot\mathbf{r} + \omega_k t)}$ that satisfy

$$\left[\omega^2 - \left(\hat{\mathbf{k}}\cdot\hat{\mathbf{b}}\right)^2 V_A^2\right]\boldsymbol{\xi} = \left[\left(C_S^2 + V_A^2\right)\left(\hat{\mathbf{k}}\cdot\boldsymbol{\xi}\right) - V_A^2\left(\boldsymbol{\xi}\cdot\hat{\mathbf{b}}\right)\left(\hat{\mathbf{k}}\cdot\hat{\mathbf{b}}\right)\right]\hat{\mathbf{k}}$$
$$- V_A^2\left(\hat{\mathbf{k}}\cdot\boldsymbol{\xi}\right)\left(\hat{\mathbf{k}}\cdot\hat{\mathbf{b}}\right)\hat{\mathbf{b}}. \tag{24.1}$$

There we examined several special cases of propagation both perpendicular and parallel to the magnetic field. Here we examine the more general case of propagation at an arbitrary angle $\theta = \cos^{-1}\left(\hat{\mathbf{k}}\cdot\hat{\mathbf{b}}\right)$ to the magnetic field.

Equation (24.1) is a 3×3 system, so its characteristic equation will yield three roots for ω^2: ω_0^2, ω_1^2, and ω_2^2. This results in six possible waves:

- Two shear waves, with $\omega = \pm\omega_0$;
- Two MA waves, with $\omega = \pm\omega_1$; and
- Two sound waves, with $\omega = \pm\omega_2$.

To be specific, we let $\hat{\mathbf{b}} = \hat{\mathbf{e}}_z$, $\mathbf{k} = k_\perp\hat{\mathbf{e}}_x + k_\parallel\hat{\mathbf{e}}_z$, $\boldsymbol{\xi} = \xi_x\hat{\mathbf{e}}_x + \xi_y\hat{\mathbf{e}}_y + \xi_z\hat{\mathbf{e}}_z$, and $k^2 = k_\parallel^2 + k_\perp^2$. Using this in Eq. (24.1), we find

$$x\text{-component: } \left(V_A^2 k^2 + C_S^2 k_\perp^2\right)\xi_x + C_S^2 k_\parallel k_\perp \xi_z = \omega^2 \xi_x, \tag{24.2}$$

$$y\text{-component: } V_A^2 k_\parallel^2 \xi_y = \omega^2 \xi_y, \tag{24.3}$$

$$z\text{-component: } C_S^2 k_\parallel k_\perp \xi_x + C_S^2 k_\parallel^2 \xi_z = \omega^2 \xi_z. \tag{24.4}$$

Notice that the y-component decouples from the x- and z-components. This immediately gives the eigenvalue

Schnack, D.D.: *Waves in a Uniform Medium: Arbitrary Angle of Propagation*. Lect. Notes Phys. **780**, 147–152 (2009)
DOI 10.1007/978-3-642-00688-3_24

$$\omega_0^2 = k_\parallel^2 V_A^2 \tag{24.5}$$

or $\omega_0^2 = k^2 V_A^2 \cos^2 \theta$, and the eigenvector

$$\xi_0 = \begin{pmatrix} 0 \\ 1 \\ 0 \end{pmatrix}, \tag{24.6}$$

i.e., $\xi_x = \xi_z = 0$, $\xi_y = 1$. This is the *shear Alfvén wave* found in Lecture 23. The polarization is shown in Fig. 24.1.

Fig. 24.1 Displacement vector (ξ), the propagation vector ($\hat{k}$), and the direction of the magnetic field ($\hat{b}$) for transverse waves

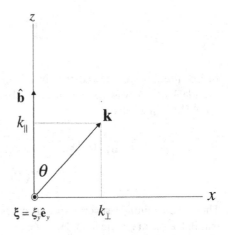

The two remaining eigenvectors have $\xi_x \neq 0$, $\xi_z \neq 0$, and $\xi_y = 0$. The characteristic equation for the coupled x- and z-components (24.2) and (24.3) is

$$\left(\frac{\omega}{k}\right)^4 - (C_S^2 + V_A^2)\left(\frac{\omega}{k}\right)^2 + C_S^2 V_A^2 \cos^2 \theta = 0. \tag{24.7}$$

The eigenvalues are

$$\left(\frac{\omega}{k}\right)_{1,2}^2 = \frac{1}{2}(C_S^2 + V_A^2)\left[1 \pm \sqrt{1 - \frac{4 C_S^2 V_A^2 \cos^2 \theta}{(C_S^2 + V_A^2)}}\right]. \tag{24.8}$$

Estimates of these solutions can be made for the interesting special case of $C_S^2 / V_A^2 = \Gamma \beta / 2 \ll 1$ (strong magnetic field). Then to the lowest order in this parameter, the eigenvalue corresponding to the ($+$) sign is

$$\left(\frac{\omega}{k}\right)_1^2 \approx V_A^2 \left(1 + \frac{C_S^2}{V_A^2} \sin^2 \theta\right). \tag{24.9}$$

For $\theta = \pi/2$ (**k** perpendicular to $\hat{\mathbf{b}}$, or propagation across the field), this mode has phase velocity $\pm\sqrt{C_S^2 + V_A^2}$. This is just the *MA wave* we found in Lecture 23. The polarization is shown in Fig. 24.2.

Fig. 24.2 Polarization of the MA wave

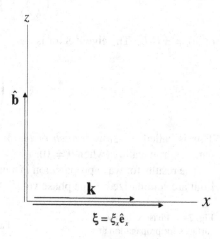

For $\theta = 0$ (propagation parallel to the field), the phase velocity is $\pm V_A$. Note, however, that this is *not* the shear Alfvén wave. The eigenvector corresponding to this frequency is found by substituting $\omega = \omega_1 = k_\parallel V_A$ (for $\theta = 0$) into Eqs. (24.2) and (24.4). The result is $\left(C_S^2 - V_A^2\right)\xi_z = 0$, so that $\xi_x = 1$, $\xi_z = 0$ is a non-trivial solution. The eigenvector is

$$\boldsymbol{\xi}_1 = \begin{pmatrix} 1 \\ 0 \\ 0 \end{pmatrix}. \tag{24.10}$$

This is different from the shear Alfvén wave eigenvector given by Eq. (24.6). This is sometimes called the *pseudo-mode*. The polarization is shown in Fig. 24.3.

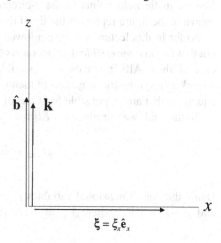

Fig. 24.3 Polarization of the "pseudo-mode"

The eigenvalue corresponding to the $(-)$ sign in Eq. (24.8) is, again in the strong field limit,

$$\left(\frac{\omega}{k}\right)^2_2 \approx C_S^2 \cos^2\theta \qquad (24.11)$$

or $\omega_2 = \pm C_s k_\parallel$. The eigenvector is

$$\boldsymbol{\xi}_2 = \begin{pmatrix} 0 \\ 0 \\ 1 \end{pmatrix}. \qquad (24.12)$$

This is called the *slow branch of the MA wave*. It becomes the sound wave for parallel propagation (when $\theta = 0$).

The results for wave propagation in a uniform, infinite medium in the strong field limit are summarized in the phase velocity diagram, shown in Fig. 24.4.

Fig. 24.4 Phase velocity surfaces for propagation at arbitrary angle to the magnetic field in a uniform medium

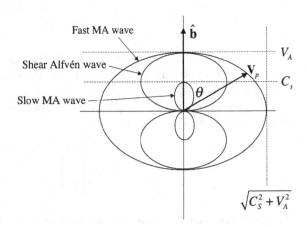

The magnetic field points in the z-direction (upward in Fig. 24.4). The surfaces shown in the figure represent the tip of the phase velocity vector $\mathbf{V}_p = (\omega/k)\hat{\mathbf{k}}$.

So far in this lecture we have allowed for compressible displacements. It turns out that for *incompressible* displacements ($\nabla \cdot \mathbf{V} = 0$), which includes the important case of shear Alfvén waves, it is possible to formulate the MHD equations in a remarkably symmetric way, and to thereby illuminate the symmetries in the MHD equations that are responsible for some of the waves we have just discussed.

To this end, we introduce the *Elsässer variables*

$$\mathbf{z}^\pm = \mathbf{V} \pm \frac{\mathbf{B}}{\sqrt{\mu_0 \rho_0}}. \qquad (24.13a,b)$$

Using the non-dimensional variables $\rho \to \rho_0$, $t \to t/\tau_A$, $\mathbf{x} \to \mathbf{x}/L$, $\mathbf{B} \to \mathbf{B}/B_0$, and $p \to p/\rho V_A^2$, where $V_A^2 = B_0^2/\mu_0\rho_0$ is the square of the Afvén speed and

$\tau_A = L/V_A$ is the Alfvén time, the incompressible MHD equations (with constant and uniform dissipation) are

$$\frac{\partial \mathbf{V}}{\partial t} + \mathbf{V} \cdot \nabla \mathbf{V} = -\nabla P + \mathbf{B} \cdot \nabla \mathbf{B} + \frac{1}{S_\nu} \nabla^2 \mathbf{V}, \qquad (24.14)$$

$$\frac{\partial \mathbf{B}}{\partial t} + \mathbf{V} \cdot \nabla \mathbf{B} = \mathbf{B} \cdot \nabla \mathbf{V} + \frac{1}{S} \nabla^2 \mathbf{B}, \qquad (24.15)$$

and

$$\nabla \cdot \mathbf{V} = 0, \qquad (24.16)$$

where $P = p + \frac{1}{2}B^2$ is the total pressure, S is the Lundquist number, and S_ν is the viscous Lundquist number (see Lecture 11). The non-dimensional Elsässer variables become

$$\mathbf{z}^\pm = \mathbf{V} \pm \mathbf{B}, \qquad (24.17\text{a,b})$$

so that $\mathbf{V} = (\mathbf{z}^+ + \mathbf{z}^-)/2$ and $\mathbf{B} = (\mathbf{z}^+ - \mathbf{z}^-)/2$.

If we first add Eqs. (24.14) and (24.15), then subtract them, and use Eq. (24.17a,b), we find that the Elsässer variables satisfy the coupled equations

$$\frac{\partial \mathbf{z}^\pm}{\partial t} + \mathbf{z}^\mp \cdot \nabla \mathbf{z}^\pm = -\nabla P + \frac{1}{2}\left(\frac{1}{S_\nu} + \frac{1}{S}\right)\nabla^2 \mathbf{z}^\pm + \frac{1}{2}\left(\frac{1}{S_\nu} - \frac{1}{S}\right)\nabla^2 \mathbf{z}^\mp, \quad (24.18\text{a,b})$$

along with

$$\nabla \cdot \mathbf{z}^\pm = 0. \qquad (24.18\text{c,d})$$

Equations (24.18 a–d) are completely equivalent to the incompressible MHD equations, Eqs. (24.14), (24.15) and (24.16). The coupled system is nonlinear. Note, however, that each individual equation is linear in either $\mathbf{z}^+$ or $\mathbf{z}^-$. The total pressure is determined by taking the divergence of either of Eq. (24.18a,b) and using Eq. (24.18c,d). The result is a Poisson equation

$$\nabla^2 P = -\nabla \mathbf{z}^\mp \cdot \nabla \mathbf{z}^\pm, \qquad (24.19\text{a,b})$$

which can be solved once the boundary conditions on P are specified. [Equation (24.19a,b) are completely symmetric in $\mathbf{z}^\pm$ and $\mathbf{z}^\mp$ and represent the same relationship, so it did not matter which of Eq. (24.18a,b) we used.]

So, how does this relate to waves, which are the subject of this lecture? If we linearize Eq. (24.18a,b) about a state with a uniform magnetic field $\mathbf{B}_0$, the equilibrium

Elsässer variables are $z_0^\pm = \pm B_0$. If we ignore dissipation, the perturbed Elsässer variables satisfy

$$\frac{\partial z_1^\pm}{\partial t} \mp B_0 \cdot \nabla z_1^\pm = -\nabla P_1 \qquad (24.20a,b)$$

and

$$\nabla^2 P_1 = \pm B_0 \cdot \nabla z_1^\pm. \qquad (24.20c)$$

The pressure clearly responds to the variation of $z_1^\mp$ along the equilibrium field. If pressure forces can be ignored (as in a low-β plasma, for example), the solution of Eq. (24.20a,b) is $z_1^\pm = z_1^\pm(x \mp B_0 t)$, where $z_1^\pm(x)$ is the initial condition. (Recall that in our non-dimensional variables, B_0 is equivalent to the Alfvén velocity.) The solution $z^+(x - B_0 t)$ is a disturbance propagating without distortion in the $+B_0$ direction, while the solution $z^-(x + B_0 t)$ is a disturbance propagating the $-B_0$ direction. Both solutions propagate at the Alfvén velocity V_A, so that $z_1^\pm$ are the right- and left-propagating *shear Alfvén waves* that we have just discussed.

The symmetry displayed by the Elsässer variables in incompressible MHD reflects the symmetric roles played by the velocity and the magnetic field in the propagation of shear Alfvén waves. Their fundamental nature can be further illustrated by writing the invariant quantities energy and cross-helicity as

$$E = \frac{1}{2} \int dV \left(B^2 + V^2 \right)$$
$$= \frac{1}{4} \int dV \left[\left(z^+ \right)^2 + \left(z^- \right)^2 \right] \qquad (24.21)$$

and

$$H_C = \int dV \mathbf{V} \cdot \mathbf{B}$$
$$= \frac{1}{4} \int dV \left[\left(z^+ \right)^2 - \left(z^- \right)^2 \right]. \qquad (24.22)$$

Because of this symmetry, the Elsässer variables play an important role in the theory of incompressible MHD turbulence. We will see them again in Lecture 36.

Lecture 25
The Calculus of Variations and the Ideal MHD Energy Principle

> *All generalizations are dangerous, even this one.*
> Alexandre Dumas

In Lecture 22, we showed that the ideal MHD force operator is self-adjoint and suggested that this allowed a formulation in which the stability of a system could be determined without solving a differential equation. Going further requires a little background in the calculus of variations. In the lecture we begin this discussion,[1] and formulate the ideal MHD energy principle.

We have seen that an equilibrium state is an extremum of the potential energy of the system; i.e., $\partial W / \partial r_i = 0$, where the r_i represent symbolically all possible local displacements of the system. Systems seek the lowest accessible state of potential energy. Therefore, if the extremum of the energy is a local minimum, the equilibrium is stable; if it is a local maximum, or an inflection point, the system is unstable (because other, lower-energy states are accessible).

In ordinary calculus, the extrema of a function $F(x)$ are determined by the roots of $F'(x) = 0$. These roots are numbers (values of x) at which F has a local maximum, minimum, or inflection point.

Now consider the expression

$$J\{y\} = \int_{x_0}^{x_1} F\left(x, y, y'\right) dx, \qquad (25.1)$$

where $y(x)$ is a function of x, $y' = dy/dx$, x_0, x_1, $y(x_0)$, and $y(x_1)$ are known numbers, and the functional form $F(x, y, z)$ is given. The problem is to determine the function $y(x)$ that "extremizes" $J\{y\}$. The formalism for solving this problem is called the *calculus of variations*. $J\{y\}$ is called a *functional* of y, because its value depends on the particular functional form $y(x)$. Once the functional form $y(x)$ is known, $J\{y\}$ is just a number.

---

[1] The discussion of the calculus of variations follows R. Courant and D. Hilbert, *Methods of Mathematic Physics*, Vol. 1, Interscience, New York (1953).

Schnack, D.D.: *The Calculus of Variations and the Ideal MHD Energy Principle*. Lect. Notes Phys. **780**, 153–158 (2009)
DOI 10.1007/978-3-642-00688-3_25 © Springer-Verlag Berlin Heidelberg 2009

Consider a function $\eta(x)$ defined in the interval $x_0 \leq x \leq x$, with $\eta(x_0) = \eta(x_1) = 0$, and let ε be a small number, i.e., $\varepsilon \ll 1$. Then the function $\bar{y} = y + \varepsilon\eta$ satisfies the same boundary conditions at x_0 and x_1 as y. If $y(x)$ is the extremizing function for $J\{y\}$, then for a local minimum $\Phi(\varepsilon) \equiv J\{y + \varepsilon\eta\} > J\{y\}$, and for a local maximum $\Phi(\varepsilon) \equiv J\{y + \varepsilon\eta\} < J\{y\}$, except when $\varepsilon = 0$. (We will not deal with inflection points.) The condition for an extremum is therefore $d\Phi/d\varepsilon = 0$ at $\varepsilon = 0$.

More specifically,

$$\Phi(\varepsilon) = \int_{x_0}^{x_1} F\left(x, y + \varepsilon\eta, y' + \varepsilon\eta'\right) dx. \tag{25.2}$$

Then

$$\frac{d\Phi}{d\varepsilon} = \int_{x_0}^{x_1} \left[\frac{\partial F}{\partial(y + \varepsilon\eta)} \frac{d}{d\varepsilon}(y + \varepsilon\eta) + \frac{\partial F}{\partial(y' + \varepsilon\eta')} \frac{d}{d\varepsilon}(y' + \varepsilon\eta') \right] dx \tag{25.3}$$

or, at $\varepsilon = 0$, the condition for an extremum

$$\left.\frac{d\Phi}{d\varepsilon}\right|_{e=0} = \int_{x_0}^{x_1} \left[\eta\frac{\partial F}{\partial y} + \eta'\frac{\partial F}{\partial y'} \right] dx = 0 \tag{25.4}$$

must hold for all functions $\eta(x)$ satisfying $\eta(x_1) = \eta(x_2) = 0$. Now

$$\frac{d}{dx}\left(\eta\frac{\partial F}{\partial y'} \right) = \eta'\frac{\partial F}{\partial y'} + \eta\frac{d}{dx}\left(\frac{\partial F}{\partial y'} \right). \tag{25.5}$$

Therefore

$$
\begin{aligned}
0 &= \int_{x_0}^{x_1} \left[\eta\frac{\partial F}{\partial y} + \frac{d}{dx}\left(\eta\frac{\partial F}{\partial y'} \right) - \eta\frac{d}{dx}\left(\frac{\partial F}{\partial y'} \right) \right] dx \\
&= \int_{x_0}^{x_1} \eta\left[\frac{\partial F}{\partial y} - \frac{d}{dx}\left(\frac{\partial F}{\partial y'} \right) \right] dx + \left.\eta\frac{\partial F}{\partial y'}\right|_{x_0}^{x_1} \\
&= \int_{x_0}^{x_1} \eta\left[\frac{\partial F}{\partial y} - \frac{d}{dx}\left(\frac{\partial F}{\partial y'} \right) \right] dx. \tag{25.6}
\end{aligned}
$$

The integrated term vanishes because of the boundary conditions on η.

Equation (25.6) must hold for all admissible functions $\eta(x)$. The *fundamental lemma of the calculus of variations* is that this can only occur if

$$\frac{\partial F}{\partial y} - \frac{d}{dx}\left(\frac{\partial F}{\partial y'}\right) = 0. \tag{25.7}$$

This is called the *Euler equation*. Using the notation $\partial F/\partial y' \equiv F_{y'}$, it is written as

$$\frac{d}{dx}F_{y'} - F_y = 0 \tag{25.8}$$

or, expanding the first term,

$$y''F_{y'y'} + y'F_{y'y} + F_{y'x} - F_y = 0. \tag{25.9}$$

As seen explicitly from Eq. (25.9), the Euler equation is a second-order ordinary differential equation for the function $y(x)$, with boundary conditions $y(x_0)$ and $y(x_1)$, which extremizes $J\{y\}$.

The function $\delta y = \varepsilon \eta$ is called the *first variation of y*. The *first variation of J* is defined as

$$\begin{aligned}\delta J &= J\{y + \varepsilon\eta\} - J\{y\}\\ &= \Phi(\varepsilon) - \Phi(0)\\ &\approx \left.\frac{d\Phi}{d\varepsilon}\right|_{\varepsilon=0}\varepsilon\end{aligned}$$

or, from the equations preceding (25.6),

$$\delta J = \int_{x_0}^{x_1} \delta y\left[\frac{\partial F}{\partial y} - \frac{d}{dx}\left(\frac{\partial F}{\partial y'}\right)\right]dx + \left.\delta y\frac{\partial F}{\partial y'}\right|_{x_0}^{x_1}. \tag{25.10}$$

The condition for an extremum is therefore $\delta J = 0$, i.e., the first variation of J must vanish. Note that, in general, the resulting expression depends explicitly on the boundary conditions through the last term in Eq. (25.10). Boundary conditions for which

$$\left.\delta y\frac{\partial F}{\partial y'}\right|_{x_0}^{x_1} = 0 \tag{25.11}$$

are called the *natural boundary conditions* for the problem $\delta J = 0$. This means that any solution of the variational problem

$$\delta J = 0 = \int_{x_0}^{x_1} \delta y\left[\frac{\partial F}{\partial y} - \frac{d}{dx}\left(\frac{\partial F}{\partial y'}\right)\right]dx$$

will *automatically* satisfy the natural boundary conditions.

Therefore, a necessary condition for $y(x)$ to be an extremum of $J\{y\}$ is the vanishing of the first variation of J, $\delta J = 0$, for all $y + \delta y$ that satisfy the boundary conditions.

Now, it is fair to ask how all this is related to ideal MHD. Recall that the ideal MHD wave equation is

$$-\rho_0 \omega^2 \boldsymbol{\xi} = \mathbf{F}\{\boldsymbol{\xi}\}. \qquad (25.12)$$

(The reason for the $\{\ldots\}$ notation is now apparent; $\mathbf{F}$ is a functional of $\boldsymbol{\xi}$.) Dotting Eq. (25.12) with $\boldsymbol{\xi}^*$ and integrating over all space, we can write

$$\omega^2\{\boldsymbol{\xi}\} = \frac{\delta W\{\boldsymbol{\xi}^*, \boldsymbol{\xi}\}}{K\{\boldsymbol{\xi}^*, \boldsymbol{\xi}\}}, \qquad (25.13)$$

where

$$\delta W\{\boldsymbol{\xi}^*, \boldsymbol{\xi}\} = -\frac{1}{2}\int \boldsymbol{\xi}^* \cdot \mathbf{F}\{\boldsymbol{\xi}\}\, dV \qquad (25.14)$$

and

$$K\{\boldsymbol{\xi}^*, \boldsymbol{\xi}\} = \frac{1}{2}\int \boldsymbol{\xi}^* \cdot \boldsymbol{\xi}\, dV. \qquad (25.15)$$

Therefore, ω^2 is an integral functional of $\boldsymbol{\xi}$.

We now take the first variation of Eq. (25.13). We will use the notation Δ instead of δ for the variation to avoid confusion with the standard notation δW for the potential energy. This procedure is facilitated by writing Eq. (25.13) as

$$K\omega^2 = -\frac{1}{2}\int \boldsymbol{\xi}^* \cdot \mathbf{F}\{\boldsymbol{\xi}\}\, dV. \qquad (25.16)$$

Taking the first variation,

$$\Delta K\omega^2 + K\Delta\omega^2 = -\frac{1}{2}\int \left(\Delta\boldsymbol{\xi}^* \cdot \mathbf{F}\{\boldsymbol{\xi}\} + \boldsymbol{\xi}^* \cdot \mathbf{F}\{\Delta\boldsymbol{\xi}\}\right) dV. \qquad (25.17)$$

Now

$$\Delta K = \int \rho_0 \Delta\boldsymbol{\xi}^* \cdot \boldsymbol{\xi}\, dV \qquad (25.18)$$

and

$$\int \boldsymbol{\xi}^* \cdot \mathbf{F}\{\Delta\boldsymbol{\xi}\}\, dV = \int \Delta\boldsymbol{\xi}^* \cdot \mathbf{F}\{\boldsymbol{\xi}\}\, dV, \qquad (25.19)$$

since $\mathbf{F}$ is self-adjoint. Then,

$$\Delta\omega^2 = -\frac{1}{K}\int \Delta\boldsymbol{\xi}^* \cdot \left(\mathbf{F}\{\boldsymbol{\xi}\} + \rho_0\omega^2\boldsymbol{\xi}\right) dV. \qquad (25.20)$$

Therefore, the condition for the vanishing of the first variation of ω^2, $\Delta\omega^2 = 0$, is the same as requiring $-\rho_0\omega^2\boldsymbol{\xi} = \mathbf{F}\{\boldsymbol{\xi}\}$, i.e., *the ideal MHD wave equation is the Euler equation for the variational problem $\Delta\omega^2 = 0$.* It is a consequence of $\mathbf{F}$ being self-adjoint.

From Eqs. (25.13) and (25.14), we see that if there exists a displacement $\boldsymbol{\xi}(\mathbf{r})$ for which $\delta W < 0$, then $\omega^2 < 0$ and the system is unstable. This leads to the *ideal MHD energy principle*: if $\delta W \geq 0$ for *all* allowable displacements, then the system is stable. If $\delta W < 0$ for *any* displacement, the system is unstable. These conditions can be shown to be both necessary and sufficient for determining the stability of an ideal MHD system.

Through the energy principle, the determination of stability is reduced to finding if there is a possible displacement for which $\delta W < 0$. This is a great simplification. However, in order to apply it in specific situations, we need to extend the energy principle to include contributions not only from the plasma, but also from the vacuum and the plasma/vacuum interface.

We can write the total potential energy of the perturbed system as[2]

$$\delta W = \delta W_F + \delta W_S + \delta W_V, \qquad (25.21)$$

where $\delta W_F = -(1/2)\int \boldsymbol{\xi}^* \cdot \mathbf{F}\{\boldsymbol{\xi}\} dV$ is the contribution from the plasma (fluid), δW_S is the contribution from the plasma/vacuum interface, and δW_V is the contribution from the vacuum. The explicit separation of δW as indicated in Eq. (25.21) requires *many* integrations by parts. After "a tedious but straightforward calculation," it turns out that

$$\delta W_F = \frac{1}{2}\int_F dV\left[\frac{|\mathbf{Q}|^2}{\mu_0} - \boldsymbol{\xi}_\perp^* \cdot (\mathbf{J} \times \mathbf{Q}) + \Gamma p_0 |\nabla \cdot \boldsymbol{\xi}|^2 + (\boldsymbol{\xi}_\perp \cdot \nabla p_0)\nabla \cdot \boldsymbol{\xi}_\perp^*\right],$$
$$\qquad (25.22)$$

$$\delta W_S = \frac{1}{2}\int_S dS\, |\hat{\mathbf{n}} \cdot \boldsymbol{\xi}_\perp|^2\, \hat{\mathbf{n}} \cdot \nabla\left[\!\!\left[p + \frac{B^2}{2\mu_0}\right]\!\!\right], \qquad (25.23)$$

and

$$\delta W_V = \frac{1}{2}\int_V dV\frac{1}{\mu_0}\left|\nabla \times \hat{\mathbf{A}}\right|^2, \qquad (25.24)$$

---

[2] See Jeffrey P. Freidberg, *Ideal Magnetohydrodynamics*, Plenum Press, New York (1987).

where F denotes the volume of the fluid, V denotes the volume of the vacuum, and S denotes the surface of the plasma/vacuum interface. We note that the vacuum contribution is always stabilizing, i.e., $\delta W_V > 0$.

As we have seen, the displacement and perturbed fields are subject to the conditions:

$$\hat{\mathbf{n}} \times \hat{\mathbf{A}} = 0 \tag{25.25}$$

for the vacuum solution, or

$$\hat{\mathbf{n}} \cdot \boldsymbol{\xi} = 0 \tag{25.26}$$

for the plasma solution, at the outer conducting boundary; and

$$\hat{\mathbf{n}} \times \hat{\mathbf{A}} = -(\hat{\mathbf{n}} \cdot \boldsymbol{\xi}) \hat{\mathbf{B}}_0 \tag{25.27}$$

and

$$-\Gamma p_0 \nabla \cdot \boldsymbol{\xi} + \frac{1}{\mu_0} (\boldsymbol{\xi} \cdot \nabla \mathbf{B}_0 + \mathbf{B}_1) \cdot \mathbf{B}_0 = \frac{1}{\mu_0} \left(\boldsymbol{\xi} \cdot \nabla \hat{\mathbf{B}}_0 + \hat{\mathbf{B}}_1 \right) \cdot \hat{\mathbf{B}}_0, \tag{25.28}$$

at the plasma/vacuum interface.

Since we only need to show that there is *some* displacement that makes $\delta W < 0$ and satisfies the boundary conditions, the usual procedure is to *guess* a functional form for $\boldsymbol{\xi}$, called a *trial function*, substitute it into the energy principle, and compute δW. (This procedure is not as random as it appears because of the Rayleigh–Ritz technique, which will be discussed in Lecture 27.) Of course, the trial function must satisfy the conditions of Eqs. (25.25, 25.26, 25.27, 25.28). It turns out that the pressure balance boundary condition, Eq. (22.29), is difficult to enforce. However, it can be shown that for every displacement $\boldsymbol{\xi}$ that does not satisfy pressure balance on S but makes $\delta W < 0$, there is a "neighboring" displacement $\tilde{\boldsymbol{\xi}}$ that satisfies pressure balance whose energy $\delta \tilde{W}$ differs from δW by an arbitrarily small amount. Therefore, it is not necessary to choose a trial function that satisfies the pressure balance condition. If a system is unstable without satisfying pressure balance, it will also be unstable if pressure balance is enforced.

This leads to what is called the *extended energy principle*: A necessary and sufficient condition for instability is that one can find a $\boldsymbol{\xi}$ and an $\hat{\mathbf{A}}$ that satisfy $\hat{\mathbf{n}} \times \hat{\mathbf{A}} = -(\hat{\mathbf{n}} \cdot \boldsymbol{\xi}) \hat{\mathbf{B}}_0$ at the fluid/vacuum interface, and have either $\hat{\mathbf{n}} \times \hat{\mathbf{A}} = 0$ or $\hat{\mathbf{n}} \cdot \boldsymbol{\xi} = 0$ at conducting boundaries, which make $\delta W < 0$.

The extended energy principle forms the basis for much of linear stability theory, especially for fusion plasmas.

Lecture 26
Examples of the Application of the Energy Principle

To state a theorem and then show examples is literally to teach backwards.

E. Kim Neubets

In this lecture, we will present several examples of the application of the energy principle for determining the stability of a magneto-fluid system that illustrate the power of the method. These examples are all based on the simple case of a fluid/vacuum interface, as shown in Fig. 26.1.

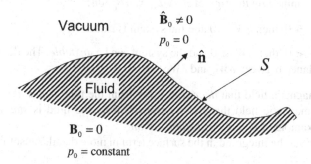

Fig. 26.1 The interface between an unmagnetized fluid and a vacuum region with a magnetic field

In the fluid, the magnetic field vanishes, $\mathbf{B}_0 = 0$, and the pressure is constant. In the vacuum, the magnetic field is finite, $\hat{\mathbf{B}}_0 \neq 0$, and the pressure vanishes.

The potential energy in the fluid is

$$\delta W_{\mathrm{F}} = \frac{1}{2} \int_F dV \Gamma p_0 \, |\nabla \cdot \boldsymbol{\xi}|^2 \geq 0, \tag{26.1}$$

since all other terms vanish; according to the energy principle, this term is stabilizing. However, δW_{F} can be minimized by choosing a displacement for which $\nabla \cdot \boldsymbol{\xi} = 0$. We conclude that $\nabla \cdot \boldsymbol{\xi} \neq 0$ *is always stabilizing when* $\nabla p_0 = 0$.

Schnack, D.D.: *Examples of the Application of the Energy Principle*. Lect. Notes Phys. **780**, 159–165 (2009)
DOI 10.1007/978-3-642-00688-3_26

The jump in total pressure across the interface S is

$$\left[\!\left[p_0 + \frac{B_0^2}{2\mu_0}\right]\!\right] = \frac{\hat{B}_0^2}{2\mu_0} - p_0. \tag{26.2}$$

The first term on the right-hand side comes from the vacuum and the second term comes from the fluid. Therefore

$$\nabla\left[\!\left[p_0 + \frac{B_0^2}{2\mu_0}\right]\!\right] = \nabla\left(\frac{\hat{B}_0^2}{2\mu_0}\right), \tag{26.3}$$

since $p_0 = $ constant. Then, if $\nabla \cdot \boldsymbol{\xi} = 0$, the total potential energy of the system is

$$\delta W = \delta W_S + \delta W_V$$

$$= \frac{1}{2}\int_S dS\,|\hat{\mathbf{n}} \cdot \boldsymbol{\xi}|^2\,\hat{\mathbf{n}} \cdot \nabla\left(\frac{\hat{B}_0^2}{2\mu_0}\right) + \frac{1}{2\mu_0}\int_S dV\,\left|\nabla \times \hat{\mathbf{A}}\right|^2. \tag{26.4}$$

As noted in Lecture 25, the vacuum contribution is always stabilizing. Stability is therefore determined by the *sign of* $\hat{\mathbf{n}} \cdot \nabla \hat{B}_0^2 \equiv \partial \hat{B}_0^2/\partial n$:

- If $\partial \hat{B}_0^2/\partial n > 0$, then $\delta W > 0$ and the system is *stable*.

- If $\partial \hat{B}_0^2/\partial n < 0$, then $\delta W < 0$ and the system can be *unstable*. The details depend on the balance between δW_S and δW_V.

A vacuum magnetic field that *decreases* away from the fluid is *destabilizing*. Conversely, the magnetic field that *increases* away from the fluid is *stabilizing*. The latter is an example of *minimum-B stabilization*.

Now consider the integrand in the surface term in more detail. A useful identity is

$$\nabla\left(\frac{1}{2}\hat{B}^2\right) = \hat{\mathbf{B}} \cdot \nabla\hat{\mathbf{B}} + \hat{\mathbf{B}} \times \left(\nabla \times \hat{\mathbf{B}}\right).$$

In the vacuum, $\mu_0\hat{\mathbf{J}} = \nabla \times \hat{\mathbf{B}} = 0$, so $\nabla\left(\hat{B}^2/2\right) = \hat{\mathbf{B}} \cdot \nabla\hat{\mathbf{B}}$. We want to compute $\nabla\hat{\mathbf{B}}$ at the interface S. The geometry is shown in Fig. 26.2.

Point C is the center of curvature of the surface. The vector $\mathbf{R}$ points from the surface to the center of curvature and $\hat{\mathbf{a}}_r = -\mathbf{R}/R$. In this geometry,

$$\nabla\hat{\mathbf{B}} = \left(\hat{\mathbf{a}}_r\frac{\partial}{\partial r} + \frac{\hat{\mathbf{b}}}{R}\frac{\partial}{\partial \theta}\right)\hat{B}\hat{\mathbf{b}}$$

$$= \hat{\mathbf{a}}_r\hat{\mathbf{b}}\frac{\partial \hat{B}}{\partial r} + \frac{\hat{\mathbf{b}}\hat{\mathbf{b}}}{R}\frac{\partial \hat{B}}{\partial \theta} - \frac{\hat{\mathbf{b}}\hat{\mathbf{a}}_r}{R}\hat{B},$$

Fig. 26.2 Details of the curvature at the fluid–vacuum interface

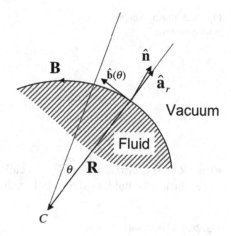

since $\partial \hat{\mathbf{b}}/\partial \theta = -\hat{\mathbf{a}}_r$. Then

$$\hat{\mathbf{B}} \cdot \nabla \hat{\mathbf{B}} = \hat{B}\hat{\mathbf{b}} \cdot \left(\hat{\mathbf{a}}_r \hat{\mathbf{b}} \frac{\partial \hat{B}}{\partial r} + \frac{\hat{\mathbf{b}}\hat{\mathbf{b}}}{R} \frac{\partial \hat{B}}{\partial \theta} - \frac{\hat{\mathbf{b}}\hat{\mathbf{a}}_r}{R} \hat{B} \right)$$

$$= \frac{\hat{\mathbf{b}}}{R} \hat{B} \frac{\partial \hat{B}}{\partial \theta} - \hat{\mathbf{a}}_r \frac{\hat{B}^2}{R},$$

so that

$$\frac{1}{2} \hat{\mathbf{n}} \cdot \nabla \hat{B}^2 = \hat{\mathbf{n}} \cdot \left(\hat{\mathbf{B}} \cdot \nabla \hat{\mathbf{B}} \right)$$

$$= \frac{\hat{\mathbf{n}} \cdot \hat{\mathbf{b}}}{R} \hat{B} \frac{\partial \hat{B}}{\partial \theta} - \hat{\mathbf{n}} \cdot \hat{\mathbf{a}}_r \frac{\hat{B}^2}{R}. \qquad (26.5)$$

But $\hat{\mathbf{n}} \cdot \hat{\mathbf{b}} = 0$ because $\hat{\mathbf{B}}$ lies in the surface S, and $\hat{\mathbf{n}} \cdot \hat{\mathbf{a}}_r = -\hat{\mathbf{n}} \cdot \mathbf{R}/R$. Therefore

$$\frac{1}{2} \hat{\mathbf{n}} \cdot \nabla \hat{B}^2 = (\hat{\mathbf{n}} \cdot \mathbf{R}) \frac{\hat{B}^2}{R^2}. \qquad (26.6a)$$

The surface energy δW_S is then

$$\delta W_S = \frac{1}{4\mu_0} \int_S dS \, |\hat{\mathbf{n}} \cdot \boldsymbol{\xi}|^2 \, (\hat{\mathbf{n}} \cdot \mathbf{R}) \frac{\hat{B}^2}{R^2}, \qquad (26.6b)$$

so that stability may be determined by the sign of $\hat{\mathbf{n}} \cdot \mathbf{R}$:

- $\hat{\mathbf{n}} \cdot \mathbf{R} > 0$ is *stabilizing*: This is called (creatively!) *good curvature*; the center of curvature lies inside the vacuum, as shown in Fig. 26.3.

Fig. 26.3 Illustration of a
good curvature

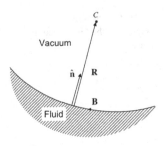

- $\hat{\mathbf{n}} \cdot \mathbf{R} < 0$ is *destabilizing*: This is called *bad curvature*; the center of curvature lies inside the fluid, as shown in Fig. 26.4.

Fig. 26.4 Illustration of bad
curvature

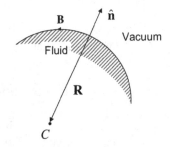

In order to find the total perturbed potential energy, we need to determine δW_V. This requires that we consider the effect of displacements of the surface S when the equilibrium surface is *flat* (i.e., $R \to \infty$). We work in a Cartesian coordinate system with x normal to the surface, and y and z in the plane of the surface. The magnetic field in the vacuum is in the z-direction. As before, the field vanishes in the fluid and the pressure is constant. The configuration is shown in Fig. 26.5.

Fig. 26.5 An expanded view
of the fluid–vacuum interface

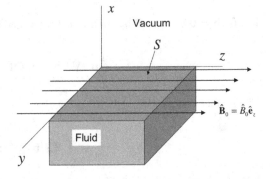

In this case, $\hat{\mathbf{n}} = \hat{\mathbf{e}}_x$, $\hat{\mathbf{B}}_0 = \hat{B}_0 \hat{\mathbf{e}}_z$, and $\hat{\mathbf{n}} \cdot \boldsymbol{\xi} = \xi_x$. We also assume that the system extends infinitely far in the y- and z-directions, so that on S we can write $\xi_x = \xi_0 e^{i(k_y y + k_z z)}$. This is the vertical displacement of the surface.

In the vacuum $\mathbf{J} = 0$, so the perturbed vector potential in the vacuum satisfies

$$\nabla \times \nabla \times \hat{\mathbf{A}} = 0. \tag{26.7}$$

With $\hat{\mathbf{A}} = \tilde{\mathbf{A}}e^{i\mathbf{k}\cdot\mathbf{r}}$, $\mathbf{k} = k_x\hat{\mathbf{e}}_x + k_y\hat{\mathbf{e}}_y + k_z\hat{\mathbf{e}}_z$, $\nabla \to i\mathbf{k}$, and using vector identities, this becomes

$$k^2\tilde{\mathbf{A}} - \mathbf{k}\left(\mathbf{k}\cdot\tilde{\mathbf{A}}\right) = 0. \tag{26.8}$$

The perturbed vacuum field is $\hat{\mathbf{B}}_1 = \tilde{\mathbf{B}}e^{i\mathbf{k}\cdot\mathbf{r}}$, where $\tilde{\mathbf{B}} = i\mathbf{k} \times \tilde{\mathbf{A}}$.

On S ($x = 0$), $\tilde{\mathbf{A}}$ must satisfy the boundary condition

$$\hat{\mathbf{n}} \times \tilde{\mathbf{A}} = -(\hat{\mathbf{n}}\cdot\boldsymbol{\xi})\hat{\mathbf{B}}_0. \tag{26.9}$$

In light of the above discussion, this becomes

$$\tilde{A}_z = 0 \tag{26.10}$$

and

$$\tilde{A}_y = -\xi_x\hat{B}_{0S}, \tag{26.11}$$

where $\hat{B}_{0S}$ is the equilibrium vacuum magnetic at the surface. Since $\tilde{\mathbf{B}} = i\mathbf{k} \times \tilde{\mathbf{A}}$, the perturbed magnetic field only depends on the component of $\tilde{\mathbf{A}}$ that is parallel to $\mathbf{k}$. We are therefore free to set $\mathbf{k}\cdot\tilde{\mathbf{A}} = 0$. From Eq. (26.10), this requires

$$\tilde{A}_x = -\frac{k_y}{k_x}\tilde{A}_y, \tag{26.12}$$

and Eq. (26.8) becomes

$$k^2\tilde{\mathbf{A}} = 0. \tag{26.13}$$

A non-trivial $\tilde{\mathbf{A}}$ therefore requires $k^2 = 0$ or $k_x = \pm i\kappa$, where $\kappa^2 = k_y^2 + k_z^2$. After some algebra, the solution for the perturbed vacuum field that is bounded at infinity is found to be

$$\hat{\mathbf{B}}_1 = \tilde{\mathbf{B}}e^{-\kappa x}e^{i\left(k_y y + k_z z\right)}, \tag{26.14}$$

where

$$\tilde{\mathbf{B}} = -\frac{i\tilde{A}_y}{k_x}\left(\hat{\mathbf{e}}_x k_z k_x + \hat{\mathbf{e}}_y k_z k_y - \hat{\mathbf{e}}_z \kappa^2\right), \tag{26.15}$$

where $k_x = \pm i\kappa$.

The magnetic energy density in the perturbed vacuum field is therefore, again after some algebra,

$$w_V = \frac{1}{2\mu_0} \hat{\mathbf{B}}_1 \cdot \hat{\mathbf{B}}_1^* e^{-2\kappa x}$$

$$= \frac{1}{\mu_0} |\tilde{A}_y|^2 k_z^2 e^{-2\kappa x}$$

or, using Eq. (26.11),

$$w_V = \frac{1}{\mu_0} k_z^2 \xi_x^2 \hat{B}_{0s}^2 e^{-2\kappa x}. \tag{26.16}$$

The vacuum contribution to the potential energy is therefore

$$\delta W_V = \int_V w_V dV$$

$$= \frac{1}{\mu_0} \iint k_z^2 \xi_x^2 \hat{B}_{0s}^2 dy dz \int_0^{\infty} dx e^{-2\kappa x}$$

$$= \frac{1}{2\mu_0} \int_S \frac{k_z^2 \xi_x^2 \hat{B}_{0s}^2}{\kappa} dS. \tag{26.17}$$

If $\nabla \cdot \boldsymbol{\xi} = 0$, $\delta W_F = 0$ [see Eq. (26.1)], and the total perturbed potential energy of the system is

$$\delta W = \delta W_S + \delta W_V$$

$$= \frac{1}{4\mu_0} \int_S dS \xi_x^2 \frac{\partial \hat{B}_0^2}{\partial x} + \frac{1}{2\mu_0} \int_S \frac{k_z^2 \xi_x^2 \hat{B}_{0s}^2}{\kappa} dS. \tag{26.18}$$

Therefore, instability occurs if $\partial \hat{B}_0^2/\partial x < 0$ and $k_z^2/\kappa \rightarrow 0$. The latter condition is equivalent to $k_z = 0$, i.e., $\mathbf{k} \cdot \hat{\mathbf{B}}_{0s} = 0$. This means that, on S, the wavefronts of the perturbation are parallel to the equilibrium magnetic field. These perturbations *do not bend the field lines*. The displacements to the surface can slip through the magnetic field without increasing the magnetic energy. The situation is sketched in Fig. 26.6.

If $\partial \hat{B}_0^2/\partial x < 0$, the upward moving tip of the perturbation enters a region where $\hat{B}_0^2/2\mu_0 < p_0$. The fluid in the tip thus expands and feels a "restoring force" in the direction of ξ_x. Instability ensues.

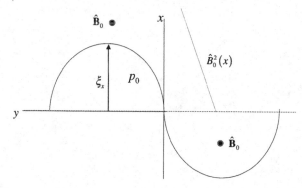

Fig. 26.6 Displacement of the surface seen in the plane perpendicular to the direction of the magnetic field, and the variation of the vacuum magnetic pressure with distance from the unperturbed surface

While the conclusions drawn here regarding stability (or lack thereof) derive from the simple problem of an unmagnetized, uniform pressure plasma in contact with a unidirectional vacuum magnetic field, the concepts of incompressible perturbations, minimum-B, good and bad curvature, and field-line bending are sufficiently general that they are applicable in other more complex situations.

Lecture 27
The Rayleigh–Ritz Technique for Estimating Eigenvalues

> *It is the mark of an educated mind to rest satisfied with the*
> *degree of precision which the nature of the subject admits and*
> *not to seek exactness where only an approximation is possible.*
> Aristotle

The energy principle provides a powerful technique for determining the stability or instability of a magneto-fluid system without resorting to the solution of a differential equation. Instead, one makes an educated guess at the minimizing displacement and then examines the sign of the resulting eigenvalue. This approach is made even more powerful, and put on a solid theoretical footing, by application of the Rayleigh–Ritz technique for estimating the eigenvalues of a self-adjoint operator. This is briefly discussed in this lecture.[1]

Let $\boldsymbol{\xi}_i, i = 0, 1, 2, \dots$ be the eigenvectors of the ideal MHD force operator $\mathbf{F}$, with eigenvalues $\omega_i^2, i = 0, 1, 2, \dots$ ordered such that $\omega_0^2 < \omega_1^2 < \omega_2^2 < \cdots$. If $\omega_0^2 < 0$ the system is unstable. If we know the eigenvector $\boldsymbol{\xi}_0$, we can compute the perturbed potential and kinetic energies, δW_0 and K_0, and thereby determine the eigenvalue $\omega_0^2 = \delta W / K$. However, finding $\boldsymbol{\xi}_0$ requires solving the ideal MHD wave equation, which may be quite tedious, difficult, or otherwise unpleasant.

Instead, we take a *guess* at $\boldsymbol{\xi}_0$, say

$$\underbrace{\boldsymbol{\xi}}_{\text{Guessed eigenvector}} = \underbrace{\boldsymbol{\xi}_0}_{\text{Exact eigenvector}} + \underbrace{\delta\boldsymbol{\xi}}_{\text{Error}} . \tag{27.1}$$

Now, *by definition*, the error $\delta\boldsymbol{\xi}$ is orthogonal to (i.e., has no projection along) $\boldsymbol{\xi}_0$ (it contains all the parts of $\boldsymbol{\xi}$ that differ from $\boldsymbol{\xi}_0$), so that we can write

$$\boldsymbol{\xi} = \boldsymbol{\xi}_0 + \sum_{i=1}^{\infty} a_i \boldsymbol{\xi}_i, \tag{27.2}$$

---

[1] We have generalized the discussion found in George Arfkin, *Mathematical Methods for Physicists*, 2nd Ed., Academic Press, New York (1970).

Schnack, D.D.: *The Rayleigh–Ritz Technique for Estimating Eigenvalues*. Lect. Notes Phys. **780**, 167–170 (2009)
DOI 10.1007/978-3-642-00688-3_27

where $i = 0$ has been excluded from the summation. Then an *estimate* of the eigenvalue ω_0^2 is

$$
\begin{aligned}
\omega^2 &= \frac{\delta W \left\{ \boldsymbol{\xi}^*, \boldsymbol{\xi} \right\}}{K \left\{ \boldsymbol{\xi}^*, \boldsymbol{\xi} \right\}} \\
&= \frac{\delta W \left\{ \boldsymbol{\xi}_0^* + \delta\boldsymbol{\xi}^*, \boldsymbol{\xi}_0 + \delta\boldsymbol{\xi} \right\}}{K \left\{ \boldsymbol{\xi}_0^* + \delta\boldsymbol{\xi}^*, \boldsymbol{\xi}_0 + \delta\boldsymbol{\xi} \right\}},
\end{aligned}
\tag{27.3}
$$

where we explicitly display the complex conjugate in the arguments.

The numerator in Eq. (27.3) is evaluated by expanding δW. We have

$$
\begin{aligned}
\delta W \left\{ \boldsymbol{\xi}_0^* + \delta\boldsymbol{\xi}^*, \boldsymbol{\xi}_0 + \delta\boldsymbol{\xi} \right\} = \delta W \left\{ \boldsymbol{\xi}_0^*, \boldsymbol{\xi}_0 \right\} + \delta W \left\{ \boldsymbol{\xi}_0^*, \delta\boldsymbol{\xi} \right\} \\
+ \delta W \left\{ \delta\boldsymbol{\xi}^*, \boldsymbol{\xi}_0 \right\} + \delta W \left\{ \delta\boldsymbol{\xi}^*, \delta\boldsymbol{\xi} \right\},
\end{aligned}
\tag{27.4}
$$

since $\mathbf{F} \left\{ \boldsymbol{\xi} \right\}$ is a linear functional. We need to evaluate the individual terms on the right-hand side of this equation. Since $\mathbf{F}$ is self-adjoint, we know that $\delta W \left\{ \boldsymbol{\xi}_0^*, \delta\boldsymbol{\xi} \right\} = \delta W \left\{ \delta\boldsymbol{\xi}^*, \boldsymbol{\xi}_0 \right\}$. Further,

$$
\delta W \left\{ \delta\boldsymbol{\xi}^*, \boldsymbol{\xi}_0 \right\} = -\frac{1}{2} \int \left(\sum_{i=1}^{N} a_i \boldsymbol{\xi}_i \right)^* \cdot \mathbf{F} \left\{ \boldsymbol{\xi}_0 \right\} dV = 0,
\tag{27.5}
$$

because the eigenvectors are orthogonal and $i = 0$ is excluded from the sum. Also,

$$
K \left\{ \boldsymbol{\xi}_i^*, \boldsymbol{\xi}_j \right\} = \frac{1}{2} \rho_0 \int \boldsymbol{\xi}_i^* \cdot \boldsymbol{\xi}_j dV = \frac{1}{2} \rho_0 \delta_{ij},
\tag{27.6}
$$

so that we can write

$$
\delta W \left\{ \boldsymbol{\xi}_i^*, \boldsymbol{\xi}_j \right\} = \omega^2 K \left\{ \boldsymbol{\xi}_i^*, \boldsymbol{\xi}_j \right\} = \frac{1}{2} \rho_0 \omega_i^2 \delta_{ij}.
\tag{27.7}
$$

Therefore

$$
\begin{aligned}
\delta W \left\{ \delta\boldsymbol{\xi}^*, \delta\boldsymbol{\xi} \right\} &= \sum_{i=1}^{\infty} \sum_{j=1}^{\infty} a_i^* a_j \delta W \left\{ \boldsymbol{\xi}_i^*, \boldsymbol{\xi}_j \right\} \\
&= \sum_{i=1}^{\infty} \sum_{j=1}^{\infty} a_i^* a_j \frac{1}{2} \rho_0 \omega_j^2 \delta_{ij} \\
&= \frac{1}{2} \rho_0 \sum_{i=1}^{\infty} |a_i|^2 \omega_i^2.
\end{aligned}
$$

The numerator of Eq. (27.3) is then

$$\delta W \left\{ \xi_0^* + \delta \xi^*, \xi_0 + \delta \xi \right\} = \delta W \left\{ \xi_0^*, \xi_0 \right\} + \delta W \left\{ \delta \xi^*, \delta \xi \right\}$$

$$= \delta W_0 + \frac{1}{2} \rho_0 \sum_{i=1}^{\infty} |a_i|^2 \omega_i^2. \qquad (27.8)$$

Similarly, the kinetic energy in the denominator of Eq. (27.3) is

$$K \left\{ \xi_0^* + \delta \xi^*, \xi_0 + \delta \xi \right\} = K \left\{ \xi_0^*, \xi_0 \right\} + K \left\{ \delta \xi^*, \delta \xi \right\}$$

$$= K_0 + \frac{1}{2} \rho_0 \sum_{i=1}^{\infty} |a_i|^2. \qquad (27.9)$$

Then, since by Eq. (27.6), $K_0 = \rho_0/2$, Eq. (27.3) becomes

$$\omega^2 = \frac{\delta W_0/K_0 + \sum_{i=1}^{\infty} |a_i|^2 \omega_i^2}{1 + \sum_{i=1}^{\infty} |a_i|^2}. \qquad (27.10)$$

If we now assume that we have made a decent guess at ξ_0, we expect the error term in Eq. (27.10) to be small (i.e., $\sum_i |a_i|^2 \ll 1$) and the denominator can be expanded by the binomial theorem. Then Eq. (27.10) can be written approximately as

$$\omega^2 \approx \frac{\delta W_0}{K_0} + \sum_{i=1}^{\infty} |a_i|^2 \omega_i^2 - \frac{\delta W_0}{K_0} \sum_{i=1}^{\infty} |a_i|^2$$

or

$$\omega^2 \approx \omega_0^2 + \sum_{i=1}^{\infty} |a_i|^2 \left(\omega_i^2 - \omega_0^2 \right), \qquad (27.11)$$

where $\omega_0^2 = \delta W_0/K_0$ is the actual eigenvalue associated with ξ_0. The summation in Eq. (27.11) is the error in our estimate of this eigenvalue.

There are two important things to note about Eq. (27.11):

1. Even though the error $\delta \xi$ in our estimate of the eigenvector is $O(a_i)$ [see Eq. (27.2)], the error in the estimate of the eigenvalue is $O(|a_i|^2) \ll O(a_i)$. Therefore, *the error in the eigenvalue is much less than the error in the eigenvector*. A 10% error in the estimate of ξ_0 results in only a 1% error in the estimate of ω^2. Our guess does not have to be very good at all; even relatively poor guesses at ξ_0 yield relatively good estimates of ω^2.
2. Since, by definition, $\omega_i^2 > \omega_0^2$, *the error in the estimate of ω^2 is always positive* [see Eq. (27.11)]. Successively improved guesses always converge from above.

Therefore, if we find an estimate $\omega^2 < 0$, then we can be assured that $\omega_0^2 < 0$ also. This provides reliable conclusions about instability.

The procedure outlined above can be improved by writing the guess at $\boldsymbol{\xi}_0$ in terms of a number of parameters and then minimizing the resulting ω^2 with respect to these parameters. That is, we write our guess as $\boldsymbol{\xi} = \boldsymbol{\xi}(A, B, C, D, \ldots)$ and compute $\delta W \{\boldsymbol{\xi}^*, \boldsymbol{\xi}\}$ and $K \{\boldsymbol{\xi}^*, \boldsymbol{\xi}\}$. This yields an estimated eigenvalue $\omega^2 = \omega^2(A, B, C, D, \ldots)$. We then differentiate the expression for ω^2 with respect to the parameters $A, B, C, D, \ldots$ and set the derivatives to zero, i.e., $\partial \omega^2/\partial A = 0$, $\partial \omega^2/\partial B = 0$, $\partial \omega^2/\partial C = 0$, $\partial \omega^2/\partial D = 0$, etc. This yields a set of simultaneous equations that can be solved for the parameters $A, B, C, D, \ldots$ that result in a stationary (and, we hope, minimum) value of ω^2. This is called the Rayleigh–Ritz technique. It forms the basis for many important computational algorithms used to study the stability of magneto-fluid systems (called "δW codes").

Lecture 28
The Gravitational Interchange Mode or g-Mode

> *We can lick gravity, but sometimes the paperwork is overwhelming.*
>
> Wernher Von Braun

We now consider the case where the magneto-fluid system is subject to a gravitational force $\mathbf{F}_g = \rho\mathbf{g}$, where $\mathbf{g}$ is a constant gravitational acceleration vector. In a system with straight field lines, the equilibrium condition is

$$\nabla\left(p + \frac{B^2}{2\mu_0}\right) = \rho\mathbf{g}. \tag{28.1}$$

Recall that in a system with curved field lines, but no gravity, the equilibrium condition is

$$\nabla\left(p + \frac{B^2}{2\mu_0}\right) = \frac{B^2}{\mu_0}\boldsymbol{\kappa}, \tag{28.2}$$

where $\boldsymbol{\kappa} = \hat{\mathbf{b}} \cdot \nabla\hat{\mathbf{b}}$ is the field-line curvature. Therefore, by using gravity as a proxy force, it is possible to study the stability properties of systems with curved field lines while using Cartesian geometry with straight field lines. This is a great simplification. This accounts for both the importance of the gravitational problem in MHD and the richness of its solutions. The study of the stability properties of the equilibrium given by Eq. (28.1), the so-called *g-mode*, or *gravitational interchange* problem, is one of the most important problems in MHD.

Since $\mathbf{F}_g = \rho\mathbf{g}$, the gravitational force will make an additional contribution to the perturbed potential energy of

$$\delta W_g = -\frac{1}{2}\int dV \boldsymbol{\xi}^* \cdot \mathbf{F}_g\{\boldsymbol{\xi}\}. \tag{28.3}$$

The perturbed gravitational force is $\mathbf{F}_g = \rho_1\mathbf{g}$. We have previously shown that the perturbed density is related to the displacement by

Schnack, D.D.: *The Gravitational Interchange Mode or g-Mode*. Lect. Notes Phys. **780**, 171–177 (2009)
DOI 10.1007/978-3-642-00688-3_28 © Springer-Verlag Berlin Heidelberg 2009

$$\rho_1 = -\boldsymbol{\xi} \cdot \nabla \rho_0 - \rho_0 \nabla \cdot \boldsymbol{\xi}, \tag{28.4}$$

so that

$$\delta W_g = \frac{1}{2} \int dV \left(\boldsymbol{\xi}^* \cdot \mathbf{g}\right) \left(\boldsymbol{\xi} \cdot \nabla \rho_0 + \rho_0 \nabla \cdot \boldsymbol{\xi}\right). \tag{28.5}$$

The full expression for δW including gravity is

$$\delta W = \frac{1}{2} \int dV \left\{ \frac{|\mathbf{Q}|^2}{\mu_0} - \boldsymbol{\xi}^* \cdot \mathbf{J} \times \mathbf{Q} + \Gamma p_0 \left|\nabla \cdot \boldsymbol{\xi}\right|^2 + \left(\boldsymbol{\xi}^* \cdot \nabla p_0\right) \nabla \cdot \boldsymbol{\xi} \right.$$
$$\left. + \left(\boldsymbol{\xi}^* \cdot \mathbf{g}\right) \left(\boldsymbol{\xi} \cdot \nabla \rho_0 + \rho_0 \nabla \cdot \boldsymbol{\xi}\right) \right\}, \tag{28.6}$$

where $\mathbf{Q} = \nabla \times (\boldsymbol{\xi} \times \mathbf{B})$.

We now consider the equilibrium of Eq. (28.1) in Cartesian geometry. Solid walls are at $x = \pm L$, $\mathbf{B} = B(x)\hat{\mathbf{e}}_y$, and $\mathbf{g} = g\hat{\mathbf{e}}_x$. The current is

$$J_z = \frac{1}{\mu_0} \frac{dB}{dx}, \tag{28.7}$$

and the force balance condition, Eq. (28.1), is

$$\frac{dp_0}{dx} + \frac{1}{2\mu_0} \frac{dB^2}{dx} = \rho_0 g. \tag{28.8}$$

The configuration is shown in Fig. 28.1.

Fig. 28.1 Geometry for the gravitational interchange mode

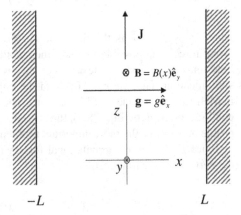

$-L$ L

We now specialize to the case $\boldsymbol{\xi} = \boldsymbol{\xi}(x, z)$, so that the displacement is independent of z. Note that now $\boldsymbol{\xi}$ is real, so $\boldsymbol{\xi}^* = \boldsymbol{\xi}$. We substitute this *ansatz* into Eq. (28.6). After considerable algebra, the result is

$$\delta W = \frac{1}{2} \int dV \left[\left(\frac{B^2}{\mu_0} + \Gamma p_0 \right) (\nabla \cdot \boldsymbol{\xi})^2 + 2\rho_0 g \xi_x \nabla \cdot \boldsymbol{\xi} + g \frac{d\rho_0}{dx} \xi_x^2 \right]. \quad (28.9)$$

First, we note that the system is stable ($\delta W > 0$) if $g = 0$.
Second, if $\nabla \cdot \boldsymbol{\xi} = 0$, stability requires

$$\int g \frac{d\rho_0}{dx} \xi_x^2 > 0. \qquad (28.10a)$$

Suppose that $g d\rho_0/dx < 0$ at a single point x_0. Then we could choose a trial function ξ_x that is zero everywhere except in a very small region around x_0, as shown in Fig. 28.2.

Fig. 28.2 Possible trial
function for the gravitational
interchange problem

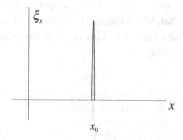

Then $\delta W < 0$ and the system is unstable. But the point x_0 is arbitrary, so the condition for stability $g d\rho_0/dx > 0$ must hold *pointwise* (i.e., at all points in $-L < x < L$), for if it is not satisfied at a single point, then $\delta W < 0$ and the system is unstable. This is called the *Rayleigh–Taylor instability*. Stable and unstable situations for incompressible perturbations are sketched in Fig. 28.3.

Fig. 28.3 Stable and unstable
density profiles for the
incompressible case

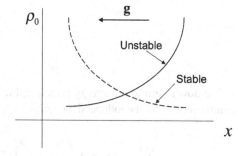

Third, $\nabla \cdot \boldsymbol{\xi} \neq 0$ is not always stabilizing when $\nabla p_0 \neq 0$; incompressible displacements are not always the most unstable. In this case, if $\xi_x \nabla \cdot \boldsymbol{\xi} > 0$, then the system is unstable if $g < 0$; if $\xi_x \nabla \cdot \boldsymbol{\xi} < 0$, then the system is unstable if $g > 0$. Note that these conclusions are true even if $d\rho_0/dx = 0$. Consider the total pressure profile shown in the left of Fig. 28.4. Let the fluid element at x be displaced downward to the new position x', as shown in the figure on the right of Fig. 28.4. The pressure at x' is greater than the pressure at x. During the displacement, the fluid element will therefore be compressed ($\nabla \cdot \boldsymbol{\xi} < 0$), its density will increase, and it will find itself heavier than its surroundings. It will continue to fall. Similarly, upwardly displaced elements will expand, find themselves lighter than their surroundings, and continue to climb. This is called *buoyancy*. It is an instability that depends on compressibility. It is often called the *Parker instability*.

In the g-mode, the system lowers its energy by interchanging fluid elements. Instabilities of this type are called, not surprisingly, *interchange modes*. The g-mode is often described as the instability of a heavier fluid being supported by a lighter one. This is the case when the fluid is incompressible and is its primary manifestation in magnetic confinement devices, such as tokamaks, where incompressibility is enforced by a strong magnetic field (as in reduced MHD). However, we have seen that, when the fluid is compressible, instability can occur even when there is no density gradient. This is often the primary manifestation of this mode in astrophysical settings.

Fig. 28.4 *Left*: Total pressure profile and gravity. *Right*: Displacement of a fluid element from x to x'

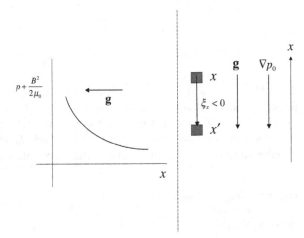

We now return to the energy principle, Eq. (28.9). Notice that the integrand is a *quadratic form* in the independent variables $\nabla \cdot \boldsymbol{\xi}$ and ξ_x, i.e., it can be written as

$$\sum_{i,j=1}^{2} x_i a_{ij} x_j = \left(\frac{B^2}{\mu_0} + \Gamma p_0 \right) (\nabla \cdot \boldsymbol{\xi})^2 + \rho_0 g \, (\nabla \cdot \boldsymbol{\xi}) \, \xi_x$$

$$+ \rho_0 g \xi_x \, (\nabla \cdot \boldsymbol{\xi}) + g \frac{d\rho_0}{dx} \xi_x^2, \tag{28.10b}$$

where $x_1 = \nabla \cdot \boldsymbol{\xi}$ and $x_2 = \xi_x$. If this is positive, the system is stable. There is a theorem stating that a necessary and sufficient condition for a quadratic form to be positive, i.e.,

$$Q = \mathbf{x}^T \cdot \mathbf{A} \cdot \mathbf{x} > 0, \tag{28.11}$$

is that the determinant of all the principal minors of $\mathbf{A}$ be greater than zero. That is,

$$\det \mathbf{A}_1 = a_{11} > 0,$$

$$\det \mathbf{A}_2 = \begin{vmatrix} a_{11} & a_{12} \\ a_{21} & a_{22} \end{vmatrix} > 0,$$

$$\det \mathbf{A}_3 = \begin{vmatrix} a_{11} & a_{12} & a_{13} \\ a_{21} & a_{22} & a_{23} \\ a_{31} & a_{32} & a_{33} \end{vmatrix} > 0,$$

etc., must hold simultaneously. Equation (28.10) is a 2×2 system, with

$$a_{11} = \frac{B^2}{\mu_0} + \Gamma p_0, \quad a_{12} = \rho_0 g,$$

$$a_{21} = \rho_0 g, \quad a_{22} = g \frac{d\rho_0}{dx}.$$

The first principal determinant is

$$a_{11} = \frac{B^2}{\mu_0} + \Gamma p_0 > 0,$$

which is always satisfied. Requiring the second principal determinant (in this case, the determinant of a_{ij}) to be positive yields

$$\left(\frac{B^2}{\mu_0} + \Gamma p \right) \left(g \frac{d\rho_0}{dx} \right) - (\rho_0 g)^2 > 0$$

or

$$g \frac{d\rho_0}{dx} > \frac{g^2}{C^2} > 0, \tag{28.12}$$

where $C^2 = C_S^2 + V_A^2$. Equation (28.12) is the condition for stability of the g-mode, including compressibility. As concluded in the discussion following Eq. (28.10), it must hold at every point in x. If it is violated at any point, the system is unstable.

There are two points:

1. The system is unstable if g and $d\rho_0/dx$ have opposite signs. This is the same conclusion as in the incompressible case (the Rayleigh–Taylor instability).

2. Even if g and $d\rho_0/dx$ have the same sign, the system will be unstable unless $g\,(d\rho_0/dx) > \rho_0 g^2/C^2$. Because of compressibility, *instability can occur even when a heavy fluid supports a light fluid*. The drive for the Parker instability can overcome the stabilization of the Rayleigh–Taylor instability (which occurs because of the alignment of the density gradient with gravity). This situation is sketched in Fig. 28.5.

Fig. 28.5 Stability diagram for the compressible gravitational interchange mode (the Parker instability)

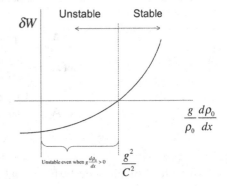

We conclude by estimating the growth rate with the Rayleigh–Ritz technique. For simplicity, we only consider the incompressible case, $\nabla \cdot \boldsymbol{\xi} = 0$, i.e., the Rayleigh–Taylor instability. (Yes, it is the same Lord Rayleigh.) We assume an exponential density profile $\rho_0 = \bar{\rho}e^{x/L_n}$. The perturbed potential energy is

$$\delta W = \frac{1}{2}\int dV g \frac{d\rho_0}{dx} g\xi_x^2. \tag{28.13}$$

We choose the trial function

$$\xi_x = \xi_0 e^{ikz}\cos\frac{\pi x}{2L}, \tag{28.14}$$

which is periodic in z and satisfies the boundary conditions $\xi_x(-L, z) = \xi_x(-L, z) = 0$, as required. Since $\nabla \cdot \boldsymbol{\xi} = 0$, the z-component of the displacement satisfies $\partial\xi_z/\partial z = -\partial\xi_x/\partial x$. Using Eq. (28.14) and integrating,

$$\xi_z = -\frac{i\pi\xi_0}{2kL}e^{ikz}\sin\frac{\pi x}{2L}. \tag{28.15}$$

The potential energy is

$$\delta W = \frac{\rho_0 g}{2L_n}\xi_0^2 \int_{-L}^{L}\cos^2\frac{\pi x}{2L}dx$$

$$= \frac{\pi\rho_0 g}{4L_n}\xi_0^2, \tag{28.16}$$

and the perturbed kinetic energy is

$$K = \frac{1}{2}\rho_0 \int\limits_{-L}^{L} \left(\xi_x^2 + \xi_z^2\right) dx$$

$$= \frac{\pi\rho_0}{4}\xi_0^2 \left(1 + \frac{\pi^2}{4k^2L^2}\right). \tag{28.17}$$

The estimated growth rate is

$$\omega^2 = \frac{\delta W}{K},$$

$$= \frac{g/L_n}{1 + \pi^2/4k^2L^2}. \tag{28.18}$$

The system is unstable if $g/L_n < 0$, as expected. For long wavelengths, $kL << 1$, we have

$$\gamma^2 \approx \frac{4}{\pi^2}(kL)^2, \tag{28.19}$$

and for short wavelengths, $kL >> 1$,

$$\gamma^2 \approx g/L_n, \tag{28.20}$$

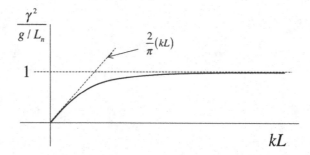

Fig. 28.6 Growth rate versus normalized wave number for the incompressible gravitational interchange mode

where $\gamma^2 = -\omega^2$ is the square of the growth rate. When the wavelength is long, the mode can "feel" the presence of the conducting wall and the growth rate is reduced. However, when the wavelength is short, the mode can evolve as though the wall were absent, and the growth rate is independent of k. The full dependence of the growth rate on wave number is given by Eq. (18.9). This is sketched in Fig. 28.6.

Lecture 29
Comments on the Energy Principle and the Minimizing Eigenfunction[1]

In this lecture, we make some comments on the MHD energy principle and the process of minimization. These remarks are quite general. Some of them require a considerable amount of vector algebra and integration by parts. Most of these details have been omitted for clarity of presentation.

After a formidable calculation, the ideal MHD energy principle (excluding gravity) can be written as

$$\delta W = \frac{1}{2} \int dV \left\{ \frac{|\mathbf{Q}_\perp^2|}{\mu_0} + \frac{B^2}{\mu_0} \left| \nabla \cdot \boldsymbol{\xi}_\perp + 2\boldsymbol{\xi}_\perp \cdot \boldsymbol{\kappa} \right|^2 + \Gamma p_0 |\nabla \cdot \boldsymbol{\xi}|^2 \right.$$

$$\left. -2 \left(\boldsymbol{\xi}_\perp \cdot \nabla p_0 \right) \left(\boldsymbol{\kappa} \cdot \nabla \boldsymbol{\xi}_\perp^* \right) - J_\parallel \left(\boldsymbol{\xi}_\perp^* \times \hat{\mathbf{b}} \right) \cdot \mathbf{Q}_\perp \right\}, \tag{29.1}$$

where

$$\mathbf{Q} = \mathbf{Q}_\perp + Q_\parallel \hat{\mathbf{b}}, \tag{29.2}$$

$$Q_\parallel = -B \left(\nabla \cdot \boldsymbol{\xi}_\perp + 2\boldsymbol{\xi}_\perp \cdot \boldsymbol{\kappa} \right) + \frac{\mu_0}{B} \boldsymbol{\xi}_\perp \cdot \nabla p_0, \tag{29.3}$$

$$\mathbf{J} = \mathbf{J}_\perp + J_\parallel \hat{\mathbf{b}}, \tag{29.4}$$

$$\boldsymbol{\xi} = \boldsymbol{\xi}_\perp + \xi_\parallel \hat{\mathbf{b}}, \tag{29.5}$$

and

$$\boldsymbol{\kappa} = \hat{\mathbf{b}} \cdot \nabla \hat{\mathbf{b}}. \tag{29.6}$$

Instability occurs if $\delta W < 0$.

---

[1] For the gory details, see Jeffrey P. Freidberg, *Ideal Magnetohydrodynamics*, Plenum Press, New York (1987).

Schnack, D.D.: *Comments on the Energy Principle and the Minimizing Eigenfunction.* Lect. Notes Phys. **780**, 179–182 (2009)
DOI 10.1007/978-3-642-00688-3_29

We remark on the terms in Eq. (29.1) as follows:

1. $\left|\mathbf{Q}_\perp^2\right|/\mu_0 > 0$ is *stabilizing*. This is the energy required to bend the field lines. It gives rise to the shear Alfvén wave.
2. $B^2\left|\nabla\cdot\boldsymbol{\xi}_\perp + 2\boldsymbol{\xi}_\perp\cdot\boldsymbol{\kappa}\right|^2/\mu_0 > 0$ is *stabilizing*. It is the energy required to compress the magnetic field. It gives rise to magneto-acoustic waves.
3. $\Gamma p_0\left|\nabla\cdot\boldsymbol{\xi}\right|^2 > 0$ is *stabilizing*. It is the energy required to compress the fluid. It gives rise to sound waves.
4. $-2\left(\boldsymbol{\xi}_\perp\cdot\nabla p_0\right)\left(\boldsymbol{\kappa}\cdot\nabla\boldsymbol{\xi}_\perp^*\right)$ can be either stabilizing or destabilizing. When negative, it gives rise to *pressure-driven instabilities*. Since $\nabla p = \mathbf{J}_\perp\times\mathbf{B}$, it contains the effects of perpendicular current.
5. $-J_\parallel\left(\boldsymbol{\xi}_\perp^*\times\hat{\mathbf{b}}\right)\cdot\mathbf{Q}_\perp$ can be either stabilizing or destabilizing. When negative, it can give rise to *current-driven instabilities* or, more accurately, instabilities driven by the *parallel* current.

Note that $\xi_\parallel$ enters δW only through the term $\nabla\cdot\boldsymbol{\xi}$. It is therefore possible to minimize δW once and for all with respect to $\xi_\parallel$. We will then be left to deal only with $\delta W\{\boldsymbol{\xi}_\perp,\boldsymbol{\xi}_\perp\}$. This minimization is carried out by letting $\xi_\parallel \to \xi_\parallel + \Delta\xi_\parallel$ in Eq. (29.1), where $\Delta\xi_\parallel$ is the variation of ξ. (The Δ notation is used to avoid confusion with δW.) Assuming that $\hat{\mathbf{b}}\cdot\hat{\mathbf{n}} = 0$ on the boundary, and after much algebra, we finally arrive at an expression for the variation of δW as

$$\Delta\left(\delta W\right) = 0 = -\mathrm{Re}\int dV\,\Delta\xi_\parallel^*\hat{\mathbf{b}}\cdot\nabla\left(\Gamma p_0\nabla\cdot\boldsymbol{\xi}\right). \tag{29.7}$$

Since this must hold for arbitrary $\Delta\xi_\parallel$, we must require

$$\hat{\mathbf{b}}\cdot\nabla\left(\Gamma p_0\nabla\cdot\boldsymbol{\xi}\right) = 0. \tag{29.8}$$

Any $\boldsymbol{\xi}$ that satisfies Eq. (29.8) minimizes δW with respect to $\xi_\parallel$. Now

$$\hat{\mathbf{b}}\cdot\nabla\left(p_0\nabla\cdot\boldsymbol{\xi}\right) = p_0\hat{\mathbf{b}}\cdot\nabla\left(\nabla\cdot\boldsymbol{\xi}\right) + \nabla\cdot\boldsymbol{\xi}\,\hat{\mathbf{b}}\cdot\nabla p_0. \tag{29.9}$$

But in equilibrium, $\hat{\mathbf{b}}\cdot\nabla p_0 = 0$, so that Eq. (29.8) becomes

$$\hat{\mathbf{b}}\cdot\nabla\left(\nabla\cdot\boldsymbol{\xi}\right) = 0. \tag{29.10}$$

This is the final form of the minimizing condition.

Equation (29.10) has a form similar to the homogeneous algebraic equation $Ax = 0$. We know that if $A \neq 0$, then the only solution is $x = 0$. The condition $A \neq 0$ is equivalent to saying that A is invertible, i.e., $A^{-1} \equiv 1/A$ exists. Conversely, if $A = 0$, A is not invertible and solutions $x \neq 0$ are possible.

By analogy, the properties of the constraint (29.10) depend on whether the operator $\hat{\mathbf{b}}\cdot\nabla$ is invertible or not. Suppose it is invertible everywhere. Then the solution of Eq. (29.10) is $\nabla\cdot\boldsymbol{\xi} = 0$; this is the minimizing condition. In light of Eq. (29.5),

$$\nabla \cdot \left(\boldsymbol{\xi}_\perp + \xi_\parallel \hat{\mathbf{b}} \right) = 0. \tag{29.11}$$

With $\hat{\mathbf{b}} = \mathbf{B}/B$ and $\nabla \cdot \mathbf{B} = 0$, this can be written as

$$\mathbf{B} \cdot \nabla \left(\frac{\xi_\parallel}{B} \right) = -\nabla \cdot \boldsymbol{\xi}_\perp. \tag{29.12}$$

Choosing $\xi_\parallel$ to satisfy Eq. (29.12) will minimize δW if $\hat{\mathbf{b}} \cdot \nabla$ is invertible everywhere.

Now suppose that $\hat{\mathbf{b}} \cdot \nabla$ is invertible everywhere *except* at some isolated locations where $\hat{\mathbf{b}} \cdot \nabla = 0$. Then for $\nabla \cdot \boldsymbol{\xi}_\perp$ to remain finite, $\xi_\parallel/B$ must go to infinity at these locations, i.e., $\xi_\parallel$ is *singular* there. This is not an acceptable trial function. Let the location of the singularity be $\mathbf{x}_0$, and let

$$\boldsymbol{\xi} = \tilde{\boldsymbol{\xi}} + \varepsilon \boldsymbol{\eta}, \tag{29.13}$$

where $\boldsymbol{\eta}(\mathbf{x})$ is zero everywhere except in a small region of size ε around $\mathbf{x} = \mathbf{x}_0$, and $\varepsilon \ll 1$. We assume that $\boldsymbol{\xi}$ is the "real" minimizing trial function that satisfies Eq. (29.12) (i.e., has $\nabla \cdot \boldsymbol{\xi} = 0$ everywhere and $\xi_\parallel$ singular at $\mathbf{x} = \mathbf{x}_0$), and that $\tilde{\boldsymbol{\xi}}$ is some "neighboring" trial function (with $\nabla \cdot \tilde{\boldsymbol{\xi}} \neq 0$ and $\xi_\parallel$ well behaved at $\mathbf{x} = \mathbf{x}_0$). Then $\nabla \cdot \boldsymbol{\xi} = \nabla \cdot \tilde{\boldsymbol{\xi}} + \varepsilon \nabla \cdot \boldsymbol{\eta}$, so that $\nabla \cdot \boldsymbol{\xi}$ and $\nabla \cdot \tilde{\boldsymbol{\xi}}$ can be chosen to be equal to each other (and therefore zero) away from $\mathbf{x}_0$, and differ by an arbitrarily small amount near $\mathbf{x}_0$. Then

$$\begin{aligned} \delta W \{\boldsymbol{\xi}, \boldsymbol{\xi}\} &= \delta W \{ \tilde{\boldsymbol{\xi}} + \varepsilon \boldsymbol{\eta}, \tilde{\boldsymbol{\xi}} + \varepsilon \boldsymbol{\eta} \} \\ &= \delta \tilde{W} + O(\varepsilon), \end{aligned}$$

where $\delta \tilde{W} = \delta W \{ \tilde{\boldsymbol{\xi}}, \tilde{\boldsymbol{\xi}} \}$ is a "neighboring" δW computed with a trial function that has $\xi_\parallel$ well behaved at $\mathbf{x}_0$. Now suppose $\delta \tilde{W} < 0$, i.e., we find an instability with the trial function $\tilde{\boldsymbol{\xi}}$. Then we can choose ε so small that δW (the "real" δW) is also negative. We therefore conclude that for every well-behaved trial function $\tilde{\boldsymbol{\xi}}$ that makes $\delta \tilde{W} < 0$, there is a "neighboring" trial function $\boldsymbol{\xi}$ (with $\xi_\parallel$ singular at $\mathbf{x}_0$) whose potential energy δW differs from $\delta \tilde{W}$ by an arbitrarily small amount, and which satisfies $\nabla \cdot \boldsymbol{\xi} = 0$.

The procedure when $\hat{\mathbf{b}} \cdot \nabla \neq 0$ at $\mathbf{x} = \mathbf{x}_0$ is as follows:

1. Choose a trial function $\tilde{\boldsymbol{\xi}}$ that is well behaved at $\mathbf{x}_0$.
2. Compute $\delta \tilde{W}$.
3. If $\delta \tilde{W} < 0$, the system is unstable.
4. It will also be unstable with the singular trial function $\boldsymbol{\xi}$.

Therefore, any conclusions drawn about instability with a non-singular trial function will be valid.

Now suppose that $\hat{\mathbf{b}} \cdot \nabla = 0$ everywhere. This is the case for the g-mode (with $\mathbf{B} = B(x)\hat{\mathbf{e}}_y$ and $\nabla = \hat{\mathbf{e}}_x \partial_x + \hat{\mathbf{e}}_z \partial_z$) and for axially symmetric perturbations to a pure Z-pinch (with $\mathbf{B} = B_\theta(r)\hat{\mathbf{e}}_\theta$ and $\nabla = \hat{\mathbf{e}}_r \partial_r + \hat{\mathbf{e}}_z \partial_z$). Then $\nabla \cdot \boldsymbol{\xi} = \nabla \cdot \boldsymbol{\xi}_\perp$, i.e., $\xi_\parallel$

does not appear in δW and the term $\Gamma p_0 \left| \nabla \cdot \boldsymbol{\xi}_\perp \right|^2$ must be retained (as was seen in the stability analysis of the g-mode). In this case, $\nabla \cdot \boldsymbol{\xi} = 0$ is not the most unstable displacement.

A further special case occurs if the field lines are all closed. Then the periodicity constraint $\xi_\parallel(l) = \xi_\parallel(l + L)$, where L is the length of a field line and l is the distance along a field line, must be imposed on each field line. It can be shown that this implies the condition

$$\int \Gamma p_0 \left| \nabla \cdot \boldsymbol{\xi} \right|^2 dV = \int \Gamma p_0 \left| \langle \nabla \cdot \boldsymbol{\xi}_\perp \rangle \right|^2 dV,$$

where

$$\langle f \rangle \equiv \frac{\int \frac{dl}{B} f}{\int \frac{dl}{B}}$$

denotes the average taken along a field line. Again, the term $\Gamma p_0 \left| \nabla \cdot \boldsymbol{\xi}_\perp \right|^2$ must be included in the analysis.

We remark on the significance of the condition $\hat{\mathbf{b}} \cdot \nabla = 0$, which occurs at locations $\mathbf{x} = \mathbf{x}_0$. In the analysis of axisymmetric equilibria, the $\mathbf{x}_0$ correspond to flux surfaces. These special surfaces are called *singular surfaces*. If we use the analogy $\nabla \rightarrow i\mathbf{k}$, then the singularity condition is $\mathbf{k} \cdot \mathbf{B} = 0$, i.e., on these surfaces the wavefronts of the perturbation are aligned with the magnetic field, so that the perturbation does not bend the field lines. We have already seen in Lecture 26 that this is destabilizing. Singular surfaces play a special role in the stability analysis of axisymmetric systems.

We will now consider several examples of the role played by the issues raised in this lecture in determining the stability properties of some specific equilibria.

Lecture 30
Examples of the Application of the Energy Principle to Cylindrical Equilibria[1]

> *Few things are harder to put up with than the annoyance of a good example.*
>
> Mark Twain

We now use the energy principle to analyze the stability properties of the cylindrical θ-pinch, the Z-pinch, and the general screw pinch. Since these equilibria depend only on the radial coordinate r, are periodic in the θ- and z-coordinates, and the fields are independent of the periodic coordinates, the displacement can be written as in terms of the Fourier decomposition

$$\boldsymbol{\xi}(\mathbf{r}) = \boldsymbol{\xi}(r)\, e^{i(m\theta + kz)}. \tag{30.1}$$

Since the equilibria have no radial component of the magnetic field,

$$\mathbf{B} \cdot \nabla = \frac{im}{r} B_\theta + ik B_z \ . \tag{30.2}$$

In Eq. (30.1), m is an integer in the range $-\infty \le m \le \infty$. It is called the *poloidal mode number*. If the system is infinitely long, k is a continuous variable. However, it is quantized if the cylinder has finite length. For example, if a torus with circular cross-section a and major radius R (has *aspect ratio* R/a) is cut and straightened into a cylinder of length $L = 2\pi R$, then periodicity requires $k = n/R$, where n is an integer in the range $-\infty \le n \le \infty$. It is called the *axial*, or *toroidal*, *mode number*.

Typical displacements of the plasma column for different values of m and k are shown in Figs. 30.1, 30.2, and 30.3.

The mode with $m = 0, k \ne 0$ is colorfully called the "sausage mode." Since $m = 0$, the displacement is azimuthally symmetric (i.e., independent of θ). The mode with $m = 1, k = 0$ is just a shift of the column with respect to the axis. The mode with $m = 1, k \ne 0$ is called a "kink mode." It distorts the column helically.

---

[1] Again, we closely follow Jeffrey P. Freidberg, *Ideal Magnetohydrodynamics*, Plenum Press, New York (1987).

Schnack, D.D.: *Examples of the Application of the Energy Principle to Cylindrical Equilibria.*
Lect. Notes Phys. **780**, 183–192 (2009)
DOI 10.1007/978-3-642-00688-3_30

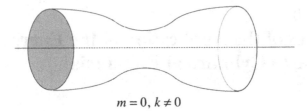

$$m = 0, \ k \neq 0$$

Fig. 30.1 The sausage mode

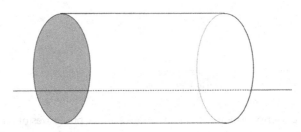

$$m = 1, \ k = 0$$

Fig. 30.2 The shift mode

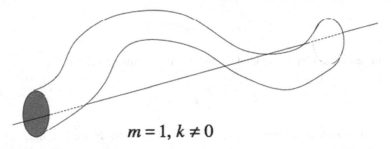

$$m = 1, \ k \neq 0$$

Fig. 30.3 The kink mode

We now consider the three cylindrical equilibria.

The θ-pinch

The equilibrium for the θ-pinch is

$$p(r) + \frac{B_z^2}{2\mu_0} = \frac{B_0^2}{2\mu_0}, \tag{30.3}$$

where B_0 is the magnetic field strength outside the fluid. It is produced externally.

Since $B_\theta = 0$, Eq. (30.2) becomes $\mathbf{B} \cdot \nabla = ikB_z$, so that this operator can be inverted as long as $k \neq 0$. In that case the minimizing condition for δW is $\nabla \cdot \boldsymbol{\xi} = 0$ or

$$\frac{1}{r} \frac{d}{dr} (r\xi_r) + \frac{im}{r} \xi_\theta + ik\xi_z = 0. \tag{30.4}$$

We can therefore use Eq. (30.4) to eliminate $\xi_\parallel = \xi_z$ from δW according to

$$\xi_z = \frac{i}{kr}\left[(r\xi_r)' + im\xi_\theta\right],$$ (30.5)

where $(\ldots)'$ denotes differentiation with respect to r. This is valid as long as $k \neq 0$. In this case we have

$$\mathbf{Q}_\perp = ikB_z\boldsymbol{\xi}_\perp = ikB_z\,(\xi_r\hat{\mathbf{e}}_r + \xi_\theta\hat{\mathbf{e}}_\theta),$$ (30.6)

$$\nabla\cdot\boldsymbol{\xi}_\perp = \frac{1}{r}(r\xi_r)' + \frac{im\xi_\theta}{r},$$ (30.7)

$$\boldsymbol{\kappa} = \hat{\mathbf{b}}\cdot\nabla\hat{\mathbf{b}} = 0,$$ (30.8)

and

$$J_\parallel = 0.$$ (30.9)

In cylindrical geometry, the potential energy per unit length is

$$\frac{\delta W}{L} = \frac{\pi}{\mu_0}\int_0^a W(r)r\,dr.$$ (30.10)

For the case $k \neq 0$, using Eqs. (30.5, 30.6, 30.7, 30.8, 30.9),

$$W(r) = B_z^2\left[k^2\left(|\xi_r|^2 + |\xi_r|^2\right) + \frac{1}{r^2}\left|(r\xi_r)'\right|^2 + \frac{m^2}{r^2}|\xi_\theta|^2\right.$$
$$\left. + \frac{im}{r^2}(r\xi_r^*)'\,\xi_\theta - \frac{im}{r^2}(r\xi_r)'\,\xi_\theta^*\right].$$ (30.11)

Then after a considerable amount of algebra, Eq. (30.10) can be re-written as

$$\frac{\delta W}{L} = \frac{\pi}{\mu_0}\int_0^a r\,dr\,B_z^2\left\{\left|k_0\xi_\theta - \frac{im}{k_0r^2}(r\xi_r)'\right|^2 + \frac{k^2}{k_0^2r^2}\left(\left|(r\xi_r)'\right|^2 + r^2k_0^2\,|\xi_r|^2\right)\right\},$$ (30.12)

where $k_0^2 = m^2/r^2 + k^2$. We note that ξ_θ appears only in the first term in the integrand. Setting this term to zero, we find that he minimizing trial function must have

$$\xi_\theta = \frac{im}{m^2 + k^2r^2}(r\xi_r)',$$ (30.13)

so that Eq. (30.12) becomes

$$\frac{\delta W}{L} = \frac{\pi}{\mu_0} \int\limits_0^a r\, dr \frac{k^2 B_z^2}{m^2 + k^2 r^2} \left\{ \left| (r\xi_r)' \right|^2 + \left(m^2 + k^2 r^2 \right) |\xi_r|^2 \right\}. \tag{30.14}$$

We can draw two conclusions from Eq. (30.14):

- $\delta W > 0$ for $k^2 \neq 0$, so that the θ-pinch is *stable* for all finite k.
- $\delta W \to 0$ for $k^2 \to \infty$, so that the stability becomes marginal for very long wavelengths.

The MHD stability of the θ-pinch is explained as follows:

- $J_\parallel = 0$ ($\mathbf{J} = J_\theta \hat{\mathbf{e}}_\theta$, $\mathbf{B} = B\hat{\mathbf{e}}_z$), so that there are no current driven modes.
- $\kappa = 0$ (no field-line curvature), so $(\boldsymbol{\xi}_\perp \cdot \nabla p_0)(\kappa \cdot \boldsymbol{\xi}_\perp) = 0$ and there are no pressure driven modes.
- Therefore, there is no MHD source of "free energy" to drive instability.

However, real θ-pinch have finite length and therefore have field-line curvature, so this can drive instability in the laboratory. This is the reason we have not considered the case $k = 0$ here. Also, there are also several non-MHD instability drives that have rendered this concept problematical for magnetic confinement fusion.
The Z-pinch
For the Z-pinch, $\mathbf{B} = B_\theta(r)\hat{\mathbf{e}}_\theta$ and $\mathbf{J} = J_z(r)\hat{\mathbf{e}}_z$, and the equilibrium condition is

$$\frac{dp_0}{dr} + \frac{B_\theta}{\mu_0 r} \frac{d}{dr} (r B_\theta) = 0. \tag{30.15}$$

Therefore $J_\parallel = 0$ and $\xi_\parallel = \xi_\theta$. For $m \neq 0$, the operator $\mathbf{B} \cdot \nabla = im B_\theta/r$ is well behaved and can be inverted. In the case the minimizing condition is $\nabla \cdot \boldsymbol{\xi} = 0$, so that

$$\xi_\theta = \frac{i}{m} \left[(r\xi_r)' + ik\xi_z \right], \tag{30.16}$$

which is valid as long as $m \neq 0$. If $m = 0$, then $\mathbf{B} \cdot \nabla = 0$ everywhere and we must consider $\nabla \cdot \boldsymbol{\xi}_\perp$ in the minimization.
When $m \neq 0$, we have

$$\mathbf{Q}_\perp = \frac{im B_\theta}{r} (\xi_r \hat{\mathbf{e}}_r + \xi_\theta \hat{\mathbf{e}}_\theta), \tag{30.17}$$

$$\nabla \cdot \boldsymbol{\xi}_\perp + 2\boldsymbol{\xi}_\perp \cdot \kappa = r \left(\frac{\xi_r}{r} \right)' + ik\xi_z, \tag{30.18}$$

$$\kappa = -\frac{\hat{\mathbf{e}}_r}{r}, \tag{30.19}$$

and

$$J_\parallel = 0. \tag{30.20}$$

Again, after much algebra, we find

$$\frac{\delta W}{L} = \frac{\pi}{\mu_0} \int\limits_0^a r\, dr \left\{ (2\mu_0 r p' + m^2 B_\theta^2) \frac{|\xi_r|^2}{r} + \frac{m^2 r^2 B_\theta^2}{m^2 + r^2 k^2} \left| \left(\frac{\xi_r}{r} \right)' \right| \right\}. \quad (30.21)$$

The last term is minimized for $k^2 \to \infty$, so the stability of the Z-pinch is determined by the sign of

$$\frac{\delta W}{L} = \frac{\pi}{\mu_0} \int\limits_0^a r\, dr \, (2\mu_0 r p' + m^2 B_\theta^2) \frac{|\xi_r|^2}{r}. \quad (30.22)$$

Suppose the integrand of Eq. (30.22) is negative in some interval $r_0 < r < r_1$. Then we can choose the trial function such that $\xi_r \neq 0$ inside this interval, and $\xi_r = 0$ outside it. Since this interval is arbitrary, we conclude that *a necessary and sufficient condition for the stability of the Z-pinch when $m \neq 0$ is*

$$2\mu_0 r \frac{dp_0}{dr} + m^2 B_\theta^2 > 0 \quad (30.23)$$

at all points in the fluid.

We can use the equilibrium condition, Eq. (30.15), to eliminate the pressure gradient from the Eq. (30.23). The result is

$$2B_\theta \frac{d}{dr}(r B_\theta) < m^2 B_\theta^2. \quad (30.24)$$

This can be re-written in either of two forms. The first is

$$\frac{r^2}{B_\theta} \frac{d}{dr} \left(\frac{B_\theta}{r} \right) < \frac{1}{2}(m^2 - 4). \quad (30.25)$$

For $r \to 0$, $B_\theta \sim r$, so $B_\theta/r \sim$ constant. In this limit, the stability condition is therefore $m^2 > 4$, so that the interior of the Z-pinch is stable for $|m| > 2$ and marginal for $|m| = 2$. For $r \to \infty$, $B_\theta \sim 1/r$ and $d(B_\theta/r)/dr \sim 1/r^3 \to 0$, so the same conclusion holds in this limit.

The second form of the stability condition is

$$\frac{1}{B_\theta^2} \frac{d}{dr}(r B_\theta^2) < m^2 - 1. \quad (30.26)$$

For $r \to \infty$, $d(r B_\theta^2)/dr \sim -1/r^2 \to 0$, and the stability condition is $m^2 > 1$. Therefore, all $|m| > 1$ are stable and $|m| = 1$ is marginal, in this limit. For $r \to 0$,

$r B_\theta^2 \sim r^3$ and the left-hand side of Eq. (30.26) $\sim r > 0$, so that the core is always unstable to the $|m| = 1$ mode.

Note that, since $J_\| = 0$, this $|m| = 1$ mode is not a current-driven mode. Rather, stability is determined by a competition between field-line bending (stabilizing) and unfavorable curvature (destabilizing). The latter wins out in the core of the Z-pinch for the mode with $|m| = 1$.

We now consider the case $m = 0$. Here $\mathbf{B} \cdot \nabla = 0$ everywhere, and so $\nabla \cdot \boldsymbol{\xi} \neq 0$. After a formidable calculation, δW is found to be

$$\frac{\delta W}{L} = \frac{\pi}{\mu_0} \int_0^a r\, dr \left\{ \frac{r \Gamma p_0 B_\theta^2 / \mu_0}{\Gamma p_0 + B_\theta^2 / \mu_0} + 2r p_0' \right\} \frac{|\xi_r|^2}{r^2}, \tag{30.27}$$

with

$$\xi_z = \frac{i}{\Gamma p_0 + B_\theta^2 / \mu_0} \left[\frac{r B_\theta^2}{\mu_0} \left(\frac{\xi_r}{r} \right)' + \frac{\Gamma p_0}{r} (r \xi_r)' \right] \tag{30.28}$$

for the minimizing perturbation. Again, stability requires that the integrand of Eq. (30.27) be positive for all r, which leads to the stability condition for $m = 0$ modes:

$$\frac{-r}{p_0} \frac{dp_0}{dr} = -\frac{d \ln p_0}{d \ln r} < \frac{4\Gamma}{2 + \Gamma \beta_\theta}, \tag{30.29}$$

where the poloidal beta is $\beta_\theta = 2 \mu_0 p_0 / B_\theta^2$. The Z-pinch can support a pressure gradient as long as it is not too large.

We remark that, since the condition (30.29) depends on the adiabatic index Γ, it requires that the fluid satisfy the adiabatic law. This is seldom the case for real plasmas. In that case, all bets are off as far as $m = 0$ stability is concerned.

The general screw pinch

You may have observed that the stability calculations become increasingly formidable as the equilibrium becomes more complex. In this regard, the case of the general screw pinch inherits some of the worst elements of the calculations for the θ-pinch and the Z-pinch. It will therefore be treated here with even more informality.

For the general screw pinch, $\mathbf{B} = B_\theta(r)\hat{\mathbf{e}}_\theta + B_z(r)\hat{\mathbf{e}}_z$ and the equilibrium condition is

$$\frac{d}{dr} \left(p_0 + \frac{B_\theta^2 + B_z^2}{2\mu_0} \right) + \frac{B_\theta^2}{\mu_0 r} = 0. \tag{30.30}$$

The minimizing perturbation has

$$\mathbf{B} \cdot \nabla \left(\frac{\xi_\|}{B} \right) = -\nabla \cdot \boldsymbol{\xi}_\perp. \tag{30.31}$$

In this case we can write $\mathbf{B} \cdot \nabla = i F(r)$, where $F(r) \equiv \mathbf{k} \cdot \mathbf{B} = m B_\theta / r + k B_z$. Therefore, $\mathbf{B} \cdot \nabla$ is well behaved everywhere that $F \neq 0$. In that case the minimizing perturbation has

$$\xi_\parallel = \frac{iB}{F} \nabla \cdot \boldsymbol{\xi}_\perp. \tag{30.32}$$

The roots r_0 of the equation $F = 0$ are called singular surfaces (again associating the radial coordinate with flux surfaces), for at these points $\mathbf{B} \cdot \nabla$ is singular and cannot be inverted. From our discussion in Lecture 29, we can still choose a "well-behaved" $\xi_\parallel$ that still minimizes δW, so that we can draw reliable conclusions from the energy principle. Physically, the surfaces r_0 are associated with $k_\parallel = 0$, so that the field-line bending term is minimized. We may expect these surfaces to play an important role in determining stability.

Under these circumstances, and "after some algebra," one finds

$$\frac{\delta W}{L} = \frac{\pi}{\mu_0} \int_0^a \left(f \left| \xi_r' \right|^2 + g \left| \xi_r \right|^2 \right) dr, \tag{30.33}$$

where

$$f = \frac{r F^2}{k_0^2} \tag{30.34}$$

and

$$g = 2 \frac{k^2}{k_0^2} \left(\mu_0 p_0' \right) + \left(\frac{k_0^2 r^2 - 1}{k_0^2 r^2} \right) r F^2 + \frac{2 k^2}{r k_0^4} \left(k B_z - \frac{m B_\theta}{r} \right) F. \tag{30.35}$$

Some general remarks can be made as follows:

1. First, $f \geq 0$ for all r, so the term $f \left| \xi_r' \right|^2$ is stabilizing. However, it vanishes at the singular surfaces r_0 where $\mathbf{k} \cdot \mathbf{B} = 0$, so that we may expect instability to be associated with these radii.
2. The sign of the term $g \left| \xi_r \right|^2$ is determined by the sign of g. The minimizing displacement therefore should have $\left| \xi_r \right|^2 > 0$, where $g < 0$, and $\left| \xi_r \right|^2 = 0$ where $g > 0$.

It turns out that the analysis of pressure-driven modes arising from the p_0' term requires a detailed analysis of the behavior of the solutions of the Euler equation (the ideal MHD wave equation) near its "regular singular points." This will be briefly outlined in Lecture 31.

For current-driven modes ($p_0' = 0$), the sign of g is determined by the sign of F. For the case of a "straight torus" when k is quantized, we can write

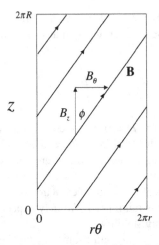

Fig. 30.4 An "unwrapped" cylindrical flux surface showing the field lines

$$F = \frac{B_\theta}{r}(m + nq),\tag{30.36}$$

where

$$q(r) = \frac{rB_z(r)}{RB_\theta(r)}\tag{30.37}$$

is called the *safety factor*. Consider an unwrapped "flux surface," i.e., a cylinder of radius r unwrapped and flattened. The magnetic field lines lie completely within such surfaces, as sketched in Fig. 30.4. The wrapping angle is $\phi = \tan^{-1} B_\theta/B_z$. We define the *pitch* of the field lines in the surface as $P(r) = rB_z/B_\theta$. This is the distance that one would travel axially (in the z-direction) by following a field line through one circuit from $\theta = 0$ to $\theta = 2\pi$. The pitch is a function of radius, meaning that the wrapping angle varies from surface to surface. The safety factor is therefore the normalized pitch, $q = P/R$. The quantity $rq'/q \equiv d\ln q/d\ln r$ is called the *magnetic shear*.

From Eq. (30.36) we see that when $F = 0$, $q = -m/n$, i.e., q is a rational number. This is why these surfaces are also called *rational surfaces*. The field lines close upon themselves after m turns in the poloidal (θ) direction and n turns in the toroidal (axial, or z) direction, within the surface.

The safety factor at the outer boundary $r = a$ is $q(a) = aB_z(a)/RB_\theta(a)$. Since $B_\theta(a) \sim I/a$, where I is the total toroidal (axial) current, $q(a) \sim a^2 B_z(a)/I \sim$ (total toroidal flux)/(total toroidal current). The more current for a given flux, the smaller $q(a)$.

The configuration is unstable if $g < 0$, which can only occur if $F < 0$ or $q(r) < -m/n$. (In the tokamak literature, it is customary to write $n \to -n$, so that $q < m/n$. Since it is only the relative orientation of $k_\theta = m/r$ and $k_z = k = n/R$

that enter the theory, it is also customary to consider only $m \geq 0$. Here we employ the standard mathematical notation for the Fourier decomposition and live with the minus sign.) Therefore, restricting the discussion to $m > 0$, the configuration is stable if $n > 0$ and may be unstable if $n < 0$ and $q < m/|n|$.

The $m = 1$ mode may be unstable if $q(r) < 1/|n|$. We therefore must require $q(r) > 1$ everywhere to assure stability. In particular, at $r = a$ we require $B_z(a)/B_\theta(a) > R/a$. This is the *Kruskal–Shafranov stability condition*. It means that the ratio of the toroidal (axial) flux to the toroidal (axial) current cannot be too small. It implies the necessity of a strong toroidal magnetic field for stability. This field, which does not contribute to confinement, must be supplied by external means. The Kruskal–Shafranov condition forms the basis for the design of the tokamak.

We have seen that the minimizing displacement must have $|\xi_r|^2 > 0$ when $g < 0$ and $|\xi_r|^2 = 0$ when $g > 0$. For the case when g is a monotonically increasing function of r, the situation is sketched in Fig. 30.5.

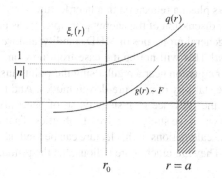

Fig. 30.5 The safety factor $q(r)$, the function $g(r)$, and the minimizing trial function $\xi(r)$ for a tokamak

This is the general case for a *tokamak*. The "top hat" shape of the displacement is typical of the $m = 1$ mode in a tokamak. Note that the displacement vanishes outside the rational surface, so that it does not "feel" the wall. The most unstable $m = 1$ displacement in a tokamak will consist (approximately) of a rigid shift of the flux surfaces inside the rational surface and no displacement outside the rational surface. We emphasize that this discussion, and the figure, are heuristic and approximate because of the stabilizing term in g proportional to F^2; the root of $g = 0$ does not exactly correspond to the root of $F = 0$. However, the picture serves as a reasonable guide to what happens when a more detailed analysis is attempted.

If instead g is a monotonically decreasing function of r, then the situation is like that shown in Fig. 30.6.

The minimizing displacement now must vanish inside the rational surface, and be non-zero outside of t. However, it must also satisfy the boundary condition $\xi_r(a) = 0$. Therefore $\xi' \neq 0$ and the first term in Eq. (30.34) can contribute to stabilization. The details depend on the relative location of the wall and the rational

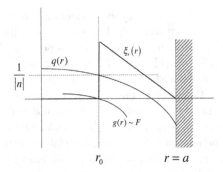

Fig. 30.6 The safety factor $q(r)$, the function $g(r)$, and the minimizing trial function $\xi(r)$ for an RFP

surface. This type of stabilization is called *wall stabilization*. A monotonically decreasing profile is characteristic of an RFP. We will describe this concept in more detail when we discuss plasma relaxation in a later lecture.

This concludes our discussion of the energy principle. It has provided a means of deducing some very general properties of MHD stability, and *we never had to solve a differential equation!* This will not be the case from now on. As mentioned, the properties of the Euler equation near a regular singular point must be investigated in order to determine the stability of pressure-driven modes. And the energy principle is no longer valid in the presence of resistivity, so of necessity we must address differential equations when we study resistive instabilities. That is what follows.

Of course, all of the calculations of this lecture can be (and have been) carried out in toroidal geometry. They are much more tedious, but the primary concepts remain unchanged.

Lecture 31
A Very Brief and General Tour of Suydam Analysis for Localized Interchange Instabilities

If stupidity got us into this mess, then why can't it get us out?

Will Rogers

We have seen that, for the general screw pinch, the perturbed potential energy is

$$\delta W = \int_0^a L\left(r, \xi, \xi'\right) dr, \tag{31.1}$$

where L is a Lagrangian density proportional to $f\xi'^2 + g\xi^2$. The Euler equation corresponding to Eq. (31.1) is

$$\frac{d}{dx}\left(\frac{\partial L}{\partial \xi'}\right) - \frac{\partial L}{\partial \xi} = 0. \tag{31.2}$$

If one uses the equilibrium condition to eliminate J_z in terms of p_0', Eq. (31.2) can be written in the standard form

$$\xi'' - P\xi' - Q\xi = 0, \tag{31.3}$$

where

$$P = \frac{3}{r} + \frac{2F'}{F} - \frac{2k^2 r}{m^2 - k^2 r^2}, \tag{31.4}$$

$$Q = \frac{1}{r^2}\left[k^2 r^2 + \left(m^2 - 1\right)\right] - \frac{2k^2 g(r)}{F\left[m^2 + k^2 r^2\right]^2} - \frac{2\mu_0 k^2 p_0'}{r F^2}, \tag{31.5}$$

and $F(r) = \mathbf{k} \cdot \mathbf{B}$, and $g(r)$ was defined in Eq. (30.35). Equation (31.3) is a second-order ordinary differential equation for the radial component of the minimizing displacement subject to the boundary conditions: $\xi(0)$ is finite and $\xi(a) = 0$.

Schnack, D.D.: *A Very Brief and General Tour of Suydam Analysis for Localized Interchange Instabilities*. Lect. Notes Phys. **780**, 193–195 (2009)
DOI 10.1007/978-3-642-00688-3_31 © Springer-Verlag Berlin Heidelberg 2009

Any point r_0 where $P \to \infty$ is a *regular singular point* of Eq. (31.3). We see from Eq. (31.4) that this occurs when $F = 0$, so that the regular singular points of the Euler equation correspond to the rational surfaces of the stability analysis. We know from the theory of ordinary differential equations (the "method of Frobenius") that the solution near a regular singular point behaves like

$$\xi \sim (r - r_0)^n \sum_j a_j (r - r_0)^j, \tag{31.6}$$

where n is a root of the *indicial equation*

$$n^2 + n + M^2 = 0, \tag{31.7}$$

which is obtained from an expansion of P and Q about $r = r_0$. If this procedure is carried out for Eqs. (31.3, 31.4, 31.5), one finds

$$M^2 = \frac{2\mu_0 p_0'}{r B_z^2} \left(\frac{\mu}{\mu'}\right)^2 \Bigg|_{r=r_0} \tag{31.8}$$

and

$$\mu = \frac{B_\theta}{r B_z} = \frac{1}{Rq}. \tag{31.9}$$

The roots of Eq. (31.7) are

$$n_{1,2} = -\frac{1}{2}\left[1 \pm \left(1 - 4M^2\right)^{1/2}\right], \tag{31.10}$$

so that near r_0,

$$\xi \sim a (r - r_0)^{n_1} + b (r - r_0)^{n_2}. \tag{31.11}$$

If $M^2 < 1/4$, then n_1 and n_2 are *real*. That means that there is at least one solution [for example, the + sign in Eq. (31.10)] that behaves like $(r - r_0)^{-|n|}$ near $r = r_0$; this solution is *singular*.

If $M^2 > 1/4$, then n_1 and n_2 are complex. In that case, the solutions always behave at least like $(r - r_0)^{-1/2}$ near $r = r_0$; they are also singular.

Therefore there is *always* a singular solution near $r = r_0$. This is not acceptable behavior for a displacement. As we saw in Lecture 29, this can always be resolved by choosing a neighboring displacement $\tilde{\xi}$ that equals ξ away from r_0, but is well behaved in the vicinity of r_0; it differs from ξ only in a small region of width 2ε, $\varepsilon \ll 1$, about r_0.

If the roots of the indicial equation are real, then ε can be chosen so small that the power series solution is dominated by the term $(r - r_0)^n$ [i.e., the first term in the expansion (31.6)]. It turns out that, when this is substituted into the energy principle, the result is $\delta W > 0$, i.e., stability.

If the roots are complex, then $n_{1,2} = -(1 \pm i\beta)/2$, where $\beta = (4M^2 - 1)^{1/2}$. Then the displacement behaves like

$$
\begin{aligned}
\xi &\sim (r - r_0)^{-\frac{1}{2} \pm i\frac{\beta}{2}} \\
&= (r - r_0)^{-1/2} (r - r_0)^{\pm i\beta/2} \\
&= (r - r_0)^{-1/2} e^{\pm i(\beta/2)\ln(r-r_0)} \\
&\approx (r - r_0)^{-1/2} \cos\left[\frac{1}{2}\beta \ln |r - r_0| + \phi\right],
\end{aligned}
\tag{31.12}
$$

where ϕ is a phase determined by the boundary conditions. This solution oscillates increasingly rapidly as $r \to r_0$. In this case, the energy principle gives

$$
\begin{aligned}
\delta W_< &= \int_0^{r_0} W(r) r\, dr \\
&= \frac{r_0^3 B_z^2 \mu'^2}{2\left[1 + (r_0\mu)^2\right]} \left[1 - 2M^2 + \left(1 - 2M^2\right)\cos 2\psi + \beta \sin 2\psi\right],
\end{aligned}
\tag{31.13}
$$

where

$$
\psi = \frac{1}{2}\beta \ln \varepsilon + \phi.
\tag{31.14}
$$

We can make ψ anything we like by a suitable choice of ε.

Now, as ψ varies between 0 and π, the term in brackets in Eq. (31.13) varies between 1 and $1 - 4M^2$. Therefore, $\delta W_< < 0$ if $1 - 4M^2 < 0$. A similar analysis holds for $r > r_0$. Therefore, the condition for *stability* is $M^2 < 1/4$, which is the same as the condition for real roots of the indicial equation. Therefore, from Eq. (31.8), the stability condition for pressure-driven modes in the general screw pinch is

$$
\frac{r}{4}\left(\frac{\mu}{\mu'}\right)^2 + \frac{2\mu_0 p_0'}{B_z^2} \geq 0
\tag{31.15}
$$

or

$$
-2\mu_0 p_0' < \frac{B_z^2}{4}\left(\frac{d \ln \mu}{d \ln r}\right)^2,
\tag{31.16}
$$

at every point in the system. This is called the Suydam criterion for stability. Again, we see that the pinch can support a negative pressure gradient, as long as it is not too large.

Lecture 32
Magnetic Reconnection

We'll meet again, don't know where, don't know when
Ross Parker and Charlie Hughes

We now begin discussions of the dynamics of the magneto-fluid system when resistivity is included in the model. This is called *resistive MHD*. There are several important differences from ideal MHD.

First, the ideal Ohm's law is no longer valid, so all the things we have been talking about regarding MHD stability, etc., are also no longer valid. In resistive MHD, the force operator is no longer self-adjoint. We will have to resort to solving differential equations.

Second, in ideal MHD the topology, or "connectedness," of the magnetic field is fixed for all time, and the magnetic field lines are co-moving with the fluid. This is not the case in resistive MHD. Now the fluid can slip through the field, and the field lines no longer have integrity in time. Their topology can change. This process is called *magnetic reconnection*, and it plays a fundamental role in the behavior of real plasmas in both laboratory and astrophysical settings, even when the resistivity, when measured by the magnetic Reynolds' number or the Lundquist number, is extremely small.

Consider the configuration shown in Fig. 32.1.

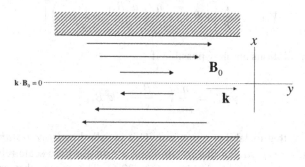

Fig. 32.1 A sheared magnetic field in slab geometry

The magnetic field is $\mathbf{B}_0 = B(x)\hat{\mathbf{e}}_y + B_{z0}\hat{\mathbf{e}}_z$, with $B_{z0} = $ constant, so the current is $\mathbf{J} = J_z(x)\hat{\mathbf{e}}_z$, and the equilibrium condition is $dp/dx = J_z B$. There are conducting walls at $x = \pm L$. The system extends to infinity at $\pm y$. This equilibrium is called a

Schnack, D.D.: *Magnetic Reconnection*. Lect. Notes Phys. **780**, 197–200 (2009)
DOI 10.1007/978-3-642-00688-3_32

sheet pinch. We choose $B(x)$ such that $B(0) = 0$. Then if $\mathbf{k} = k\hat{\mathbf{e}}_y$, $F(x) = \mathbf{k} \cdot \mathbf{B}_0 = 0$ at $x = 0$, so $x = 0$ is a singular surface.

We now consider the dynamics of the sheet pinch in ideal MHD. The transverse (x) magnetic field evolves according to Faraday's law. We assume instability and write the time dependence as $e^{\gamma t}$. Then $\gamma B_{x1} = -ikE_{z1}$ or

$$E_{z1} = \frac{i\gamma B_{x1}}{k}. \tag{32.1}$$

From the ideal MHD Ohm's law, $E_{z1} = -V_{x1}B(x)$ so that

$$V_{x1}(x) = \frac{i\gamma B_{x1}}{F(x)}. \tag{32.2}$$

If $\gamma \neq 0$, $V_{x1} \to \infty$ at $x = 0$ (where $F = 0$). Well-behaved solutions require $\gamma = 0$, so that the sheet pinch is stable in ideal MHD.

Now include resistivity. If the resistivity is constant, the perturbed field evolves according to

$$\frac{\partial \mathbf{B}_1}{\partial t} = \nabla \times (\mathbf{V}_1 \times \mathbf{B}_0) + \frac{\eta}{\mu_0}\nabla^2\mathbf{B}_1. \tag{32.3}$$

Again assuming instability, the x-component is

$$\gamma B_{x1} = iFV_{x1} + \frac{\eta}{\mu_0}\left(\frac{d^2 B_{x1}}{dx^2} - k^2 B_{x1}\right), \tag{32.4}$$

so that

$$V_{x1} = -\frac{i}{F}\left[\gamma B_{x1} - \frac{\eta}{\mu_0}\left(\frac{d^2 B_{x1}}{dx^2} - k^2 B_{x1}\right)\right]. \tag{32.5}$$

Well-behaved solutions are now possible if

$$\gamma B_{x1} = \frac{\eta}{\mu_0}\left(\frac{d^2 B_{x1}}{dx^2} - k^2 B_{x1}\right) \tag{32.6}$$

near $F = 0$, so that instability can occur. This is called *resistive instability*. Note, however, that $\gamma \sim \eta$, so that the growth is on a resistive time scale relative to the scale lengths on the right-hand side of Eq. (32.6). This means that any unstable growth will be on a time scale $\gamma^{-1} \sim \tau_R$ much slower than the Alfvén time τ_A, especially if the Lundquist number $S = LV_A/(\eta/\mu_0) = \tau_R/\tau_A \gg 1$; we expect $\gamma\tau_A \ll 1$. Recall that Alfvén waves incorporate the effects of inertia. This implies that for time scales much longer than the Alfvén time, we can neglect inertia in the region away from $F = 0$. Further, since the resistivity only affects the solutions near $F = 0$, we can ignore resistivity as well in this "outer region"; this region is therefore always in a state of MHD equilibrium.

A possible configuration of the sheet pinch including the effects of resistivity is shown in Fig. 32.2.

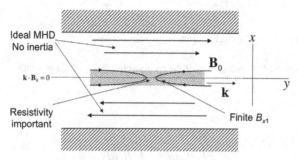

Fig. 32.2 A "reconnected" magnetic field line resulting from resistivity

The resistivity is important only in a small layer about the singular surface, where $F = 0$. Equation (32.6) allows finite B_{x1} at $x = 0$. Field lines that were originally straight can now break, change their topology, and "reconnect" within this small layer. This is called *magnetic reconnection*. The region outside the small layer is governed by ideal MHD and inertia is ignored.

If there is periodicity in the y-direction, as implied by $\mathbf{k} = k\hat{\mathbf{e}}_y$, then magnetic reconnection results in the formation of *magnetic islands*, as shown in Fig. 32.3.

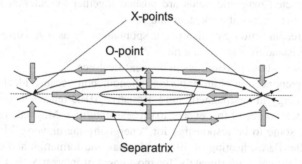

Fig. 32.3 Flow pattern in the vicinity of a magnetic island

The magnetic island has a *separatrix*, which separates field lines of different topologies (open outside the island and closed inside the island). The magnetic reconnection occurs at the *X-points*. The center of the island is called the *O-point*. Typical flows associated with magnetic islands are also shown.

The width of magnetic island can be defined by means of flux conservation within the separatrix (see Fig. 32.4).

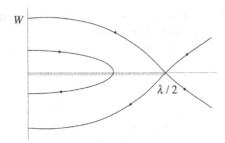

Fig. 32.4 Illustration of the calculation of the reconnected flux and island width

We require

$$\int\limits_{0}^{W} B_y dx = \int\limits_{0}^{\lambda/2} B_{x1} dy. \tag{32.7}$$

Near $x = 0$, $B_y = B'_{y0} x \approx B_0 x / L$, so that the left-hand side is $B_0 W^2 / 2L$. With $B_{x1} = B_1 \sin ky$, the right-hand side is just $2B_1/k$. Solving for the width, we find

$$W = \left(\frac{4B_1 L}{B_0 k}\right)^{1/2}, \tag{32.8}$$

where B_1 is the amplitude of the perturbed magnetic field.

Magnetic reconnection can occur as a *steady-state process* in which two oppositely directed magnetic fields are pushed together by external means. The reconnection then occurs at a constant rate γ.

Magnetic reconnection can also occur spontaneously as a *resistive instability*. The magnetic island then grows at a rate $e^{\gamma t}$.

We will discuss both possibilities in the next Lectures.

Magnetic reconnection is an important phenomenon because ideal MHD constraints trap energy in the magnetic field. Resistive MHD relaxes those constraints and allows a new source of free energy to drive instabilities. Magnetic reconnection is thought by some to be responsible for "energizing the universe," by means of solar and stellar flares, heating of diffuse plasmas, the formation and evolution of astrophysical jets, etc. Unfortunately, for most cases of interest $S \gg 1$ (for a tokamak $S \sim 10^{7-10}$, and it is even larger in astrophysical settings). Since $\gamma \to 0$ when $\eta \to 0$, we expect (and will find) that $\gamma \sim S^{-\alpha}$, $0 \le \alpha \le 1$. It is difficult to account for the observed rate of energy release with these slow growth rates. The quest for a cause of "fast" magnetic reconnection has been alive for five decades, and it still goes on. Undoubtedly it will continue for many more.

Lecture 33
Steady Reconnection: The Sweet–Parker Problem

> *Most of the basic truths of life sound absurd at first hearing.*
>
> Elizabeth Goudge

In this lecture, we discuss the problem of steady-state magnetic reconnection. We consider two adjacent flux systems of oppositely directed magnetic field, as in the sheet pinch of Lecture 32. They meet at $x = 0$, where $B = 0$. The walls are taken to be far away, i.e., $L \to \infty$.

We now imagine that these flux systems are pushed together by some external means with velocity u_0, i.e., $V_x (x \to \pm\infty) \to \pm u$. As the systems are forced together, the current density J_z near $x = 0$ will intensify and will form an extended current sheet in y of length 2Δ. In order to conserve mass, the fluid will be forced out from the center of the current sheet with velocity V_0 in the $\pm y$-directions. In accordance with the discussions of Lecture 32, we expect magnetic reconnection to occur in some small region of width 2δ. The problem is to find a steady-state solution and determine the rate of magnetic reconnection. This problem was first posed and solved by Sweet and Parker,[1] and it is called the *Sweet–Parker problem*. It is one of the most famous and important problems in resistive MHD. The situation is sketched in Fig. 33.1.

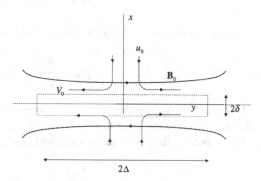

Fig. 33.1 Basic geometry for the Sweet–Parker problem

---

[1] P. A. Sweet, *Electromagnetic Phenomena in Cosmical Physics*, Cambridge University Press, Cambridge, UK, 1958; E. N. Parker, J. Geophys. Res. **62**, 509 (1957).

Schnack, D.D.: *Steady Reconnection: The Sweet–Parker Problem*. Lect. Notes Phys. **780**, 201–204 (2009)
DOI 10.1007/978-3-642-00688-3_33

We picture[2] a steady state in which fluid enters the current sheet at velocity u_0. This is the rate at which magnetic flux enters the inner layer. This magnetic flux is then acted upon by resistivity, gives up its hold on the fluid, and reconnects. The fluid then flow out along the current sheet. No more fluid can enter the layer and carry in magnetic flux until the previous bit of fluid leaves. The ratio of the inflow velocity to the outflow velocity, u_0/V_0, therefore determines the rate at which magnetic flux can be destroyed by reconnection; this is the reconnection rate.

If we assume incompressibility, then the mass coming into the inner layer must equal the mass exiting the inner layer. Since $\nabla \cdot \mathbf{V} = 0$,

$$u_0 L = V_0 \delta. \tag{33.1}$$

Away from the inner layer, the flow and field are governed by ideal MHD, so

$$E_z = u_0 B_0. \tag{33.2}$$

Within the inner layer, resistivity dominates, so

$$E_z = \eta J_z = \frac{\eta B_0}{\mu_0 \delta}. \tag{33.3}$$

In steady state, $E_z \sim$ constant, so equating (33.2) and (33.3),

$$u_0 \approx \frac{\eta}{\mu_0} \delta. \tag{33.4}$$

We have seen that we can ignore inertia away from the inner layer, so that in the outer region the x-component of force balance is

$$\frac{\partial}{\partial x} \left(p + \frac{B^2}{2\mu_0} \right) = 0. \tag{33.5}$$

Integrating this outward from the current sheet (from $x = 0$ to ∞), we have

$$p_\infty + \frac{B_\infty^2}{2\mu_0} = p_{max}, \tag{33.6}$$

where p_∞ and B_∞ are the pressure and magnetic fields far from the current sheet and p_{max} is the pressure at the center of the current sheet; we identify p_∞ and B_∞ with p_0 and B_0, the upstream pressure and field. Then

$$\frac{1}{2} B_0^2 = p_{max} - p_0. \tag{33.7}$$

---

[2] This discussion follows that of Dieter Biskamp, *Magnetic Reconnection in Plasmas*, Cambridge University Press, Cambridge, UK (2000).

Force balance along the sheet, in the y-direction, is $\rho \mathbf{V} \cdot \nabla \mathbf{V} = -\nabla p$ or

$$\frac{\partial}{\partial y}\left(\frac{1}{2}\rho V_y^2\right) = -\frac{\partial p}{\partial y}. \tag{33.8}$$

Integrating from 0 to Δ, we have

$$\frac{1}{2}\rho V_0^2 = p_{max} - p_0. \tag{33.9}$$

Equating Eqs. (33.7) and (33.8), we find

$$V_0 = \frac{B_0}{\sqrt{\mu_0 \rho}} \equiv V_{A0}. \tag{33.10}$$

The downstream (outflow) velocity is therefore equal to the upstream Alfvén velocity. This is a limit to how fast the fluid can get out of the layer so that more reconnection can occur. It is set by the upstream conditions, far from the current sheet. It acts as a throttle on the reconnection rate.

Equations (33.1), (33.4), and (33.10) are three equations in the five unknowns u_0, V_0, δ, Δ, and η. We can determine u_0 and δ in terms of V_0, Δ, and η. One result is

$$\left(\frac{\delta}{\Delta}\right)^2 = \frac{(\eta/\mu_0)\,\Delta}{V_A} \equiv S_0 \tag{33.11}$$

or

$$\frac{\Delta}{\delta} = S_0^{1/2}, \tag{33.12}$$

where S_0 is the Lundquist number based on the upstream magnetic field and the length of the current layer. Generally $S_0 \gg 1$, so that the current layer becomes very long and thin. The other result is

$$\frac{u_0}{V_0} = \frac{u_0}{V_{A0}} \equiv M_0 = S_0^{-1/2}, \tag{33.13}$$

where M_0 is the upstream Alfvén Mach number. Therefore, the steady-state reconnection rate scales like $S_0^{-1/2} \ll 1$. This is called the Sweet–Parker rate. It is one of the most famous results in MHD.

If you push harder from the outside, so that u_0 increases, then for a given S_0, $M_0 = $ constant, so that $V_0 = V_{A0}$ must also increase to maintain the same ratio. This occurs by compressing the field and increasing B_0; the rate of reconnection remains constant and is set by the resistivity. Pushing harder at fixed S_0 just compresses the magnetic field, it does not increase the reconnection rate.

Unfortunately, for $S \sim 10^{10}$, $M_0 \sim 10^{-5}$ is not large enough to account for the inferred reconnection rate associated with solar flares, coronal mass ejections, tokamak sawtooth crashes, and other experiments. To date, nobody has found a way within the resistive MHD model to make the reconnection rate scale faster than $S^{-1/2}$; one is stuck with Sweet–Parker. Over the past several years, there has been considerable research that indicates that so-called fast reconnection (i.e., faster than Sweet–Parker) can occur if effects outside of resistive MHD are accounted for in Ohm's law. This is not surprising in light of Eq. (33.12); at large S, the width of the reconnection layer δ can become small enough that the fundamental MHD assumptions break down, and other physics may dominate the reconnection process. However, theories based on these equations are quite complicated and are difficult to analyze and solve in more general geometry. Today, one still relies primarily on resistive MHD to study the global dynamics, with appeals to artificially enhanced "anomalous" resistivity to obtain sufficient rates of reconnection.

Lecture 34
Resistive Instabilities: The Tearing Mode

> *I have yet to see any problem, which, when you looked at it*
> *in the right way, did not become still more complicated.*
>
> Poul Anderson

We now discuss the situation in which magnetic reconnection arises spontaneously
in the form of an exponentially growing instability. This instability was originally
called the *tearing mode*, because it "tears" the magnetic field apart, something that
is forbidden in ideal MHD.

We again consider the sheet-pinch geometry of Lectures 32 and 33. A sketch of
$F(x) = \mathbf{k} \cdot \mathbf{B}$ versus x is shown in Fig. 34.1.

Fig. 34.1 Magnetic field
profile for the sheet pinch
configuration

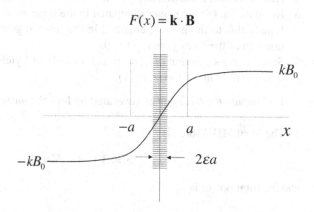

The half width of the current sheet (the region of non-uniform magnetic field) is a.
The current density is $\mu_0 J_z = dB_0/dx = B_0'$. We anticipate from the discussion
of Lecture 32 that the effects of resistivity are felt primarily in a small region about
$x = 0$; we take its half width to be εa, where $\varepsilon \ll 1$. Outside this layer the system
is in ideal MHD equilibrium. Inside the layer the effects of resistivity and inertia
must be accounted for. The flow is assumed to be incompressible. We expect that
any instability will have $\gamma \sim \eta^\alpha$ within the layer. Our approach will be to make use
of the separation of scales ($\varepsilon \ll 1$, $\gamma \tau_A \ll 1$) to obtain separate solutions inside
and outside the layer. These must then be matched at the boundary of the layer.
The matching condition will provide expressions for the growth rate γ and the layer
width εa as functions of the resistivity η, the wave number k, and the equilibrium
parameters a and B_0.

Schnack, D.D.: *Resistive Instabilities: The Tearing Mode*. Lect. Notes Phys. **780**, 205–211 (2009)
DOI 10.1007/978-3-642-00688-3_34   © Springer-Verlag Berlin Heidelberg 2009

The presentation given in this lecture is heuristic, although it leads to the correct conclusions. Of course, it can all be made thoroughly rigorous (although less transparent). The seminal paper on the topic is Furth, Killeen, and Rosenbluth, *Phys. Fluids* **6**, 459 (1963), often abbreviated at "FKR." The presentation given here follows that of Manheimer and Lashmore-Davies.[1]

The goal is to find expressions for the growth rate γ and the inner layer width εa in terms of the resistivity η and the wave number k. The plan of attack is as follows:

1. Compute the *total power* $P = \mathbf{V} \cdot \mathbf{F}$ generated by the MHD force $\mathbf{F}$. If $P > 0$, then $d(\text{kinetic energy})/dt > 0$ and the system is unstable; if $P < 0$, then $d(\text{kinetic energy})/dt < 0$ and the system is stable.
2. Make a *separation of scales argument* ($\varepsilon \ll 1$, $\gamma \tau_A \ll 1$) to conclude that all of the energy is dissipated (i.e., turns into kinetic energy) in the inner region.
3. Use the equation of motion and the induction equation in the inner layer to obtain an ordinary differential equations relating V_{x1} and B_{x1}.
4. Reduce this to an inhomogeneous ODE for V_{x1}.
5. Estimate the *characteristic length scales* associated with the solutions of this ODE to obtain a relationship $f_1(\varepsilon a, \gamma) = 0$ with η and k as parameters.
6. Estimate the *Ohmic power* dissipated in the inner region.
7. Equate this to the power generated in the outer region to obtain a second relationship of the form $f_2(\varepsilon a, \gamma) = 0$.
8. Simultaneous solution of $f_1 = 0$ and $f_2 = 0$ will yield the desired expressions $\gamma = g_1(\eta, k)$ and $\varepsilon a = g_2(\eta, k)$.

1. Compute the total power generated by $\mathbf{F}$ *in the outer region*

The ideal MHD force is

$$\mathbf{F} = \mathbf{J} \times \mathbf{B} - \nabla p, \tag{34.1}$$

and the total power is

$$P = \int dV \mathbf{V} \cdot \mathbf{F}. \tag{34.2}$$

Since $\nabla \cdot \mathbf{V} = 0$, we can write $\mathbf{V} = \nabla \times \mathbf{R}$, where $\mathbf{R} = R\hat{\mathbf{e}}_z$. For reference, the velocity components are

$$V_{x1} = ikR \tag{34.3}$$

and

$$V_{y1} = -\frac{dR}{dx}. \tag{34.4}$$

---

[1] Wallace M. Manheimer and Chris Lashmore-Davies, *MHD Instabilities in Simple Plasma Configurations*, Naval Research Laboratory, Washington, DC (1984).

Therefore,

$$P = \int dV \nabla \times \mathbf{R} \cdot \mathbf{F}$$

$$= \int dV \left[\nabla \cdot (\mathbf{R} \times \mathbf{F}) + \mathbf{R} \cdot \nabla \times \mathbf{F} \right]$$

$$= \int dV \mathbf{R} \cdot \nabla \times \mathbf{F} + \int_S dS \, (\mathbf{R} \times \mathbf{F}) \cdot \hat{\mathbf{n}}$$

$$= \int dV \mathbf{R} \cdot \nabla \times \mathbf{F}. \tag{34.5}$$

The surface integral has been dropped because the boundary is far from the current sheet and the solution is "well behaved" at infinity.

The curl of the perturbed force is

$$\nabla \times \mathbf{F}_1 = \nabla \times (\mathbf{J}_0 \times \mathbf{B}_1) + \nabla \times (\mathbf{J}_1 \times \mathbf{B}_0). \tag{34.6}$$

In this geometry Eq. (34.6) has only a z-component, which is

$$\hat{\mathbf{e}}_z \cdot \nabla \times \mathbf{F}_1 = \frac{1}{\mu_0} B_0' \frac{d B_{x1}}{dx} + \frac{1}{\mu_0} B_0'' B_{x1} + \frac{ik}{\mu_0} B_0' B_{y1} + \frac{ik}{\mu_0} B_0 \left(\frac{d B_{y1}}{dx} - ik B_{x1} \right). \tag{34.7}$$

Since $\nabla \cdot \mathbf{B}_1 = 0$, $B_{y1} = (i/k) d B_{x1}/dx$, and Eq. (34.7) becomes

$$\hat{\mathbf{e}}_z \cdot \nabla \times \mathbf{F}_1 = -\frac{B_0}{\mu_0} \left[\frac{d^2 B_{x1}}{dx^2} - \left(k^2 + \frac{B_0''}{B_0} \right) B_{x1} \right]. \tag{34.8}$$

The power, Eq. (34.5), is then

$$P = -\int_{-\infty}^{\infty} dx \, R \frac{B_0}{\mu_0} \left[\frac{d^2 B_{x1}}{dx^2} - \left(k^2 + \frac{B_0''}{B_0} \right) B_{x1} \right], \tag{34.9}$$

Now, from Eq. (34.3), $R = -(i/k) V_{x1}$, and from the ideal MHD induction equation, $V_{x1} = (\gamma/ik B_0) B_{x1}$, so that

$$P = \int_{-\infty}^{\infty} dx \frac{\gamma}{\mu_0 k^2} B_{x1} \left[\frac{d^2 B_{x1}}{dx^2} - \left(k^2 + \frac{B_0''}{B_0} \right) B_{x1} \right]. \tag{34.10}$$

2. Separation of scales

Since we are ignoring inertia in the outer region, the only contribution from γ in Eq. (34.10) can come from the inner region, i.e.,

$$P = \int\limits_{-\varepsilon a}^{\varepsilon a} dx \frac{\gamma}{\mu_0 k^2} B_{x1} \left[\frac{d^2 B_{x1}}{dx^2} - \left(k^2 + \frac{B_0''}{B_0} \right) B_{x1} \right], \tag{34.11}$$

i.e., *all the power must be generated* (and energy released) *in the inner region.*

Recall that $x = 0$ (where $F = 0$) is a regular singular point of the ideal MHD wave equation, and the ideal MHD solutions are not well behaved there. They are shown in Fig. 34.2.

Fig. 34.2 Illustration of the constant-ψ approximation

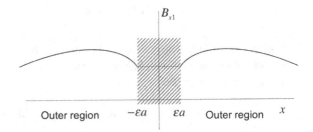

Since $\hat{\mathbf{n}} \cdot \mathbf{B}_1$ must be continuous everywhere, B_{x1} is continuous at $x = 0$, but there may be a discontinuity in dB_{x1}/dx. In order to obtain a smooth solution, we make the *approximation* that $B_{x1} = $ constant *within the inner layer*, i.e., $B_{x1} \approx B_{x1}(0)$. This is called the "constant-ψ approximation" because of notation used in FKR. Also, since we expect the inner layer to be very narrow, we also expect $\left| d^2 B_{x1}/dx^2 \right| \gg \left| k^2 B_{x1} \right|$ across this region, as can be seen heuristically from Fig. 34.2.

With these assumptions, Eq. (34.11) becomes, approximately,

$$\begin{aligned} P &\approx \frac{\gamma}{\mu_0 k^2} \int\limits_{-\varepsilon a}^{\varepsilon a} dx\, B_{x1}(0) \frac{d^2 B_{x1}}{dx^2} \\ &\approx \frac{\gamma B_{x1}(0)}{\mu_0 k^2} \left(\left. \frac{d B_{x1}}{dx} \right|_{\varepsilon a} - \left. \frac{d B_{x1}}{dx} \right|_{-\varepsilon a} \right). \end{aligned} \tag{34.12}$$

We now define the quantity

$$\Delta' \equiv \frac{1}{B_{x1}(0)} \left(\left. \frac{d B_{x1}}{dx} \right|_{\varepsilon a} - \left. \frac{d B_{x1}}{dx} \right|_{-\varepsilon a} \right), \tag{34.13}$$

which depends only on the solution in the outer region, i.e., the region in ideal MHD equilibrium. The total power is then

$$P = \frac{\gamma}{\mu_0 k^2} B_{x1}^2(0) \Delta'. \tag{34.14}$$

The sign of P depends on the sign of Δ'. If $P > 0$, then d(kinetic energy)$/dt > 0$ and the system is unstable, and vice versa. Therefore, the *criterion for instability* is

$$\Delta' > 0. \tag{34.15}$$

This is a famous result first obtained in FKR. It is worth a few remarks given as follows:

1. While all the power is dissipated in the inner region, the stability properties are completely determined by the ideal MHD solution in the outer region.
2. The interplay between the inner and outer regions is important and is what makes the tearing mode problem both interesting and difficult.
3. $\Delta' > 0$ is an important result because *you do not need to do the whole resistive MHD problem to determine stability*. Just look at ideal MHD in the outer regions and determine Δ' at the singular surface. This is a great simplification and is the basis for many computationally based tests of resistive MHD stability.

3. *Equations for the inner region*

In the inner region we must solve the equation of motion and the induction equation simultaneously. For the equation of motion, it proves convenient to use the z-component of its curl. Invoking $\nabla \cdot \mathbf{V} = 0$, we have

$$\frac{i\gamma\rho_0}{k}\left(\frac{d^2 V_{x1}}{dx^2} - k^2 V_{x1}\right) = -\frac{B_0}{\mu_0}\left[\frac{d^2 B_{x1}}{dx^2} - \left(k^2 + \frac{B_0''}{B_0}\right)B_{x1}\right]. \tag{34.16}$$

The induction equation in the inner layer is

$$\gamma B_{x1} = ik B_0 V_{x1} + \frac{\eta}{\mu_0}\left(\frac{d^2 B_{x1}}{dx^2} - k^2 B_{x1}\right). \tag{34.17}$$

4. *Inhomogeneous equation for V_{x1}*

Eliminating the term $d^2 B_{x1}/dx^2 - k^2 B_{x1}$ between these equations results in

$$\rho_o\left(\frac{d^2 V_{x1}}{dx^2} - k^2 V_{x1}\right) - \frac{k^2 B_0^2}{\gamma\eta}V_{x1} = \frac{ik}{\gamma\eta}\left(\gamma - \frac{\eta}{\mu_0}\frac{B_0''}{B_0}\right)B_{x1}(0), \tag{34.18}$$

where we have used the constant-ψ approximation in writing the right-hand side. This is an inhomogeneous ordinary differential equation for V_{x1}.

5. *Estimate size of inner region*

Since, in the inner layer, $d^2/dx^2 \gg k^2$, the length scale for the variation of the solution can be estimated by balancing the first and third terms on the left-hand side of Eq. (34.18), i.e.,

$$\left| \rho_o \frac{d^2 V_{x1}}{dx^2} \right| \sim \left| \frac{k^2 B_0^2}{\gamma \eta} V_{x1} \right|. \tag{34.19}$$

Using $B_0(x) \approx B_0 x/a$ and setting $x \approx \varepsilon a$, we find

$$\varepsilon a \approx \left(\frac{\rho_0 \gamma \eta a^2}{B_0^2 k^2} \right)^{1/4}. \tag{34.20}$$

This can be thought of as an equation of the form $f_1(\varepsilon a, \gamma) = 0$.

6. Estimate Ohmic power in inner region

The Ohmic power in the inner region is

$$P_\Omega \approx \eta J_{z1}^2(0) \varepsilon a, \tag{34.21}$$

where

$$\mu_0 J_{z1} \approx \frac{B_{y1}(\varepsilon a) - B_{y1}(-\varepsilon a)}{\varepsilon a}. \tag{34.22}$$

Since $\nabla \cdot \mathbf{B}_1 = 0$, $B_{y1} = (i/k)dB_{x1}/dx$ and Eq. (34.22) becomes

$$J_{z1} \approx \frac{1}{\mu_0 k \varepsilon a} \Delta' B_{x1}(0). \tag{34.23}$$

The Ohmic power in the inner region is therefore

$$P_\Omega \approx \eta \varepsilon a \left(\frac{\Delta' B_{x1}(0)}{\mu_0 k \varepsilon a} \right)^2. \tag{34.24}$$

7. Power dissipated in inner region equals power generated in outer region

Equating P_Ω from Eq. (34.24) to the power generated in the outer region, Eq. (34.14), we can find

$$\gamma \approx \frac{(\eta/\mu_0) \Delta'}{\varepsilon a}. \tag{34.25}$$

[In light of Eq. (34.15), it is heartening to find that $\gamma \sim \Delta'$.] This can be viewed as a second equation of the form $f_2(\varepsilon a, \gamma) = 0$.

8. Solve for γ and εa

The form of Eq. (34.25) indicates simple resistive diffusion within the inner region. The interesting behavior of resistive instabilities comes about because the size of the inner region εa also depends on resistivity through Eq. (34.20). From Eqs. (34.20) and (34.25), we find

$$\gamma = \left(\frac{B_0^2 k^2}{\mu_0 \rho_0 a^2} \right)^{1/5} \Delta'^{4/5} \left(\frac{\eta}{\mu_0} \right)^{3/5} \tag{34.26}$$

and

$$\varepsilon a = \left(\frac{B_0^2 k^2}{\mu_0 \rho_0 a^2}\right)^{-1/5} \Delta'^{1/5} \left(\frac{\eta}{\mu_0}\right)^{2/5}.$$ (34.27)

These are the principal results, along with the instability condition, Eq. (34.15).

The scaling of γ with a *fractional power* of η indicates that the tearing mode grows significantly faster than resistive diffusion. The equilibrium field diffuses at a rate $\gamma_R \sim \eta/a^2$, so that $\gamma/\gamma_R \sim \eta^{-2/5} \gg 1$ if η is very small. Resistive instabilities can have appreciable effect even if the background is evolving resistively.

Equations (34.26) and (34.27) can be written in non-dimensional form as

$$\gamma \tau_A = \left(\Delta' a\right)^{4/5} \alpha^{2/5} S^{-3/5}$$ (34.28)

and

$$\varepsilon = \left(\Delta' a\right)^{1/5} \alpha^{-2/5} S^{-2/5},$$ (34.29)

where $S = \tau_R/\tau_A$ is the Lundquist number and $\alpha = ka$ is a non-dimensional wave number. However, since we have assumed from the outset that $S \gg 1$, these scalings should be viewed with caution at intermediate values of S. The range of validity for Eqs. (34.28) and (34.29) depends on the particular problem under investigation. For tokamaks, this generally requires $S > 10^6$.

We conclude with two special cases. For the first case the equilibrium field is $-B_0$ for $x < \varepsilon a$, B_0 for $x > \varepsilon a$, and varies linearly between these values in the inner region.

By solving $\nabla \times (\mathbf{J}_0 \times \mathbf{B}_1 + \mathbf{J}_1 \times \mathbf{B}_0) = 0$ in the outer regions, it can be shown that

$$\Delta' = 2k \left[\frac{1 - ka - ka \tanh ka}{ka - (1 - ka) \tanh ka}\right].$$ (34.30)

Then $\Delta' > 0$ requires $ka < 1$. Similarly, if we take $B_0(x) = B_0 \tanh(x/a)$, we find

$$\Delta' = \frac{2}{a}\left(\frac{1}{ka} - ka\right),$$ (34.31)

so that, again, instability requires $ka < 1$. The tearing mode evidently prefers long wavelength. This is because long-wavelength displacements minimize the field-line bending energy, which we know to be stabilizing. With $ka \ll 1$, Eq. (34.31) yields $\Delta' \approx (2/a)(1/ka)$ or $\Delta' \approx 1/\alpha$. The non-dimensional scaling laws, Eqs. (34.28) and (34.29), become in this long-wavelength limit

$$\gamma \tau_A \approx \alpha^{-2/5} S^{-3/5}$$ (34.32)

and

$$\varepsilon \approx \alpha^{-3/5} S^{-2/5}.$$ (34.33)

We remark that the tearing mode grows more slowly than the Sweet–Parker rate of $S^{-1/2}$.

Lecture 35
Resistive Instabilities: Closing Remarks

And so it goes.

Kurt Vonnegut, *Slaughterhouse Five*

In Lecture 34, we presented a heuristic discussion of the tearing mode. The tearing mode is centered about $x = 0$, where $F(x) = \mathbf{k} \cdot \mathbf{B} = 0$. Note that this does not require that $\mathbf{B}(0) = 0$; it only requires that $\mathbf{k}$ be perpendicular to $\mathbf{B}$ there. Thus the tearing mode (and other resistive modes) can occur even when there is a large component of $\mathbf{B}$ in the z-direction. The scaling of the growth rate and resistive layer width with the Lundquist number and normalized wave number for the tearing mode are, for long wavelength,

$$\gamma \tau_{\mathrm{A}} \approx \alpha^{-2/5} S^{-3/5} \qquad (35.1)$$

and

$$\varepsilon \approx \alpha^{-3/5} S^{-2/5}. \qquad (35.2)$$

The tearing mode derives its free energy from the configuration of the magnetic field. As in ideal MHD, there are other resistive instabilities that derive their free energy from different sources.[1]

For example, if we include the gravity in the x-direction we can get an instability called the *resistive g-mode*. The growth rate and resistive layer width for this mode scale like

$$\gamma \tau_{\mathrm{A}} \approx \alpha^{2/3} S^{-1/3} G^{2/3} \qquad (35.3)$$

and

$$\varepsilon \approx \alpha^{-1/3} S^{-1/3} G^{1/6}, \qquad (35.4)$$

where $G = \tau_{\mathrm{A}}^2 A_1$. The form of A_1 depends on the origin of the gravitational force. If it is from a true gravitational field, then

---

[1] H. P. Furth, J. Killeen, and M. N. Rosenbluth, Phys. Fluids **6**, 459 (1963)

Schnack, D.D.: *Resistive Instabilities: Closing Remarks*. Lect. Notes Phys. **780**, 213–218 (2009)
DOI 10.1007/978-3-642-00688-3_35 © Springer-Verlag Berlin Heidelberg 2009

$$A_1 \sim -\frac{g}{\rho_0}\frac{d\rho_0}{dx};$$ (35.5)

if it is from an accelerating frame of reference, then

$$A_1 \sim -\frac{1}{\rho_0}\frac{d}{dx}\left(\rho_0 \dot{V}_0\right);$$ (35.6)

and, if it is from field line curvature, then

$$A_1 \sim -\frac{1}{\tau_A^2}\left(\frac{a^2}{4R_c}\right)\frac{d\beta_0}{dx},$$ (35.7)

where R_c is the radius of curvature of the field lines.

If we include a resistivity gradient $d\eta/dx$, then we can get an instability called the *rippling mode*, which scales like

$$\gamma\tau_A \approx \alpha^{2/5}S^{-3/5}$$ (35.8)

and

$$\varepsilon \approx \alpha^{-2/5}S^{-2/5}.$$ (35.9)

There are some resistive instabilities that do not satisfy the "constant-ψ" approximation. This can occur in a general screw pinch with a monotonically increasing q-profile, when the displacement has the form of the "top hat" trial function introduced in Lecture 30. Then the perturbation must vary rapidly across the resistive layer centered about the singular surface. (This type of instability is thought to be responsible for the "crash phase" of tokamak sawtooth oscillations.) These *non-constant-ψ modes* can have a larger growth rate, i.e., $\gamma\tau_A \sim S^{-1/2}$, but still do not exceed the Sweet–Parker rate.

Resistive instabilities can be *wall stabilized*. The magnetic field must remain parallel to any conducting boundary. Recall that the drive, or free energy, for the tearing mode comes from the outer region. If conducting boundaries are placed at $x = \pm L$, they will inhibit the bending of the field lines in the outer region that accompanies the formation of magnetic islands. If they are placed close enough to the singular surface, an unstable mode with wave number k can be completely stabilized. This effect is obviously minimized at long wavelength.

The transition from a state with no magnetic islands to one containing magnetic islands can be thought of as an example of *neighboring equilibria*. Consider two equilibrium states that satisfy the boundary conditions, one with magnetic islands and one without. Let the state without magnetic islands have energy W_1, and the state with magnetic islands have W_2, and let the system initially be in the state without islands. The system wants to be in the minimum energy state. If $W_1 > W_2$, the system would prefer to be in state 2, with islands. However, this transition is forbidden by the ideal MHD constraints on the invariance of the magnetic topology. In resistive MHD, such a transition is possible. This situation has been analyzed, and it turns out that the energy of such equilibria depends on the parameter Δ'. When $\Delta' < 0$, then $W_1 < W_2$ and the state without islands is the preferred state; when $\Delta' > 0$, then $W_1 > W_2$ and the state with islands is preferred. But $\Delta' > 0$ is just the instability criterion for the tearing mode derived in Lecture 34, so the problems of stability to small perturbations and the energy of neighboring equilibria are equivalent in this regard.

The mathematical theory of resistive instabilities relies on the method of *matched asymptotic expansions*. This is a method for obtaining an approximate solution to a boundary-value problem with an ordinary differential equation whose highest derivative is multiplied by a small parameter.[2] For example, consider the equation

$$\varepsilon \frac{d^2 u}{dx^2} + (1 + \varepsilon) \frac{du}{dx} + u = 0, \tag{35.10}$$

with $\varepsilon \ll 1$, subject to the boundary conditions $u(0) = 0$ and $u(1) = 1$. One is tempted to set $\varepsilon = 0$ and solve the simpler equation

$$\frac{du}{dx} + u = 0. \tag{35.11}$$

However, Eq. (35.11) is of lower order than Eq. (35.10), and therefore requires only a single-boundary condition, whereas the solution of Eq. (35.10) requires two boundary conditions. For Eq. (35.11), the boundary condition corresponding to a non-trivial solution is $u(1) = 1$. Therefore, the condition $u(0) = 0$ imposed on $u(x)$ *cannot be satisfied* by the solution of Eq. (35.11). Put in another way, if $u(x, \varepsilon)$ is a solution Eq. (35.10) and $u(x, 0)$ is a solution of Eq. (35.11), then

$$\lim_{\varepsilon \to 0} u(x, \varepsilon) \neq u(x, 0). \tag{35.12}$$

This is an example of non-uniform convergence and is illustrated in Fig. 35.1.

In Lecture 34, we showed that the perturbed velocity satisfies Eq. (34.18), which can be rewritten as

$$\eta \left(\frac{d^2 V_{x1}}{dx^2} - k^2 V_{x1} \right) - \frac{k^2 B_0^2}{\gamma \rho_o} V_{x1} = \frac{ik}{\gamma \rho_o} \left(\gamma - \frac{\eta}{\mu_0} \frac{B_0''}{B_0} \right) B_{x1}(0). \tag{35.13}$$

The small parameter η multiplies the highest derivative; if $\eta = 0$ we have ideal MHD. In light of Eq. (35.12), we see that the solution of the resistive MHD equations in the limit $\eta \to 0$ does *not* approach the solution of the ideal MHD equations.

---

[2] For an in-depth presentation of these techniques, including the example given here, see C. M. Bender and S. A. Orszag, *Advanced Mathematical Methods for Scientists and Engineers*, Springer, New York (1999).

Fig. 35.1 Illustration that the
limit of $u(x, \varepsilon)$ as $\varepsilon \to 0$ does
not equal to $u(x, 0)$

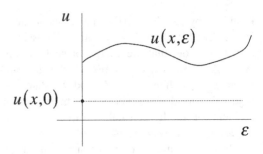

The ideal MHD equations require fewer boundary conditions. The tearing mode still exists in the limit $\eta \to 0$, but it does not exist in ideal MHD.

We now return to the boundary-value problem given by Eq. (35.10) and following. Since $\varepsilon \ll 1$, we expect solutions of Eqs. (35.10) and (35.11) to be almost equal except where $|u''| \sim 1/\varepsilon \gg 1$. From Fig. 35.2, we see that this will occur in a small region near $x = 0$, where the solutions must rapidly diverge in order for $u(x, \varepsilon)$ to satisfy the boundary condition at $u(0) = 0$. We anticipate the region of disagreement to scale like $\Delta x \sim \varepsilon^\alpha$. Since this occurs near a boundary, the region Δx is called a *boundary layer*.

We can obtain an approximate solution of Eq. (35.10) by dividing the problem into two regions:

1. An outer region, $x > \varepsilon^\alpha$, where $u = u(x, 0)$, which satisfies $u' + u = 0$ with $u(1, 0) = 1$.
2. An inner region, $x < \varepsilon^\alpha$, where $u = u(x, \varepsilon)$, which satisfies $\varepsilon u'' + (1 + \varepsilon) u' + u = 0$ with $u(0, \varepsilon) = 0$.

These solution must be matched at $x \sim \varepsilon^\alpha$.

The solution in the outer region is $u(x, 0) = e^{1-x}$. In the inner region, it is useful to rescale the independent variable as $\xi = x/\varepsilon$. Then Eq. (35.10) becomes

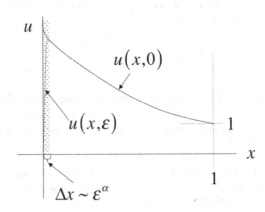

Fig. 35.2 Anticipated
behavior of the solution of
Eq. (35.10), showing the thin
boundary layer near $x = 0$

$$\frac{1}{\varepsilon}\frac{d^2u_1}{d\xi^2} + \frac{1}{\varepsilon}(1+\varepsilon)\frac{du_1}{d\xi} + u_1 = 0,$$ (35.14)

If we now let $\varepsilon \to 0$ and integrate, we have

$$\frac{du_1}{d\xi} + u_1 = C,$$ (35.15)

where C is a constant of integration. The solution of Eq. (35.15) that satisfies $u_1(0) \equiv u(0, \varepsilon) = 0$ is

$$u_1(\xi) = C\left(1 - e^{-\xi}\right).$$ (35.16)

The key step in the method is to require that

$$\lim_{\xi \to \infty} u_1(\xi) = \lim_{x \to 0} u(x, 0).$$ (35.17)

This is often stated as *the outer limit of the inner solution equals the inner limit of the outer solution.* Applying Eq. (35.17) results in $C = e$, so that $u_1(x) = C\left(1 - e^{-x/\varepsilon}\right)$. The approximate solution of the boundary-value problem is then

$$\tilde{u}(x, \varepsilon) = u(x, 0) + u_1(x) - u(0, 0).$$ (35.18)

We must subtract the common value of the two solutions at $x = 0$ and $\xi \to \infty$ to assure that the solution is continuous. The result is

$$\tilde{u}(x, \varepsilon) = e\left(e^{-x} - e^{-x/\varepsilon}\right).$$ (35.19)

Actually, in this example Eq. (35.19) satisfies the differential equation (35.10) exactly. However, $\tilde{u}(1, \varepsilon) = 1 - e^{1-1/\varepsilon} \neq 1$, so that the boundary condition at $x = 1$ is not satisfied. It is therefore only an approximate solution of the boundary-value problem. Nonetheless, $\lim_{\varepsilon \to 0} \tilde{u}(1, \varepsilon) = 1$, so that the approximate solution converges to the actual solution in the proper limit.

In this case, it is possible to obtain an exact solution of the boundary-value problem as $u(x, \varepsilon) = e\left(e^{-x} - e^{-x/\varepsilon}\right) / \left(1 - e^{1-1/\varepsilon}\right)$. The difference between this exact solution and the approximate solution of Eq. (35.19) is sketched in Fig. 35.3.

Recall that in our analysis of the tearing mode, we found that the ideal MHD solution has a discontinuity Δ' in its first derivative at $x = 0$ and that this must be matched to the resistive solution in a thin resistive layer about $x = 0$. This layer is equivalent to the "boundary layer" in the above example (even though it does not occur at a boundary). Furth, Killeen, and Rosenbluth used the method of matched asymptotic expansions to solve this problem. The lower-order ideal MHD equations are valid in the outer region. The higher-order resistive equations are valid in the inner region. In the inner region the equations are rescaled and

Fig. 35.3 Exact solution
(*solid line*) and approximate
solution (*dashed line*) of
Eq. (35.10)

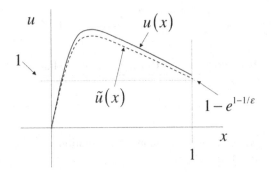

solved. The solutions in the inner and outer regions are then matched by requiring $B_{x1}(|\xi| \to \infty)_{inner} = B_{x1}(|x| \to 0)_{outer}$ and $\Delta'_{inner} = \Delta'_{outer}$. The mathematics required to solve the equations in the inner region is quite complicated, involving expansions in Hermite polynomials, etc. Nonetheless, the approach is as given here and results in the scaling laws given by Eqs. (35.1) and (35.2).

Of course, this can all be done in toroidal geometry. Then the criterion for instability becomes $\Delta' > \Delta'_C > 0$, so that toroidal geometry is actually stabilizing.

Finally, we comment on nonlinear effects on the growth of the tearing mode. In the linear regime the initial island width W, defined in a previous lecture, is much smaller than the layer width εa, and the mode grows exponentially. However, it turns out that, when $W \approx \varepsilon a$, there are nonlinear $\mathbf{J} \times \mathbf{B}$ forces that oppose the island growth, and these must compete with the drive of the linear instability. The result is that when the island width becomes comparable to or larger than the width of the resistive layer, the island width grows as

$$\frac{dW}{dt} = 1.22\eta\Delta' \tag{35.20}$$

or $W \sim \eta\Delta't$; exponential growth ceases and the island grows linearly in time with a rate proportional to the resistivity or $1/S$. Equation (35.20) is called the *Rutherford equation*.[3] Further analysis[4] reveals that

$$\frac{dW}{dt} = 1.22\eta\left[\Delta'(W) - \alpha W\right], \tag{35.21}$$

where $\Delta'(W)$ is now a function of W and α depends on equilibrium parameters. The island stops growing (saturates) when $\Delta'(W) - \alpha W = 0$, which can be solved for the saturated island width.

---

[3] P. H. Rutherford, Phys. Fluids **16**, 1903 (1958).

[4] R. B. White, D. A. Monticello, M. N. Rosenbluth, and B. V. Waddell, Phys. Fluids **20**, 800 (1977).

Lecture 36
Turbulence[1]

Abandon hope, all ye who enter here.
Dante, The Divine Comedy

Suppose we set up an experiment in which we can control all the *mean* parameters. An example might be steady flow through a pipe, where we can control the mean velocity $\bar{V}$. Now insert a probe, or some such measuring device, at a fixed location far from the boundaries and measure the flow velocity as a function of time. The result of this measurement might look something like Fig. 36.1.

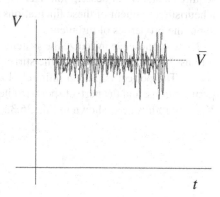

Fig. 36.1 Measurement of a fluctuating velocity field

All measurements are taken under identical conditions. However, the results of the measurements at different times are *not* the same. Instead, we find that the velocity takes on *random values*. Although $\bar{V}$ (the average) is determined precisely by the controllable conditions, the random values are not. Fluctuating motions of this kind are said to be *turbulent*.

The random fluctuations in V have a probability distribution with a mean value $\bar{V}$, as shown in Fig. 36.2.

---

[1] This lecture is greatly influenced by G. K. Batchelor, *The Theory of Homogenous Turbulence*, Cambridge University Press, Cambridge, UK (1953), and Dieter Biskamp, *Magnetohydrodynamic Turbulence*, Cambridge University Press, Cambridge, UK (2003). The thorough reading and constructive comments of P. W. Terry are gratefully acknowledged.

Schnack, D.D.: *Turbulence*. Lect. Notes Phys. **780**, 219–240 (2009)
DOI 10.1007/978-3-642-00688-3_36

Fig. 36.2 Probability
distribution for the velocity
fluctuations shown in
Fig. 36.1

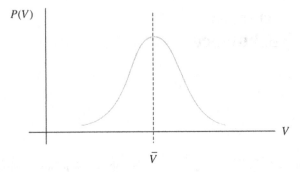

The function $P(V)$ is called the *probability distribution function* (PDF). If the fluctuations are truly random,[2] the PDF is Gaussian. If the PDF and $\bar{V}$ are independent of the position, the turbulence is said to be *homogeneous*. If the PDF is independent of arbitrary rotations of the system, and of reflections about any plane, the turbulence is said to be *isotropic*. Isotropic turbulence has no preferred direction in space. If the random flow looks the same on all spatial scales, the turbulence is said to be *self-similar* (or *scale invariant*). A rigorous theoretical study of turbulence requires a statistical description. Here we will not go that far. Rather, we will give a heuristic treatment of these fluctuations that nonetheless yields important insights about the properties of turbulence.

Now, suppose we initialize the system with long wavelength, steady, smooth conditions. For example, consider the stirring of a perfect cup of coffee with a perfect spoon. The perfect cup is an infinitely long cylinder of radius a with no boundary perturbations, and the perfect spoon excites only a single circular eddy with velocity $\mathbf{V}_{stir}$ and radius a, as shown in Fig. 36.3.

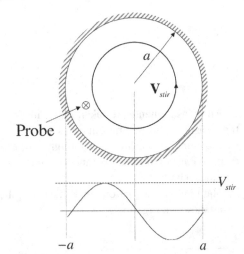

Fig. 36.3 The flow resulting
from a perfect cup stirred
perfectly by a perfect spoon

---

[2] Random fluctuations are said to be *Markovian*, meaning that there is no correlation between them; they are completely independent. In turbulence, the fluctuations may be correlated to some degree, meaning that the value of one measurement may depend in some way on the value of previous measurements. This can lead to non-Gaussian statistics.

The stirring is continued until the system reaches steady state. Then the velocity is measured at the probe position shown in Fig. 36.3. The result looks like Fig. 36.1, with $\bar{V} = V_{\text{stir}}$. The question is, how did these small-scale random fluctuations come about if only the longest wavelength is excited by the stirring of the spoon?

The answer is that the fluctuations arise because of the *nonlinearities* in the fluid equations.[3] In the absence of pressure forces, and with constant density, the evolution of the velocity is governed by the equation of motion

$$\frac{\partial V}{\partial t} + V\frac{\partial V}{\partial x} = \nu\frac{\partial^2 V}{\partial x^2}, \tag{36.1}$$

where ν is the viscosity. The second term on the left-hand side is nonlinear, containing the product of the velocity and its derivative. Consider the effect of this term on the evolution of V. At $t = 0$ the velocity is given by $V(0) = V_0 \sin k_0 x$, where $k_0 = \pi/a$ (or wavelength $\lambda_0 = 2a$). We suppose that the Reynolds' number $\text{Re} = V_0 a/\nu \gg 1$, so that we neglect the effects of viscosity at long wavelengths. Then at $t = \Delta t \ll 1/k_0 V_0$, a short time later, the velocity is

$$\begin{aligned}
V(\Delta t) &= V(0) - \Delta t\, V\frac{\partial V}{\partial x} \\
&= V_0 \sin k_0 x - \Delta t\,(V_0 \sin k_0 x)\,(k_0 V_0 \cos k_0 x) \\
&= V_0 \sin k_0 x - \frac{1}{2}k_0 \Delta t\, V_0^2 \sin 2k_0 x.
\end{aligned} \tag{36.2}$$

The new velocity has a component whose wave number is $k_1 = 2k_0$ or wavelength $\lambda_1 = a = \lambda_0/2$; it has half the wavelength of the driven input velocity. The motion now contains the original eddy plus eddies that are half the size, as shown schematically in Fig. 36.4.

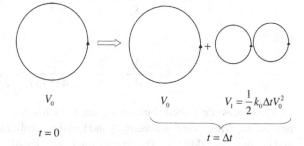

Fig. 36.4 Creation of smaller eddies after a short increment of time

---

[3] It is important to recognize that the eddies to be described here are also nonlinearly unstable. In hydrodynamics this is due to the *Kelvin—Helmholtz instability*, which derives its free energy from sheared flow. In MHD, this is modified by the presence of the magnetic field. This instability accounts for the observed dynamical variability of turbulent flow.

Note that the velocity of the smaller eddies is $V_1 = k_0 \Delta t V_0^2/2 \ll V_0$, since we have assumed that $k_0 V_0 \Delta t \ll 1$. The flow at $t = \Delta t$ is still dominated by the large eddy, but has a small component with half the wavelength superimposed on it. Similarly, at time $t = 2\Delta t$, the velocity is

$$
\begin{aligned}
V(2\Delta t) &= V(\Delta t) - \Delta t V \frac{\partial V}{\partial x} \\
&= V_0 \sin k_0 x - V_1 \sin 2k_0 x \\
&\quad - \Delta t \left(V_0 \sin k_0 x - V_1 \sin 2k_0 x\right) \frac{\partial}{\partial x}\left(V_0 \sin k_0 x - V_1 \sin 2k_0 x\right) \\
&= \left(V_0 - 3\Delta t V_1 V_0\right) \sin k_0 x \\
&\quad - \left(V_1 + k_0 \Delta t V_0^2\right) \sin 2k_0 x \\
&\quad + 3\Delta t k_0 V_1 V_0 \sin 3k_0 x \\
&\quad - 2\Delta t k_0 V_1^2 \sin 4k_0 x.
\end{aligned} \tag{36.3}
$$

The first two terms on the right-hand side are modifications to the eddies with $\lambda_0 = 2a$ and $\lambda_1 = a$, respectively. The last two terms are new eddies with smaller wavelengths. The amplitudes of eddies (i.e., the flow velocity associated with the eddies) with successively smaller wavelength are successively smaller. The flow remains dominated by the largest eddy with wavelength $\lambda_0 = 2a$, whose amplitude is slightly modified from the input value, plus a superposition of successively smaller and slower eddies with wavelengths, given by

$$
\lambda_1 = \frac{2\pi}{k_1} = \frac{2\pi}{2k_0} = \frac{2\pi}{2(\pi/a)} = a, \tag{36.4}
$$

$$
\lambda_2 = \frac{2\pi}{k_2} = \frac{2\pi}{3k_0} = \frac{2\pi}{3(\pi/a)} = \frac{2}{3}a, \tag{36.5}
$$

and

$$
\lambda_3 = \frac{2\pi}{k_3} = \frac{2\pi}{4k_0} = \frac{2\pi}{4(\pi/a)} = \frac{1}{2}a. \tag{36.6}
$$

The process described above continues at each successive time step; at each step, new eddies with shorter wavelength and smaller amplitude are generated. If continued indefinitely, eddies with arbitrarily small wavelengths (large wave numbers) will be generated.[4] This is inevitable and accounts for the small-scale random velocities that occur in the measurements. Since the amplitude of the small eddies decreases

---

[4] This analysis is meant to be heuristic, and not a recommendation for further analysis. As stated, it applies only to very short times, and we are generally interested in the time asymptotic behavior of the flow.

with wave number k, a plot of $\varepsilon(k)$, the energy in an eddy with wave number k, versus k might look something like Fig. 36.5.

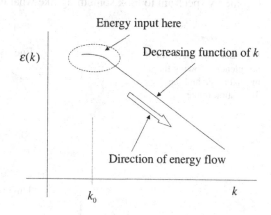

Fig. 36.5 Illustration of the cascade of energy from large to small wavelength as a result of nonlinear mode coupling

Energy is continually input at wave number k_0. It is continually spread to higher and higher k (to smaller and smaller eddies) as a result of the nonlinear interactions described above. This is called a *cascade*. The cascade of energy to higher k will continue indefinitely unless other processes intervene.

The cascade will be unable to continue when the magnitude of the viscous term [the term on the right-hand side of Eq. (36.1)] becomes comparable with the nonlinear term, for then $\partial V/\partial t \approx 0$ and the process described above can no longer operate. We therefore require $V\partial V/\partial x \approx \nu\partial^2 V/\partial x^2$. With $V \approx V_k \exp(ikx)$ and $\partial/\partial x \to ik$ (where V_k is the velocity of the eddy with wavelength $\lambda_k = 2\pi/k$), this will occur at a wave number such that $V_k \approx \nu k$ or

$$\frac{V_k}{k\nu} \approx \frac{V_k\lambda_k}{\nu} \approx 1. \qquad (36.7)$$

We recognize the expression $\mathrm{Re}_{\lambda_k} = V_k\lambda_k/\nu$ as the Reynolds' number associated with the kth eddy. The cascade will cease when $\mathrm{Re}_{\lambda_k} \approx 1$ or

$$k \approx k_D \equiv \frac{|V_k|}{\nu}; \qquad (36.8)$$

k_D is called the *dissipation wave number*.

The qualitative picture of steady-state turbulence is therefore as follows. Energy is continually input at a small wave number k_0. As a result of nonlinearities in the governing dynamical equations, this energy cascades to higher and higher wave number, producing eddies with smaller and smaller amplitude. The cascade will cease when k approaches k_D, where dissipation can compete with the nonlinearity. All the energy that is input at $k \approx k_0$ is dissipated near $k \approx k_D$; this is called the *dissipation range*. The energy simply moves through the intervening wave numbers. The range $k_0 < k < k_D$ is called the *inertial range*, for it is dominated by $dV/dt \approx 0$ (the Lagrangian derivative) and dissipation plays no role. We may therefore expect the energy spectrum to look something like what is sketched in Fig. 36.6.

Fig. 36.6 The spectrum of turbulence, showing the inertial range and the dissipation range

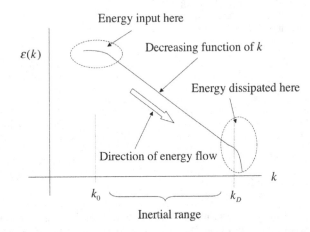

This plot is said to display the *spectrum of turbulence*.

Now, it is fair to ask why we should care about all of this. The spoon still stirs the cup, and the mean values of the flow, which are presumably what we care about, are repeatable from experiment to experiment. The answer is that these very small-scale fluctuations affect the evolution of the mean quantities. Recall that the Reynolds' number at the scale of the mean flow is much more greater than one. If we only concentrated on the mean flow, we might anticipate that there would be very little dissipation in the system. However, we know that, in steady state, *all* of the energy that is input at the large scales must be dissipated, and at the *same rate* that it is input; it is just dissipated at the small scales. The system adjusts to make this happen. *The effective dissipation rate is much larger than one might conclude from considerations of the largest eddies alone.* Therefore, we cannot understand the dissipative properties of the system without including effects of the turbulence. Further, the turbulent eddies provide a much more effective mechanism for mixing and transport within the system than might be deduced from the material properties of the fluid alone; the cream is more effectively mixed with the coffee because of the turbulence.

These properties can be demonstrated if we assume that $k_D \gg k_0$, i.e., there is a large-scale separation between the length scales at which energy is input and the length scales at which it is dissipated. We can then write any physical quantity

$$u(x,t) = \langle u(x,t) \rangle + \tilde{u}(x,t). \tag{36.9}$$

where $\langle u(x,t) \rangle$ is a long wavelength "mean" value and $\tilde{u}(x,t)$ is a random, small scale "fluctuating" component.[5] These have the properties that $\langle \langle u(x,t) \rangle \rangle = \langle u(x,t) \rangle$ and $\langle \tilde{u}(x,t) \rangle = 0$.

The incompressible hydrodynamic equations of motion (with constant density, which we arbitrarily set to 1) are

$$\frac{\partial \mathbf{V}}{\partial t} + \mathbf{V} \cdot \nabla \mathbf{V} = -\nabla p + \nu \nabla^2 \mathbf{V} \tag{36.10}$$

and

$$\nabla \cdot \mathbf{V} = 0. \tag{36.11}$$

[As an aside, we remark that Eq. (36.11) serves as a closure; the divergence of Eq. (36.10) yields

$$\nabla^2 p = -\nabla \cdot (\mathbf{V} \cdot \nabla \mathbf{V}), \tag{36.12}$$

which is a Poisson equation that determines the pressure. Solutions of Eq. (36.12) assure that the velocity remains solenoidal. Equation (36.12) is thus the "equation of state" for incompressible flows.] Substituting the *ansatz* of Eq. (36.9) into Eqs. (36.10) and (36.11), we have

$$\frac{\partial}{\partial t} \left(\langle \mathbf{V} \rangle + \tilde{\mathbf{V}} \right) + \left(\langle \mathbf{V} \rangle + \tilde{\mathbf{V}} \right) \cdot \nabla \left(\langle \mathbf{V} \rangle + \tilde{\mathbf{V}} \right) = -\nabla \left(\langle p \rangle + \tilde{p} \right) + \nu \nabla^2 \left(\langle \mathbf{V} \rangle + \tilde{\mathbf{V}} \right),$$

$$\tag{36.13}$$

$$\nabla \cdot \langle \mathbf{V} \rangle = 0, \tag{36.14}$$

and

$$\nabla \cdot \tilde{\mathbf{V}} = 0. \tag{36.15}$$

Equation (36.13) can be expanded as

$$\frac{\partial}{\partial t} \left(\langle \mathbf{V} \rangle + \tilde{\mathbf{V}} \right) + \langle \mathbf{V} \rangle \cdot \nabla \langle \mathbf{V} \rangle + \langle \mathbf{V} \rangle \cdot \nabla \tilde{\mathbf{V}} + \tilde{\mathbf{V}} \cdot \nabla \langle \mathbf{V} \rangle + \tilde{\mathbf{V}} \cdot \nabla \tilde{\mathbf{V}}$$
$$= -\nabla \left(\langle p \rangle + \tilde{p} \right) + \nu \nabla^2 \left(\langle \mathbf{V} \rangle + \tilde{\mathbf{V}} \right). \tag{36.16}$$

Equation (36.16) contains both mean and fluctuating parts. We can obtain an equation for the evolution of the mean flow alone by operating on Eq. (36.16) with the "mean operator" $\langle \ldots \rangle$, and using the properties following Eq. (36.9):

---

[5] This implies a large separation of both space and time scales between the mean and fluctuating parts of the flow.

$$\frac{\partial \langle \mathbf{V} \rangle}{\partial t} + \langle \mathbf{V} \rangle \cdot \nabla \langle \mathbf{V} \rangle + \langle \tilde{\mathbf{V}} \cdot \nabla \tilde{\mathbf{V}} \rangle = -\nabla \langle p \rangle + \nu \nabla^2 \langle \mathbf{V} \rangle . \tag{36.17}$$

Since $\nabla \cdot \tilde{\mathbf{V}} = 0$, we can write $\tilde{\mathbf{V}} \cdot \nabla \tilde{\mathbf{V}} = \nabla \cdot (\tilde{\mathbf{V}} \tilde{\mathbf{V}})$, so that Eq. (36.17) becomes

$$\frac{\partial \langle \mathbf{V} \rangle}{\partial t} + \langle \mathbf{V} \rangle \cdot \nabla \langle \mathbf{V} \rangle = -\nabla \langle p \rangle + \nabla \cdot \left[-\langle \tilde{\mathbf{V}} \tilde{\mathbf{V}} \rangle + \nu \nabla \langle \mathbf{V} \rangle \right] . \tag{36.18}$$

The mean pressure is to be determined by the condition $\nabla \cdot \langle \mathbf{V} \rangle = 0$. We note that there is an additional contribution $-\langle \tilde{\mathbf{V}} \tilde{\mathbf{V}} \rangle$ to the stress tensor for the mean flow. This term encapsulates the *enhanced dissipation* that results from the presence of the fluctuating component $\tilde{\mathbf{V}}$. This term is second order in the fluctuating velocity.

An equation for the fluctuating part of the velocity can be obtained by subtracting Eq. (36.18) from Eq. (36.16). The result is

$$\frac{\partial \tilde{\mathbf{V}}}{\partial t} + \langle \mathbf{V} \rangle \cdot \nabla \tilde{\mathbf{V}} + \tilde{\mathbf{V}} \cdot \nabla \langle \mathbf{V} \rangle + \tilde{\mathbf{V}} \cdot \nabla \tilde{\mathbf{V}} = -\nabla \langle p \rangle + \nu \nabla^2 \tilde{\mathbf{V}} . \tag{36.19}$$

The time rate of change of $\tilde{\mathbf{V}} \tilde{\mathbf{V}}$ can be found by forming the combination

$$\frac{\partial}{\partial t} \tilde{\mathbf{V}} \tilde{\mathbf{V}} = \tilde{\mathbf{V}} \frac{\partial \tilde{\mathbf{V}}}{\partial t} + \frac{\partial \tilde{\mathbf{V}}}{\partial t} \tilde{\mathbf{V}} . \tag{36.20}$$

After much algebra, the result is

$$\begin{aligned}
\frac{\partial}{\partial t} \tilde{\mathbf{V}} \tilde{\mathbf{V}} = &- \langle \mathbf{V} \rangle \left[\nabla \cdot (\tilde{\mathbf{V}} \tilde{\mathbf{V}}) \right] - \left[\nabla \cdot (\tilde{\mathbf{V}} \tilde{\mathbf{V}}) \right] \langle \mathbf{V} \rangle \\
&- \nabla \cdot \left(\langle \mathbf{V} \rangle \tilde{\mathbf{V}} \tilde{\mathbf{V}} + \tilde{\mathbf{V}} \langle \mathbf{V} \rangle \tilde{\mathbf{V}} + \tilde{\mathbf{V}} \tilde{\mathbf{V}} \langle \mathbf{V} \rangle + \tilde{\mathbf{V}} \tilde{\mathbf{V}} \tilde{\mathbf{V}} \right) \\
&- \nabla \left(\tilde{p} \tilde{\mathbf{V}} \right) + \tilde{p} \nabla \tilde{\mathbf{V}} - \left[\nabla \left(\tilde{p} \tilde{\mathbf{V}} \right) - \tilde{p} \nabla \tilde{\mathbf{V}} \right]^T \\
&+ \nu \left[\nabla^2 \tilde{\mathbf{V}} \tilde{\mathbf{V}} - 2 \left(\tilde{\mathbf{V}} \nabla \right) \cdot \left(\nabla \tilde{\mathbf{V}} \right) \right] .
\end{aligned} \tag{36.21}$$

As we can see, turbulence becomes very messy very quickly! [Here the notation is $(\tilde{\mathbf{V}} \nabla) \cdot (\nabla \tilde{\mathbf{V}}) \equiv (\partial_l \tilde{V}_i)(\partial_l \tilde{V}_j)$.] When we take the average of Eq. (36.21), we obtain an equation for $\langle \tilde{\mathbf{V}} \tilde{\mathbf{V}} \rangle$ that has the form

$$\frac{\partial}{\partial t} \langle \tilde{\mathbf{V}} \tilde{\mathbf{V}} \rangle = \left(\text{terms containing } \langle \mathbf{V} \rangle, \langle \tilde{\mathbf{V}} \tilde{\mathbf{V}} \rangle, \text{ and } \langle \tilde{p} \tilde{\mathbf{V}} \rangle \right) - \nabla \cdot \langle \tilde{\mathbf{V}} \tilde{\mathbf{V}} \tilde{\mathbf{V}} \rangle . \tag{36.22}$$

The equation for the second-order moment $\langle \tilde{\mathbf{V}} \tilde{\mathbf{V}} \rangle$ contains the third-order moment $\langle \tilde{\mathbf{V}} \tilde{\mathbf{V}} \tilde{\mathbf{V}} \rangle$! If this procedure were carried further, we would find that the equation for $\langle \tilde{\mathbf{V}} \tilde{\mathbf{V}} \tilde{\mathbf{V}} \rangle$ contains $\langle \tilde{\mathbf{V}} \tilde{\mathbf{V}} \tilde{\mathbf{V}} \tilde{\mathbf{V}} \rangle$, and so on to infinity. This is formally identical to the *closure problem* that was discussed in a Lecture 7; there are always more unknowns than equations.

A central part of turbulence theory is obtaining expressions for $\langle \tilde{\mathbf{V}}\tilde{\mathbf{V}} \rangle$ in terms of $\langle \mathbf{V} \rangle$ and $\langle p \rangle$. For example, one form of a closure relation for incompressible hydrodynamics is

$$\langle \tilde{\mathbf{V}}\tilde{\mathbf{V}} \rangle = \frac{1}{3}\langle \tilde{V}^2 \rangle \mathbf{I} - \nu_T \left(\nabla \langle \mathbf{V} \rangle + \nabla \langle \mathbf{V} \rangle^T \right). \tag{36.23}$$

Here ν_T is a *turbulent viscosity*. It is sometimes called the *eddy viscosity*, since it is due to the small-scale turbulent eddies. Then the mean flows evolve according to

$$\begin{aligned}
\frac{\partial \langle \mathbf{V} \rangle}{\partial t} + \langle \mathbf{V} \rangle \cdot \nabla \langle \mathbf{V} \rangle = &-\nabla \cdot \left(\langle p \rangle + \frac{1}{3}\langle \tilde{V}^2 \rangle \right) \mathbf{I} \\
&+ \nabla \cdot \left[(\nu + \nu_T) \left(\nabla \langle \mathbf{V} \rangle + \nabla \langle \mathbf{V} \rangle^T \right) \right].
\end{aligned} \tag{36.24}$$

The turbulence produces both an addition to the isotropic pressure force and an enhanced dissipation rate. One form of the turbulent viscosity is

$$\nu_T = \tilde{V} l_m, \tag{36.25}$$

where $\tilde{V}$ is the approximate amplitude of the fluctuating velocity and l_m is a length scale that represents the distance momentum is transported during one "eddy turnover time." i.e., the time it takes the largest scale eddy to make a single circulation; l_m is called the *mixing length*. If we assume $|\nabla \tilde{V}| \sim |\nabla \langle \mathbf{V} \rangle|$, so that $\tilde{V}/l_m \sim |d\langle V \rangle/dx|$, then $\nu_T = l_m^2 |d\langle V \rangle/dx|$. This is very approximate. In practical use, l_m is considered an "adjustable" parameter that is determined by the fit to data, etc.

As we can see, the problem of turbulent closure is very difficult. It is therefore interesting that significant insight into the properties of the turbulent spectrum can be simply obtained from a dimensional analysis. The theory is due to Kolmogorov,[6] and the results are called *Kolmogorov turbulence*.

We let l be the size of the largest eddy, u be the mean velocity, and Δu be the variation of u over a distance l. The frequency of the largest eddy is u/l; this is the inverse of the *eddy turnover time*. This frequency determines the period with which the flow pattern repeats itself when viewed from a fixed frame of reference;[7] with respect to such a system, the entire flow pattern moves with the mean velocity u.

Let λ be the size of an eddy and V_λ be the velocity associated with that eddy. Then the Reynolds' number associated with this scale of the flow is

---

[6] Our presentation of Kolmogorov turbulence follows L. D. Landau and E. M. Lifschitz, *Fluid Mechanics*, Pergamon Press, London, UK (1959).

[7] This is not to imply that anything about the turbulence is "periodic." Large eddies are observed appear and disappear dynamically. The eddy turnover time is often defined as the mean lifetime, or *de-correlation time*, of the large eddies.

$$\text{Re}_\lambda = \frac{V_\lambda \lambda}{\nu}. \tag{36.26}$$

We assume that for the large eddies, $\text{Re}_\lambda \gg 1$, so that no dissipation occurs at this scale. We define λ_0 as the scale where $\text{Re}_\lambda \approx 1$. Dissipation occurs at this scale.

As we discussed previously, while dissipation is ultimately due to viscosity, its value (or magnitude) derives from the large eddies where the energy input occurs. In steady state, the energy dissipation rate is essentially independent of viscosity and must depend only on the properties of the large eddies. These are characterized by l, Δu, and the fluid density ρ. Let ε be the mean dissipation rate per unit time per unit mass of fluid. It has units of Joules/sec/kg or L^2/T^3. Since ε can depend only on the large eddies, it can depend only on ρ, l, and Δu. We therefore write

$$\varepsilon \sim \rho^\alpha l^\beta \Delta u^\gamma \tag{36.27}$$

or

$$\frac{L^2}{T^3} = \left(\frac{M}{L^3}\right)^a L^\beta \left(\frac{L}{T}\right)^\gamma. \tag{36.28}$$

Equating exponents on each side, we find $\alpha = 0$, $\beta = -1$, and $\gamma = 3$, so that

$$\varepsilon \approx \frac{(\Delta u)^3}{l}. \tag{36.29}$$

This sets the order of magnitude of energy dissipation in turbulent flow.

We now assume that, at least in the inertial range, the turbulence is self-similar, i.e., the local properties of the turbulence are independent of l and Δu (which characterize the large eddies). Therefore, the velocity V_λ of an eddy of size λ can only depend on ρ, ε, and λ. Proceeding as before, we find

$$V_\lambda \approx (\varepsilon \lambda)^{1/3} = \Delta u \left(\frac{\lambda}{l}\right)^{1/3} \tag{36.30}$$

or, in terms of the wave number k, $V_k \approx \varepsilon^{1/3} k^{-1/3}$. This is called *Kolmogorov and Obukhov's law*: the variation of velocity over a small distance is proportional to the cube root of the distance. The energy per unit wave number is $e_k \approx dE_k/dk \approx |V_k^2|/k$, where E_k is the energy in an eddy with wave number k. Then using Eq. (36.30) and following, we have for the inertial range,

$$e_k \approx \varepsilon^{2/3} k^{-5/3}. \tag{36.31}$$

This is the famous *Kolmogorov spectrum*. If plotted using a log–log scale, the slope of the line in the inertial range in Fig. 36.6 is $-5/3$. This simple estimate turns out to be remarkably accurate for hydrodynamic turbulence.

Using Eqs. (36.29) and (36.30), the Reynolds' number, Eq. (36.26), at scale λ is $\mathrm{Re}_\lambda \approx \Delta u \lambda^{4/3}/(l^{1/3}\nu)$. But $\mathrm{Re} = l\Delta u/\nu$ is the Reynolds' number that characterizes the large-scale eddies. We can therefore write

$$\mathrm{Re}_\lambda = \left(\frac{\lambda}{l}\right)^{4/3} \mathrm{Re}. \tag{36.32}$$

Dissipation occurs at λ_0 when $\mathrm{Re}_\lambda \approx 1$ or

$$\frac{\lambda_0}{l} \approx \mathrm{Re}^{-3/4}. \tag{36.33}$$

This is very small if $\mathrm{Re} \gg 1$.

In terms of the dissipation wave number k_D, defined previously, we have $k_\mathrm{D}/k_\mathrm{in} \approx \mathrm{Re}^{3/4}$, where we have called k_in the wave number where the energy is input (i.e., where the cup is stirred). Using Eq. (36.33) and the second equality in Eq. (36.30), we have that the ratio of the energy in the smallest and largest eddies is $\left|V_\lambda^2\right|/\left|\Delta u^2\right| \approx \mathrm{Re}^{-1/2}$. Using these results, the Kolmogorov spectrum is roughly sketched in Fig. 36.7.

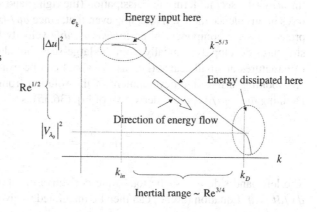

Fig. 36.7 The Kolmogorov spectrum for fully developed turbulence, showing the scaling of the various regions

Let n be the number of "degrees of freedom" per unit volume in the turbulent flow. This is approximately the number of eddies or number of spatial scales, per unit volume, associated with the flow. It has units of L^{-3}. It can depend on the density ρ, the rate of energy input ε, and the size of the smallest scale, which in turn depends on the viscosity. Dimensional analysis then yields $n \approx \left(\varepsilon/\nu^3\right)^{3/4}$. But $\varepsilon \approx \Delta u^3/l$ and $\nu = l\Delta u/\mathrm{Re}$, so that $n \approx \left(\mathrm{Re}^{3/4}/l\right)^3 = 1/\lambda_0^3$. The total number of degrees of freedom is just $N = nl^3$, so that

$$N \approx \mathrm{Re}^{9/4} \tag{36.34}$$

or about $\mathrm{Re}^{3/4}$ degrees of freedom per spatial dimension.

This last result has important consequences for numerical simulation of flows at large Reynolds' number. Based on Eq. (36.34), an approximate rule of thumb is that it requires $\sim$ Re pieces of information per spatial dimension to model turbulent flow accurately. For computer simulation, this translates into mesh points or Fourier modes. Simulation of flows with Re $\sim 10^4$ (very large by present state-of-the-art computational standards, but quite moderate physically) therefore requires approximately 10^3 mesh points in *each spatial dimension* or more than 10^9 total mesh points. Increasing the Reynolds' number by a factor of 10 requires increasing the total number of mesh points by a factor of almost 10^3 (10^4 if factors related to numerical stability are taken into account). One quickly reaches the limits of any present or foreseeable computer technology.

In this discussion, we have explicitly assumed that that the turbulence is self-similar, i.e., that the eddies look the same on all scales. That this may *not* be the case can be seen from looking at *Burger's equation*, which in one dimension is

$$\frac{\partial u}{\partial t} + u \frac{\partial u}{\partial x} = \nu \frac{\partial^2 u}{\partial x^2}. \tag{36.35}$$

This model equation [which is identical to Eq. (36.1)] contains the essential non-linearity (the second term) and dissipation (the right-hand side) that play important roles in turbulence theory. (Note, however, that, since $\partial u/\partial x \neq 0$, u cannot be interpreted as the x-component of an *incompressible* velocity field.) Since small-scale structures develop dynamically from the large-scale initial conditions, the behavior of structures at small scales (i.e., as $k \to \infty$) can be examined by setting $\nu = 0$ and looking at the time development of the solution from some initial condition. Defining $\omega = \partial u/\partial x$, the x-derivative of Eq. (36.35) is

$$\frac{\partial \omega}{\partial t} + u \frac{\partial \omega}{\partial x} = -\omega^2. \tag{36.36}$$

The left-hand side is just the total time derivative of $\omega[x(t)]$ along the trajectory $dx/dt = u$. Equation (36.36) can then be integrated to give the solution

$$\omega[x(t)] = \frac{1}{t - t_0}. \tag{36.37}$$

Therefore, the slope of the function u (analogous to the vorticity of a fluid with velocity u) becomes *infinite* in a time $\Delta t \sim 1/\omega_0$, where ω_0 represents the initial condition. This is called a *finite time singularity* (FTS): the solution becomes infinite in a finite time. This is to be contrasted with exponential behavior $\omega \sim e^{\gamma t}$, which has a large but still finite value at all (finite) times. This example *suggests* (but does not prove) that as $k \to \infty$ in fluid turbulence, the vorticity in the small eddies may become spatially concentrated, and possibly singular (at least theoretically), being drawn into thinner and thinner strands, sheets, or filaments, as the turbulent state is approached. In any case, the flow may not look the same at all scales; it may not be self-similar.

Attempts to determine whether fluid turbulence exhibits FTS behavior have proven problematical and inconclusive. Consider, for example, two-dimensional, incompressible fluid turbulence, which is governed by the equations

$$\frac{\partial \mathbf{V}}{\partial t} + \mathbf{V} \cdot \nabla \mathbf{V} = -\nabla p \tag{36.38}$$

and

$$\nabla \cdot \mathbf{V} = 0, \tag{36.39}$$

where $\mathbf{V} = V_x \hat{\mathbf{e}}_x + V_y \hat{\mathbf{e}}_y$ and $\partial/\partial z = 0$. These can be written in terms of the z-component of the vorticity, $\omega = \hat{\mathbf{e}}_z \cdot \nabla \times \mathbf{V}$, and the velocity potential, defined by $\mathbf{V} = \hat{\mathbf{e}}_z \times \nabla \phi$, as

$$\frac{\partial \omega}{\partial t} + \mathbf{V} \cdot \nabla \omega = 0 \tag{36.40}$$

and

$$\nabla^2 \phi = \omega. \tag{36.41}$$

Equation (36.40) states that the vorticity is constant along the trajectory of a fluid element. There is no term on the right-hand side to drive singular behavior as there is in the case of Burger's equation. Therefore, an FTS does *not* exist in planar, two-dimensional, incompressible fluid turbulence.

The situation is somewhat different if the two-dimensional flow has symmetry about an axis. We consider such a case in cylindrical geometry. The velocity is $\mathbf{V} = \mathbf{V}_P(r, z) + V_\theta(r, z)\hat{\mathbf{e}}_\theta$, where $\mathbf{V}_P = V_r \hat{\mathbf{e}}_r + V_z \hat{\mathbf{e}}_z$ is the poloidal velocity. Operating on Eq. (36.38) with $\hat{\mathbf{e}}_z \cdot \mathbf{r} \times$, we find

$$\left(\frac{\partial}{\partial t} + \mathbf{V}_P \cdot \nabla \right) r V_\theta = 0, \tag{36.42}$$

so that $L_z = r V_\theta$, the axial component of the angular momentum, is constant along the trajectory of a fluid particle. There is no FTS associated with angular momentum (a result that could be anticipated by the symmetry of the stress tensor). However, subtracting $\hat{\mathbf{e}}_r \cdot \partial/\partial z$ from $\hat{\mathbf{e}}_z \cdot \partial/\partial r$ of Eq. (36.38), we find, after rearranging some terms,

$$\left(\frac{\partial}{\partial t} + \mathbf{V}_P \cdot \nabla \right) \frac{\omega}{r} = -\frac{1}{r^4} \frac{\partial}{\partial z} L_z^2, \tag{36.43}$$

where now $\omega = \hat{\mathbf{e}}_\theta \cdot \nabla \times \mathbf{V}$ is the poloidal component of the vorticity. The inhomogeneous term on the right-hand side offers the *possibility* of an FTS in the vorticity of the poloidal flow. But its existence cannot be proven.

So far our approach has been analytic. Another approach to studying the properties of complex systems such as turbulent flows is to use numerical simulation. Unfortunately, numerical solutions are limited to finite values of the variables and to finite (as opposed to infinitesimal) sizes of spatial structures. Further, several types of numerical (as opposed to physical) instabilities are known to lead to solutions that behave like Eq. (36.37), i.e., they go to infinity (really, generate numbers larger than can be represented on a digital computer) in a finite (as opposed to exponential) time. So, it is not surprising that attempts to settle the question of FTS in hydrodynamic turbulence by means of direct numerical simulation of Eqs. (36.40, 36.41) have been inconclusive.[8] What seems clear is that the vorticity does tend to become concentrated in filamentary structures at the smallest scales that can be resolved.

Further, there are some *indications* (although not general conclusions) that an FTS *may* occur in three-dimensional flows. An equation for the vorticity $\omega = \nabla \times \mathbf{V}$ is found by taking the curl of Eq. (36.38). Since $\mathbf{V} \cdot \nabla \mathbf{V} = \nabla(V^2/2) - \mathbf{V} \times \omega$ and $\nabla \cdot \mathbf{V} = \nabla \cdot \omega = 0$, we have $\nabla \times (\mathbf{V} \cdot \nabla \mathbf{V}) = -\nabla \times (\mathbf{V} \times \omega) = -\omega \cdot \nabla \mathbf{V} + \mathbf{V} \cdot \nabla \omega$, so that

$$\frac{\partial \omega}{\partial t} + \mathbf{V} \cdot \nabla \omega = \omega \cdot \nabla \mathbf{V}. \tag{36.44}$$

The left-hand side is the advection of the vorticity with the motion of the fluid element. The right-hand side is called the *vorticity stretching term*. It results from the projection of the rate of strain tensor $\sim \nabla \mathbf{V}$ (see Lecture 7) in the direction of the vorticity ω. It clearly vanishes for two-dimensional planar flows [see Eq. (36.40)]. In three dimensions, it serves as a source of vorticity at small scales and may result in a singularity. This *tendency* is borne out by numerical simulation. Whether this actually occurs in finite or exponential time seems moot, since in any physical system dissipation will intervene before the singularities can develop. What *does* seem to be important is the formation of filamentary structures at small scales, where the energy is dissipated; the turbulence is *not* self-similar. This may affect the properties of the PDF and the spectrum of turbulence, and hence the associated turbulent transport. Of course, as discussed previously, in steady state the dissipation rate is, by definition, equal to the input rate and is therefore independent of the viscosity.

Now, how is all this related to MHD? The answer is that nobody really knows, although there are many ideas.[9] MHD turbulence certainly may have different properties than hydrodynamic turbulence. For one thing, the magnetic field provides a preferred direction in space, so the turbulence will no longer be isotropic. We have

[8] See Dieter Biskamp, *Magnetohydrodynamic Turbulence*, Cambridge University Press, Cambridge, UK (2003), pp. 34ff.

[9] See, for example, Dieter Biskamp, *Magnetohydrodynamic Turbulence*, Cambridge University Press, Cambridge, UK (2003), from which this discussion follows.

seen (for example, Lectures 13 and 32) that motions, or eddies, tend to stretch out along field lines, so that $k_\perp \gg k_\parallel$, as sketched in Fig. 36.8.

Fig. 36.8 In MHD turbulence, the eddies tend to be elongated in the direction of the magnetic field

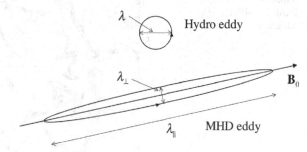

When the mean field is much larger than the fluctuating field, we might expect the turbulence to be approximately two-dimensional on the plane perpendicular to $\langle \mathbf{B} \rangle$.

From Eq. (36.29), the size of an eddy is approximately $l \sim \Delta u^3/\varepsilon$. If we take $\Delta u \approx V_A$, then $L = V_A^3/\varepsilon$ is called the "integral scale" for MHD turbulence. It turns out that, for a given eddy,

$$\frac{k_\perp}{k_\parallel} \approx (L k_\perp)^{1/3}, \tag{36.45}$$

so that MHD turbulence becomes more anisotropic as $k_\perp$ increases. The perpendicular energy spectrum remains Kolmogorov-like,

$$e_{k_\perp} \approx \left(\frac{V_A^3}{L}\right)^{2/3} k_\perp^{-5/3}. \tag{36.46}$$

However, the parallel spectrum differs from Kolmogorov:

$$e_{k_\parallel} \approx \varepsilon^{3/2} V_A^{-5/2} k_\parallel^{-5/2}. \tag{36.47}$$

MHD turbulence does not appear to be self-similar. In MHD we know that the eddies are stretched out along the magnetic field and that this anisotropy increases at smaller scales [see Eqs. (36.46, 36.47)]. Thus the small scales are more "stretched" than the large scales, and the stretching changes with increasing $k_\perp$. This is manifested in the structure of the current density $\tilde{\mathbf{J}}$ at small scales. At these scales, the magnetic flux tends to get "squeezed" by the eddies to form long, thin current filaments. An example of filamentary spatial structures in hydrodynamic turbulence is illustrated in Fig. 36.9. We might expect the current filaments in MHD turbulence to look like this.

Fig. 36.9 The spatial structure at small scales in intermittent turbulence. Note the coherent structures (http://www-vis.lbl.gov/Events/SC04/Incite3/index.html)

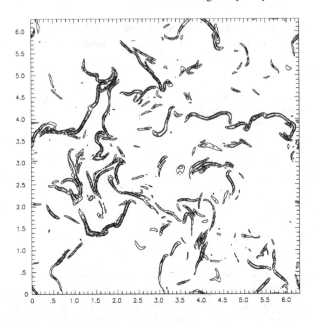

For self-similar turbulence the variation would be random. However, in MHD the current appears as semi-discrete sheets or spikes. The energy is therefore dissipated in discrete reconnection events (see Lecture 33) at small scales, rather than as a continuous process. The turbulence is said to be *intermittent*. The structures at small scales are not space filling, and the turbulence at this scale ceases to be self-similar. This affects the PDF for the current. If the fluctuations comprising the turbulence are random, as we assumed at the beginning of this lecture, then the PDF is a Gaussian, $P(\varepsilon) \sim e^{-\varepsilon}$, where ε is the energy. It is found that in intermittent turbulence, $P(\varepsilon) \sim e^{-(\ln \varepsilon)^2}$. This differs from Gaussian primarily by having a much longer tail at high energy, representing the relative abundance of these large-amplitude events. Of course, the PDF affects the validity of the closure expressions that determine the effect of the turbulence on the large-scale flows. We expect intermittent turbulence to yield different transport properties than self-similar turbulence.

The tendency in MHD toward intermittency and spiky structures at small scales can be illustrated by considering the Elsässer variables $\mathbf{z}^{\pm} = \mathbf{V} \pm \mathbf{B}$, introduced in Lecture 24. In terms of these variables, the incompressible ideal MHD equations are

$$\frac{\partial \mathbf{z}^{\pm}}{\partial t} + \mathbf{z}^{\mp} \cdot \nabla \mathbf{z}^{\pm} = -\nabla P \tag{36.48}$$

and

$$\nabla \cdot \mathbf{z}^{\pm} = 0. \tag{36.49}$$

An equation for the generalized vorticity $\boldsymbol{\omega}^{\pm} = \nabla \times \mathbf{z}^{\pm} = \boldsymbol{\omega} \pm \mathbf{J}$ is found by taking the curl of Eq. (36.48). The second term can be written as

$$\mathbf{z}^{\mp} \cdot \nabla \mathbf{z}^{\pm} = \nabla(\mathbf{z}^{\pm} \cdot \mathbf{z}^{\mp}) - \nabla \mathbf{z}^{\mp} \cdot \mathbf{z}^{\pm} - \mathbf{z}^{\mp} \times \boldsymbol{\omega}^{\pm}, \tag{36.50}$$

so that

$$\nabla \times (\mathbf{z}^{\mp} \cdot \nabla \mathbf{z}^{\pm}) = \nabla \times (\nabla \mathbf{z}^{\mp} \cdot \mathbf{z}^{\pm}) - \boldsymbol{\omega}^{\pm} \cdot \nabla \mathbf{z}^{\mp} + \mathbf{z}^{\mp} \cdot \nabla \boldsymbol{\omega}^{\pm}. \tag{36.51}$$

The first term on the right-hand side is

$$\nabla \times (\nabla \mathbf{z}^{\mp} \cdot \mathbf{z}^{\pm}) = \varepsilon_{ijk} \partial_j \left[\left(\partial_k z_l^{\mp} \right) z_l^{\pm} \right] = \varepsilon_{ijk} \left[\left(\partial_k z_l^{\mp} \right) \left(\partial_j z_l^{\pm} \right) + z_l^{\pm} \partial_j \partial_k z_l^{\mp} \right]$$

$$= \sum_{l=1}^{3} \nabla z_l^{\pm} \times \nabla z_l^{\mp} + \left(\nabla \times \nabla \mathbf{z}^{\mp} \right) \cdot \mathbf{z}^{\pm}. \tag{36.52}$$

The last term vanishes as the curl of a gradient. The generalized MHD vorticity equation is then

$$\frac{\partial \boldsymbol{\omega}^{\pm}}{\partial t} + \mathbf{z}^{\mp} \cdot \nabla \boldsymbol{\omega}^{\pm} = \boldsymbol{\omega}^{\pm} \cdot \nabla \mathbf{z}^{\mp} + \sum_{l=1}^{3} \nabla z_l^{\pm} \times \nabla z_l^{\mp}. \tag{36.53}$$

It differs in form from the hydrodynamic vorticity equation, Eq. (36.44), in that it has two terms on the right-hand side. The first is generalized vorticity stretching. The second is new in MHD. Either term may (or may not!) lead to filamentary structures and a possible FTS. Since the first term can be responsible for filamentary structure in the vorticity, we might suspect that the second term may lead to filamentary structure in the current density. Note that, for a two-dimensional planar system, the vorticity stretching term vanishes as in hydrodynamics, but the second term remains. We therefore expect MHD turbulence to differ from hydrodynamic turbulence, even in two dimensions.

The tendency toward current sheet formation in MHD (and the associated small scale structure in the current density) can be illustrated by considering a two-dimensional planar system in the vicinity of an X-point (a null in the poloidal field). In this case it is sufficient to consider the ideal reduced MHD equations [see Lecture 13, Eqs. (13.29, 13.30)],

$$\frac{\partial \psi}{\partial t} + \mathbf{V} \cdot \nabla \psi = 0 \tag{36.54}$$

and

$$\frac{\partial \omega}{\partial t} + \mathbf{V} \cdot \nabla \omega = \mathbf{B} \cdot \nabla J, \tag{36.55}$$

with $\mathbf{B} = \hat{\mathbf{e}}_z \times \nabla \psi + B_0 \hat{\mathbf{e}}_z$, $\mathbf{V} = \hat{\mathbf{e}}_z \times \nabla \phi$, $J = \nabla^2 \psi$, and $\omega = \nabla^2 \phi$. [These are equivalent to Eq. (36.53). The right-hand side of Eq. (36.55) arises from the second term on the right-hand side of Eq. (36.53).] In the vicinity of an X-point, the flux function is, for example,

$$\psi = \frac{1}{2}\left(\frac{x^2}{\xi^2(t)} - \frac{y^2}{\eta^2(t)}\right), \tag{36.56}$$

whose level curves are hyperbolae with asymptotes whose slopes are $\pm\eta/\xi$. We allow these parameters to vary with time. With this choice, $J = \nabla^2\psi = J(t)$. If we further choose

$$\phi = \Lambda(t)xy, \tag{36.57}$$

we have $\omega = \nabla^2\phi = 0$, so that Eq. (36.55) is satisfied identically. Substituting Eqs. (36.56) and (36.57) into Eq. (36.54), and recalling that x and y are independent variables, yields $\dot{\xi} = -\Lambda\xi$ and $\dot{\eta} = \Lambda\eta$. With the choice $\Lambda = $ constant, these have exponentially growing solutions, so that the current grows exponentially, i.e., $J \sim e^{\Lambda t}$. This is not an FTS, but it does indicate a tendency for the localized current density to grow to large values. This exponential growth of the current density in two-dimensional MHD turbulence has been borne out by numerical simulation. There does not appear to be a true FTS in two-dimensional MHD turbulence, although there is a tendency to form current sheets at small scale.

In three-dimensional turbulence, the properties of the structures at small scales are determined by the balance between the two terms on the right-hand side of Eq. (36.53). High-resolution numerical simulation of three-dimensional MHD turbulence indicates an exponential growth of current sheets, rather than an FTS. These fine scale structures seem to be a fundamental property of MHD flows.

A further property of MHD turbulence is the tendency of the flow and magnetic field to be aligned. If one solves the variational problem of minimizing the total energy (kinetic plus magnetic) with the constraint of constant cross-helicity (see Lecture 12; constrained variational problems will be discussed in more detail in Lecture 37), the result (in non-dimensional variables) is $\mathbf{V} = \pm\mathbf{B}$, which must be valid at each point in space; the velocity and magnetic field tend to become aligned. This is called an Alfvénic state, and the tendency toward *dynamic alignment* is called the *Alfvén effect*. In such a state, either $\mathbf{z}^+$ or $\mathbf{z}^-$ vanishes, and Eq. (36.53) indicates that the nonlinearity in the MHD equations ceases to function. This is sometimes called *nonlinear depletion*, i.e., the nonlinearities are depleted by the dynamics. According to our discussions in Lecture 24, it also means that, at least locally, and at this scale, all the Alfvén waves are either right or left traveling; they are all propagating in the same direction and hence cannot interact with each other. This must therefore be the final state of MHD turbulence. Numerical simulations indicate that the MHD fluid tends to form into small regions where the velocity and the magnetic field are either positively or negatively aligned. The details of this state depend on the amount of energy and cross-helicity in the initial conditions.

Returning to the closure problem, if we again assume a separation of scales, as in Eq. (36.9), the equations for the mean components of the velocity and magnetic field in MHD are (in non-dimensional form)

$$\frac{\partial \langle \mathbf{V} \rangle}{\partial t} + \langle \mathbf{V} \rangle \cdot \nabla \langle \mathbf{V} \rangle = -\nabla \langle p \rangle + \langle \mathbf{B} \rangle \cdot \nabla \langle \mathbf{B} \rangle + \nabla \cdot \langle \tilde{\mathbf{V}}\tilde{\mathbf{V}} - \tilde{\mathbf{B}}\tilde{\mathbf{B}} \rangle + \nu \nabla^2 \langle \mathbf{V} \rangle \quad (36.58)$$

and

$$\frac{\partial \langle \mathbf{B} \rangle}{\partial t} = \nabla \times (\langle \mathbf{V} \rangle \times \langle \mathbf{B} \rangle) + \nabla \times \langle \tilde{\mathbf{V}} \times \tilde{\mathbf{B}} \rangle + \eta \nabla^2 \langle \mathbf{B} \rangle . \quad (36.59)$$

Closure expressions are required for the turbulent stress $\langle \tilde{\mathbf{V}}\tilde{\mathbf{V}} - \tilde{\mathbf{B}}\tilde{\mathbf{B}} \rangle$ and the (negative of the) turbulent electric field $\langle \tilde{\mathbf{V}} \times \tilde{\mathbf{B}} \rangle$. One form of these closures is

$$\langle \tilde{\mathbf{V}}\tilde{\mathbf{V}} - \tilde{\mathbf{B}}\tilde{\mathbf{B}} \rangle = \frac{1}{3} \langle \tilde{V}^2 - \tilde{B}^2 \rangle \mathbf{I}$$
$$- \nu_t^V \left[\nabla \langle \mathbf{V} \rangle + (\nabla \langle \mathbf{V} \rangle)^T \right] - \nu_t^M \left[\nabla \langle \mathbf{B} \rangle + (\nabla \langle \mathbf{B} \rangle)^T \right] \quad (36.60)$$

and

$$\langle \tilde{\mathbf{V}} \times \tilde{\mathbf{B}} \rangle = \alpha_t \langle \mathbf{B} \rangle - \beta_t^M \langle \mathbf{J} \rangle + \beta_t^V \langle \boldsymbol{\omega} \rangle , \quad (36.61)$$

where $\boldsymbol{\omega} = \nabla \times \mathbf{V}$ is the vorticity, ν_t^V is an eddy viscosity due to velocity fluctuations, and ν_t^M is an eddy viscosity due to magnetic field fluctuations.

Equation (36.61) is important. If we define the turbulent electric field as $\mathbf{E} \equiv -\langle \tilde{\mathbf{V}} \times \tilde{\mathbf{B}} \rangle$, then the effect of the first term in Eq. (36.61) is to generate a mean electric field parallel to the mean magnetic field, $\mathbf{E} = -\alpha_t \langle \mathbf{B} \rangle$. This is remarkable, since in ideal MHD the electric field $\mathbf{E} = -\mathbf{V} \times \mathbf{B}$ is always perpendicular to the magnetic field. Not surprisingly, this is called the α-effect. It is of central importance in *dynamo theory* and will be discussed further in Lecture 38. The second term in Eq. (36.61) produces a mean electric field parallel to the mean current; the coefficient β_t^M enters as an additional turbulent resistivity. This is called the β-effect. The last term in Eq. (36.61) represents the effect of the fluctuating velocity field on the $\langle \mathbf{B} \rangle$.

The approach in this lecture has been theoretical. It only becomes physics when it is compared with what occurs in nature. This can be determined from experiment, as in the case of hydrodynamic flow in a pipe or wind tunnel, or from observations of astrophysical plasmas, such as the interstellar medium or the solar wind. The most striking thing about these data is that they display the general form of the Kolmogorov spectrum, i.e., an input range, an inertial range, and a dissipation range, as sketched in Fig. 36.7. An example of the spectrum fluctuations in an optical signal induced by inhomogeneities in the atmospheric refractive index is shown in Fig. 36.10. The presence of input, inertial, and dissipation ranges are apparent, and the 5/3 power law in the inertial range is in agreement with the predictions of Kolmorogov theory.

Fig. 36.10 Measured spectrum of turbulence in the earth's atmosphere due to inhomogeneities in the index of refraction induced by fluctuations in the density and temperature. The 5/3 power law in the inertial range is in agreement with Kolmogorov theory. (See etoile.berkeley.edu/~jrg/SEEING/node3.html.)

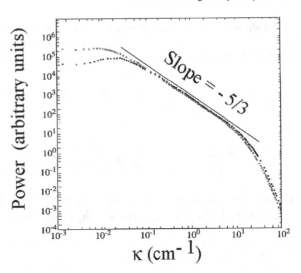

While the power law in the inertial range often appears to be close to the Kolmogorov value, closer examinations of data for both hydrodynamic and MHD turbulence show that small deviations from the value 5/3 exist and are real. This deviation is often attributed to the intermittency, or the lack of self-similarity at small scales, that we have discussed previously. A clear example of this phenomenon in MHD turbulence is shown in Fig. 36.11, which shows the spectrum of the fluctuating magnetic energy in a region of the Earth's magnetosphere as measured in situ by a constellation of satellites. (The horizontal axis is the wave number normalized to the ion Larmor radius. The separation between the satellites was about 100 km.)

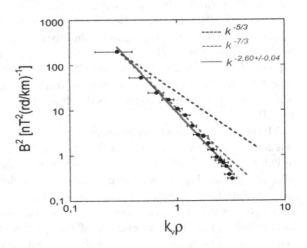

Fig. 36.11 Spectrum of fluctuations in the magnetic energy in a region of the earth's magnetosphere as measured by the CLUSTER mission.[10] (See http://sci.esa.int/science-e/www/object/index.cfm?fobjectid=38841)

---

[10] F. Sahraoui, G. Belmont, L. Rezeau, N. Cornilleau-Wehrlin, M. André, S. Buchert, and H. Rème, Phys. Rev. Letters **96**, 075002 (2006).

The exponent for the power law in the inertial range is approximately 8/3, much steeper than the Kolmogorov value.

MHD turbulent spectra are not often measured in the laboratory. An example of data from an RFP plasma is shown in Fig. 36.12. (We will discuss the RFP configuration in more detail in Lectures 37 and 38. The terms "Standard" and "PPCD" refer to different modes of experimental operation. "Right" and "Left" propagation can be thought of as referring to different "twists" in the background magnetic field. These details are not important here.) The toroidal mode number n is a proxy for the wave number $k = n/R$, where R is the major radius of the toroidal device. In this example, the plasma is continually "stirred" by a handful of unstable, long wavelength-resistive MHD modes (the modes with large amplitude near $n = 0$). This is the energy input scale. Note, however, that not only is the spectrum not Kolmogorov, but *it does not have an inertial range*. It appears that some sort of dissipation occurs across the entire range of wave numbers, even though the collisional dissipation is quite small. (There is also an anomalous heating of the ions in this experiment, which may be related to non-MHD effects that are operating on all scales.) This behavior is as yet unexplained.

Fig. 36.12 Experimentally measured magnetic energy spectrum for an RFP device. There is no inertial range. (Courtesy of P. W. Terry)

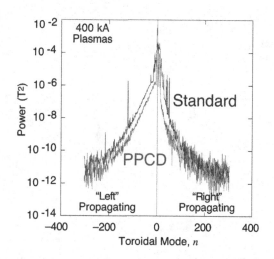

In this lecture we have implicitly maintained our fundamental assumption, introduced in Lecture 1, that the material under consideration looks the same no matter how finely it is subdivided. With regard to turbulence, this means that the material looks the same at all scales: in the input range, the inertial range, and the dissipation range; at both low k and high k. For "normal" fluids, such as air and water in the terrestrial environment, this is usually a good assumption. However, for MHD the situation is not as clear. For example, in a hot magnetized plasma the ion Larmor radius is only 10^{-2} to 10^{-3} of the largest scale length. Clearly, even in cases where the largest scales are well described by MHD, if the inertial range extends for more than a few decades beyond the large energy-producing eddies, both the high-k end of the inertial range and the dissipation range will be greatly influenced by physics that is not incorporated in the MHD model. The validity of using MHD to model turbulence in these circumstances must be questioned. This is especially clear from the example of Fig. 36.12. Incorporating non-MHD effects into comprehensive models of turbulence in magnetized plasmas is an area that needs further theoretical development.

Lecture 37
MHD Relaxation: Magnetic Self-Organization[1]

Relax! Don't do it!

Frankie Goes to Hollywood

Magnetized fluids and plasmas are observed to exist naturally in states that are relatively independent of their initial conditions or the way in which the system was prepared. Their properties are completely determined by boundary conditions and a few global parameters, such as magnetic flux, current, and applied voltage. Successive experiments carried out with the same global parameters yield the same mean state, even though they were not initiated in exactly the same way (for example, how the gas initially fills the vacuum chamber or the breakdown process). Further, if the system is disturbed it tends to return to the same state. These preferred states are called *relaxed*, or *self-organized*, states, and the dynamical process of achieving these states is called *plasma relaxation*, or *self-organization*. Relaxed states cannot result from force balance or stability considerations alone, because there may be many different stable equilibria corresponding to a given set of parameters and boundary conditions. Some other process must be at work.

The energy principle says that a system tries to achieve its state of minimum potential energy

$$W = \int_{V_0} \left(\frac{B^2}{2\mu_0} + \frac{p}{\Gamma - 1} \right) dV. \tag{37.1}$$

Taken literally, minimization of W yields the state $B = 0$, $p = 0$, which is physically irrelevant. Clearly, the minimization must be constrained in some way. Minimization with the condition that the total magnetic flux Φ be fixed yields $B = $ constant, $p = 0$, which is better but still not physically realistic. Further constraints are required.

---

[1] The seminal references and summary are contained in J. B. Taylor, Rev. Mod. Phys. **58**, 741 (1986). Much of the present discussion is a condensation of Sergio Ortolani and Dalton D. Schnack, *Magnetohydrodynamics of Plasma Relaxation*, World Scientific, Singapore (1993).

Schnack, D.D.: *MHD Relaxation: Magnetic Self-Organization.* Lect. Notes Phys. **780**, 241–259 (2009)
DOI 10.1007/978-3-642-00688-3_37

Recall that, in ideal MHD, the integrals

$$K_l = \int_{V_l} \mathbf{A} \cdot \mathbf{B} dV, \, l = 1, 2, \ldots, \tag{37.2}$$

are constant on each and every flux tube V_l in the system. These are the *Wöltjer invariants* (see Lecture 12). There are an infinite number of these constraints. In an earlier lecture we showed that the existence of the Wöltjer invariants is equivalent to the assumption of ideal MHD, $\mathbf{E} = -\mathbf{V} \times \mathbf{B}$, and vice versa. We therefore seek to minimize W subject to the constraint of ideal MHD. If we vary the magnetic field and pressure independently, then

$$\delta W = 0 = \int_{V_0} \left(\frac{\mathbf{B} \cdot \delta \mathbf{B}}{2\mu_0} + \frac{\delta p}{\Gamma - 1} \right) dV$$

$$= \frac{1}{\mu_0} \int_{V_0} \mathbf{B} \cdot \nabla \times (\boldsymbol{\xi} \times \mathbf{B}) \, dV + \frac{1}{\Gamma - 1} \int_{V_0} \delta p \, dV, \tag{37.3}$$

where $\boldsymbol{\xi}$ is the displacement. Since $\delta \mathbf{B}$ and δp are independent, the last integral is minimized by setting $\delta p = 0$. The first integrand is rewritten using the vector identities

$$\nabla \cdot [\mathbf{B} \times (\boldsymbol{\xi} \times \mathbf{B})] = \mathbf{B} \cdot \nabla \times (\boldsymbol{\xi} \times \mathbf{B}) - (\boldsymbol{\xi} \times \mathbf{B}) \cdot \nabla \times \mathbf{B} \tag{37.4}$$

and

$$\mathbf{B} \times (\boldsymbol{\xi} \times \mathbf{B}) = \boldsymbol{\xi} B^2 - \mathbf{B} \boldsymbol{\xi} \cdot \mathbf{B}, \tag{37.5}$$

so that

$$\int_{V_0} (\boldsymbol{\xi} \times \mathbf{B}) \cdot \nabla \times \mathbf{B} dV + \oint_S \hat{\mathbf{n}} \cdot \left[\boldsymbol{\xi} B^2 - \mathbf{B} \boldsymbol{\xi} \cdot \mathbf{B} \right] dS = 0. \tag{37.6}$$

The surface integral vanishes with the boundary conditions $\hat{\mathbf{n}} \cdot \boldsymbol{\xi} = \hat{\mathbf{n}} \cdot \mathbf{B} = 0$, and the volume integral becomes

$$\int_V \boldsymbol{\xi} \cdot [(\nabla \times \mathbf{B}) \times \mathbf{B}] dV = 0. \tag{37.7}$$

Since this must hold for arbitrary $\boldsymbol{\xi}$, minimization of W requires

$$(\nabla \times \mathbf{B}) \times \mathbf{B} = 0 \tag{37.8}$$

or

$$\nabla \times \mathbf{B} = \lambda(\mathbf{r}) \mathbf{B}. \tag{37.9}$$

where $\lambda(\mathbf{r}) = \mathbf{J} \cdot \mathbf{B}/B^2$ is related to the parallel current density. It is a function of space that satisfies the equation

$$\mathbf{B} \cdot \nabla \lambda = 0. \tag{37.10}$$

It is constant along field lines. The relaxed magnetic fields are *force-free* (see Lecture 17).

We remark on the minimization with respect to the pressure. Instead of setting $\delta p = 0$, we could have used the adiabatic (ideal MHD) energy equation

$$\delta p = -\Gamma p_0 \nabla \cdot \boldsymbol{\xi} - \boldsymbol{\xi} \cdot \nabla p_0. \tag{37.11}$$

Minimization then yields $\nabla p = (\nabla \times \mathbf{B}) \times \mathbf{B}$, instead of Eq. (37.8). The pressure will be constant along a flux tube, but can vary from flux tube to flux tube. In that case, the pressure distribution would be determined by the details of the way the system was prepared, and would be unrepeatable. This is *not* what is observed. Instead, if there is a small amount of resistivity the flux tubes will break. The pressure will mix and equilibrate, resulting in a state with $\nabla p = 0$. To quote Taylor[2]: "Relaxation proceeds by reconnection of lines of force, and during this reconnection plasma pressure can equalize itself so that the fully relaxed state is a state of uniform pressure. Hence, the inclusion of plasma pressure does not does not change our conclusion about the relaxed state. Of course, one may argue that pressure relaxation might be slower than field relaxation, so that the former was incomplete and some residual pressure gradients would remain . . .. However, no convincing argument for determining the residual pressure gradient has yet been given. We shall, therefore, consider ∇p to be negligible in relaxed states – which in any event is a good approximation for low-β plasmas."

Dependence on the initial conditions is also a problem for the force-free relaxed states given by Eqs. (37.9) and (37.10). The function $\lambda(\mathbf{r})$ is determined by the way the system is prepared, which is uncontrollable. We conclude that ideal MHD *over-constrains* the system.

Taylor recognized that in a slightly resistive plasma contained within a perfectly conducting boundary, one flux tube will retain its integrity, and that is the flux tube containing the entire plasma! Then only the single quantity

$$K_0 = \int_{V_o} \mathbf{A} \cdot \mathbf{B} dV \tag{37.12}$$

will remain invariant. We recognize this as the *total magnetic helicity*. Note that this is not a proof; rather it is a conjecture based upon physical insight. *Taylor's conjecture* is then that *MHD systems tend to minimize their magnetic energy subject to the constraint that the total magnetic helicity remains constant.* In order to carry out

---

[2] J. B. Taylor, Rev. Mod. Phys. **58**, 741(1986).

this calculation, we need to know something about how constrained minimization is expressed in the calculus of variations.

Constrained Variation and Lagrange Multipliers[3]

Problem I: Given a continuous function $f(x, y, \ldots)$ of N variables in a close region G, find the point $(x_0, y_0, \ldots)$ where f has an extremum.

Solution I: Set

$$\frac{\partial f}{\partial x} = 0,$$

$$\frac{\partial f}{\partial y} = 0, \qquad (37.13)$$

$$\text{etc.}$$

This yields N simultaneous equations in N unknowns whose solution is $x = x_0$, $y = y_0$, etc.

Problem II: Now suppose that the variables $(x, y, \ldots)$ are no longer independent, but are subject to the restrictions, or constraints,

$$g_1(x, y, \ldots) = 0,$$
$$g_2(x, y, \ldots) = 0,$$
$$\cdot$$
$$\cdot \qquad (37.14)$$
$$\cdot$$
$$g_h(x, y, \ldots) = 0,$$

where $h < N$. Find $(x_0, y_0, \ldots)$.

Solution IIA: Use Eq. (37.14) to algebraically eliminate h of the unknowns. Then the procedure of solution I yields $N - h$ simultaneous equations in $N - h$ unknowns, which can be solved for $(x_0, y_0, \ldots)$. This can be quite tedious.

Solution IIB: Introduce $h + 1$ new parameters $\lambda_0, \lambda_1, \ldots, \lambda_h$, and construct the function

$$F = \lambda_0 f + \lambda_1 g_1 + \lambda_2 g_2 + \ldots + \lambda_h g_h. \qquad (37.15)$$

The unknowns are now $(x, y, \ldots, \lambda_0, \lambda_1, \ldots, \lambda_h)$. There is one more unknown than equations, so we can determine $(x_0, y_0, \ldots)$ and the *ratios* of $(\lambda_0, \lambda_1, \ldots)$ from the unconstrained problem

$$\frac{\partial F}{\partial x} = 0, \ \frac{\partial F}{\partial y} = 0, \ \ldots\ldots$$
$$\frac{\partial F}{\partial \lambda_1} = g_1 = 0, \ \frac{\partial F}{\partial \lambda_2} = g_2 = 0, \ \ldots\ldots \qquad (37.16)$$

---

[3] This discussion follows R. Courant and D. Hilbert, *Methods of Mathematic Physics*, Vol. 1, Interscience, New York (1953).

If $\lambda_0 \neq 0$, we can set $\lambda_0 = 1$ since F is homogeneous in the λ_i. This procedure avoids the algebra of eliminating the unknowns from the constraints. The λ_i are called *Lagrange multipliers*, and the procedure is called the *method of Lagrange multipliers*.

We now apply this method to the constrained variational problem.

Problem III: Find $y(x)$ that makes

$$J\{y\} = \int_{x_0}^{x_1} F\left(x, y, y'\right) dx \tag{37.17}$$

stationary, has given boundary values $y(x_0) = y_0$, $y(x_1) = y_1$, and is subject to the subsidiary condition (constraint)

$$K = \int_{x_0}^{x_1} G\left(x, y, y'\right) dx = C. \tag{37.18}$$

Solution III: Let $y(x)$ be the desired extremal, and consider the neighboring curve

$$y + \delta y = y(x) + \varepsilon_1 \eta(x) + \varepsilon_2 \zeta(x), \tag{37.19}$$

with $\eta(x_0) = \eta(x_1) = \zeta(x_0) = \zeta(x_1) = 0$. Then

$$\Phi\left(\varepsilon_1, \varepsilon_2\right) = \int_{x_0}^{x_1} F\left(x, y + \varepsilon_1\eta + \varepsilon_2\zeta, y' + \varepsilon_1\eta' + \varepsilon_2\zeta'\right) dx \tag{37.20}$$

must be stationary at $\varepsilon_1 = \varepsilon_2 = 0$ with respect to all sufficiently small values of ε_1 and ε_2 for which

$$\Psi\left(\varepsilon_1, \varepsilon_2\right) = \int_{x_0}^{x_1} G\left(x, y + \varepsilon_1\eta + \varepsilon_2\zeta, y' + \varepsilon_1\eta' + \varepsilon_2\zeta'\right) dx = C. \tag{37.21}$$

Let

$$\chi = \lambda_0 \Phi\left(\varepsilon_1, \varepsilon_2\right) + \lambda \Psi\left(\varepsilon_1, \varepsilon_2\right), \tag{37.22}$$

where λ_0 and λ are Lagrange multipliers. Then for an extremum, we require

$$\left. \frac{\partial \chi}{\partial \varepsilon_1} \right|_{\substack{\varepsilon_1=0 \\ \varepsilon_2=0}} = \left. \frac{\partial}{\partial \varepsilon_1} \left[\lambda_0 \Phi + \lambda \Psi\right] \right|_{\substack{\varepsilon_1=0 \\ \varepsilon_2=0}} = 0 \tag{37.23}$$

and

$$\frac{\partial \chi}{\partial \varepsilon_2}\bigg|_{\substack{\varepsilon_1=0 \\ \varepsilon_2=0}} = \frac{\partial}{\partial \varepsilon_2} \left[\lambda_0 \Phi + \lambda \Psi\right]\bigg|_{\substack{\varepsilon_1=0 \\ \varepsilon_2=0}} = 0. \qquad (37.24)$$

Using the results from Lecture 25 on the calculus of variations, we have

$$\frac{\partial \Phi}{\partial \varepsilon_1} = \int_{x_0}^{x_1} \eta \, [F]_y \, dx, \qquad (37.25)$$

$$\frac{\partial \Phi}{\partial \varepsilon_2} = \int_{x_0}^{x_1} \zeta \, [F]_y \, dx, \qquad (37.26)$$

$$\frac{\partial \Psi}{\partial \varepsilon_1} = \int_{x_0}^{x_1} \eta \, [G]_y \, dx, \qquad (37.27)$$

and

$$\frac{\partial \Psi}{\partial \varepsilon_2} = \int_{x_0}^{x_1} \zeta \, [G]_y \, dx, \qquad (37.28)$$

where we have introduced the notation

$$[F]_y \equiv \frac{\partial F}{\partial y} - \frac{d}{dx}\left(\frac{\partial F}{\partial y'}\right). \qquad (37.29)$$

Equations (37.23) and (37.24) then become

$$\int_{x_0}^{x_1} \left\{\lambda_0 \, [F]_y + \lambda \, [G]_y\right\} \eta \, dx = 0, \qquad (37.30)$$

and

$$\int_{x_0}^{x_1} \left\{\lambda_0 \, [F]_y + \lambda \, [G]_y\right\} \zeta \, dx = 0. \qquad (37.31)$$

From Eq. (37.30), we find

$$\frac{\lambda_0}{\lambda} = -\frac{\int_{x_0}^{x_1} \eta\, [F]_y\, dx}{\int_{x_0}^{x_1} \eta\, [G]_y\, dx}, \tag{37.32}$$

so that the ratio λ_0/λ is independent of ζ. Then since ζ is arbitrary, we conclude from Eq. (37.31) that

$$\lambda_0\, [F]_y + \lambda\, [G]_y = 0 \tag{37.33}$$

or $\lambda_0/\lambda = -[G]_y\, /\, [F]_y$. If $\lambda_0 \neq 0$ (i.e., $[G]_y \neq 0$), we can set $\lambda_0 = 1$, and the minimizing condition is

$$[F]_y + \lambda\, [G]_y = 0 \tag{37.34}$$

or

$$\frac{d}{dx}\frac{\partial}{\partial y'}(F + \lambda G) - \frac{\partial}{\partial y}(F + \lambda G) = 0. \tag{37.35}$$

So, minimizing $\int F dx$ subject to the constraint $\int G dx = C$ is equivalent to minimizing $\int (F + \lambda G)\, dx$ without constraint.

Then, according to Taylor's conjecture, we should minimize the functional $I = W - \lambda' K_0$ without constraint [where λ' is a constant, and the minus sign is conventional; eventually λ' will be related to the variable λ used in Eqs. (37.9) and (37.10)], i.e., the proper variational problem is $\delta I = \delta W - \lambda' \delta K_0 = 0$. (Do not confuse δW and δK_0 with their use in the energy principle.)

Proceeding, we have

$$\delta W = \frac{1}{\mu_0} \int_{V_0} \delta \mathbf{B} \cdot \mathbf{B} dV = \frac{1}{\mu_0} \int_{V_0} (\nabla \times \delta \mathbf{A}) \cdot \mathbf{B} dV$$

$$= \frac{1}{\mu_0} \int_{V_0} [\delta \mathbf{A} \cdot \nabla \times \mathbf{B} + \nabla \cdot (\delta \mathbf{A} \times \mathbf{B})] dV$$

$$= \frac{1}{\mu_0} \int_{V_0} \delta \mathbf{A} \cdot \nabla \times \mathbf{B} dV + \frac{1}{\mu_0} \int_{S_0} (\delta \mathbf{A} \times \mathbf{B}) \cdot \hat{\mathbf{n}} dS , \tag{37.36}$$

and

$$\delta K_0 = \int_{V_0} (\delta \mathbf{A} \cdot \mathbf{B} + \mathbf{A} \cdot \delta \mathbf{B}) dV$$

$$= \int_{V_0} [\delta \mathbf{A} \cdot \mathbf{B} + \nabla \cdot (\mathbf{A} \times \delta \mathbf{A}) + \delta \mathbf{A} \cdot \nabla \times \mathbf{A}] \, dV$$

$$= 2 \int_{V_0} \delta \mathbf{A} \cdot \mathbf{B} dV + \int_{S_0} (\mathbf{A} \times \delta \mathbf{A}) \cdot \hat{\mathbf{n}} dS \ , \tag{37.37}$$

so that

$$\delta I = \int_{V_0} \delta \mathbf{A} \cdot \left[\frac{1}{\mu_0} \nabla \times \mathbf{B} - 2\lambda' \mathbf{B} \right] dV + \int_{S_0} (\hat{\mathbf{n}} \times \delta \mathbf{A}) \cdot (\mathbf{B} + \lambda' \mathbf{A}) \, dS. \tag{37.38}$$

The surface S is a perfect conductor where we require $\hat{\mathbf{n}} \times \delta \mathbf{E} = -i\omega \hat{\mathbf{n}} \times \delta \mathbf{A} = 0$, so that the surface term vanishes. Then setting $\delta I = 0$,

$$\int_{V_0} \delta \mathbf{A} \cdot \left[\frac{1}{\mu_0} \nabla \times \mathbf{B} - 2\lambda' \mathbf{B} \right] dV = 0. \tag{37.39}$$

Since this must hold for arbitrary $\delta \mathbf{A}$, we obtain the minimizing condition as

$$\nabla \times \mathbf{B} = \lambda \mathbf{B}, \tag{37.40}$$

where $\lambda \equiv 2\mu_0 \lambda'$ is a *constant*. This means that the system has lost memory of the details of how it was prepared. States that satisfy Eq. (37.40) are relaxed states. They are independent of the initial conditions, in agreement with experiment.

We will see that Taylor's conjecture leads to states that agree with experimental results over a wide range of parameters. But why should it be true? Why should the helicity be invariant while the energy is minimized? For example, consider K_0. We showed in Lecture 12 that

$$\frac{dK_0}{dt} = -2 \int \mathbf{E} \cdot \mathbf{B} dV. \tag{37.41}$$

In ideal MHD, $\mathbf{E} = -\mathbf{V} \times \mathbf{B}$ and $dK_0/dt = 0$. However, in resistive MHD, $\mathbf{E} = -\mathbf{V} \times \mathbf{B} + \eta \mathbf{J}$ and

$$\frac{dK_0}{dt} = -2\eta \int \mathbf{J} \cdot \mathbf{B} dV \approx O(\eta) \neq 0, \tag{37.42}$$

so that K_0 is *not* constant. Further,

$$\frac{dW}{dt} = -\int \mathbf{J} \cdot \mathbf{E} dV = -\eta \int J^2 dV \approx O(\eta) \neq 0, \tag{37.43}$$

so that K_0 and W formally decay at the same rate! So, in what sense does W decay while K_0 remains constant?

What matters is the *relative* rate of decay of energy with respect to helicity. The dynamical processes that are responsible for relaxation should dissipate W faster than K_0, even if they are at the same order in the resistivity. The ratio W/K_0 should be minimized.

Taylor envisioned relaxation to occur as a result of resistive MHD turbulence acting at small scales. If we measure time in units of the Alfvén time τ_A, then Eqs. (37.42) and (37.43) can be written non-dimensionally as

$$\frac{dW}{dt} = -\frac{2}{S} \int J^2 dV \qquad (37.44)$$

and

$$\frac{dK_0}{dt} = -\frac{2}{S} \int \mathbf{J} \cdot \mathbf{B} dV, \qquad (37.45)$$

where $S = \tau_R/\tau_A \gg 1$ is the Lundquist number. We write the magnetic field as $\mathbf{B} = \sum \mathbf{B_k} e^{i\mathbf{k} \cdot \mathbf{r}} \to B_k e^{ikx}$ and the current as $\mathbf{J} = \sum i\mathbf{k} \times \mathbf{B_k} e^{i\mathbf{k} \cdot \mathbf{r}} \to k B_k e^{ikx}$. Then at large k,

$$\frac{dW_k}{dt} \approx -\frac{2}{S} k^2 B_k^2 \qquad (37.46)$$

and

$$\frac{dK_{0_k}}{dt} \approx -\frac{2}{S} k B^2. \qquad (37.47)$$

Now $dW_k/dt \approx O(1)$ when $k^2 B^2/S \approx 1$ or $k_W \approx B S^{1/2} \approx O(S^{1/2})$. This is the wave number at which W is dissipated. But at this wave number,

$$\left. \frac{dK_0}{dt} \right|_{k=k_W} \approx \frac{k_W B^2}{S} = B^3 S^{-1/2} \approx O(S^{-1/2}) \ll 1. \qquad (37.48)$$

This suggests that small-scale turbulence may dissipate energy more efficiently than helicity. This process is an example of *selective decay of invariants*.

It can also be argued that K_0 is preserved by long wavelength motions. To do this, we first need to define the *helical flux*. In cylindrical geometry, the condition $\nabla \cdot \mathbf{B} = 0$ is

$$\frac{1}{r} \frac{\partial}{\partial r} (r B_r) + \frac{1}{r} \frac{\partial B_\theta}{\partial \theta} + \frac{\partial B_z}{\partial z} = 0. \qquad (37.49)$$

We define a new independent variable $\phi = m\theta + nz/R$, where m and n are poloidal and toroidal mode numbers. Then Eq. (37.49) becomes

$$\frac{1}{r} \frac{\partial}{\partial r} (r B_r) + \frac{1}{r} \frac{\partial}{\partial \phi} \left(m B_\theta + \frac{nr}{R} B_z \right) = 0. \qquad (37.50)$$

This condition will be satisfied identically if

$$r B_r = -\frac{\partial \chi_{m,n}}{\partial \phi} \tag{37.51}$$

and

$$m B_\theta + \frac{nr}{R} B_z = \frac{\partial \chi_{m,n}}{\partial r}, \tag{37.52}$$

where $\chi_{m,n}$ is the *helical flux function* associated with mode numbers (m, n). Integrating Eq. (37.52) from 0 to r, we find

$$\chi_{m,n} = m \int_0^r B_\theta dr' + \frac{n}{R} \int_0^r B_z r' dr'$$

$$= m\psi + \frac{n}{R}\Phi, \tag{37.53}$$

where ψ is the poloidal flux and Φ is the toroidal flux.

Now consider the integral

$$\hat{K} = \int F(\mathbf{r}, t)\, \mathbf{A} \cdot \mathbf{B}\, dV. \tag{37.54}$$

If $F = 1$, then $\hat{K} = K_0$. The rate of change of $\hat{K}$ is

$$\frac{d\hat{K}}{dt} = \int_{V_0} \left[\frac{\partial F}{\partial t}\mathbf{A} \cdot \mathbf{B} + F\frac{\partial \mathbf{A}}{\partial t} \cdot \mathbf{B} + F\mathbf{A} \cdot \frac{\partial \mathbf{B}}{\partial t} \right] dV \tag{37.55}$$

$$= \int_{V_0} \left[\frac{\partial F}{\partial t}\mathbf{A} \cdot \mathbf{B} - 2F\mathbf{E} \cdot \mathbf{B} + F\nabla \cdot (\mathbf{A} \times \mathbf{E}) \right] dV. \tag{37.56}$$

The second term in the integrand vanishes in ideal MHD. The remainder can be written as

$$\frac{d\hat{K}}{dt} = \int_{V_0} \left[\frac{\partial F}{\partial t}\mathbf{A} \cdot \mathbf{B} - (\mathbf{A} \times \mathbf{E}) \cdot \nabla F + \nabla \cdot (F\mathbf{A} \times \mathbf{E}) \right] dV$$

$$= \int_{V_0} \left[\frac{\partial F}{\partial t}\mathbf{A} \cdot \mathbf{B} - (\mathbf{A} \times \mathbf{E}) \cdot \nabla F \right] dV + \int_{S_0} F (\hat{\mathbf{n}} \times \mathbf{E}) \cdot \mathbf{A}\, dS. \tag{37.57}$$

The surface term vanishes because $\hat{\mathbf{n}} \times \mathbf{E} = 0$ on S_0, and in ideal MHD, $\mathbf{A} \times \mathbf{E} = \mathbf{B}(\mathbf{A} \cdot \mathbf{V}) - \mathbf{V}(\mathbf{A} \cdot \mathbf{B})$, so that, finally,

$$\frac{d\hat{K}}{dt} = \int_{V_0} \left[\left(\frac{\partial F}{\partial t} + \mathbf{V} \cdot \nabla F \right) \mathbf{A} \cdot \mathbf{B} - (\mathbf{B} \cdot \nabla F) \mathbf{A} \cdot \mathbf{V} \right] dV. \tag{37.58}$$

Then, in ideal MHD, $d\hat{K}/dt = 0$ if (a) $dF/dt \equiv \partial F/\partial t + \mathbf{V} \cdot \nabla F = 0$, so that F is co-moving with the fluid, *and* (b) $\mathbf{B} \cdot \nabla F = 0$, so that F is constant along field lines.

Both ψ, the poloidal flux, and Φ, the toroidal flux, satisfy these conditions, as does any function $F(\psi, \Phi)$. In particular, any function of the *helical* flux $\chi_{m,n}$, defined in Eq. (37.53), satisfies these conditions. Therefore, a mode with mode numbers (m, n) preserves the invariants

$$\hat{K}_\alpha(m, n) = \int_{V_0} \chi_{m,m}^\alpha \mathbf{A} \cdot \mathbf{B} dV, \alpha = 0, 1, 2, \ldots\ldots \tag{37.59}$$

However, in the realistic case where all modes are present, the only invariant preserved by *all* the modes is $\hat{K}_0 = K_0$, the global helicity invariant.

The above is a heuristic argument, as it relies on ideal MHD, and we know that resistivity is present. However, it gives more credence to the conjecture that minimizing $I = W - \lambda K_0$ is a plausible approach. It also suggests how relaxation may occur as a result of long wavelength motions, with low (m, n), rather than by small-scale turbulence. In any case, the real test is to compare the predictions of the theory with the results of experiment.

For the most part, we will restrict ourselves to doubly periodic cylindrical geometry. The curl of Eq. (37.40) is

$$\nabla^2 \mathbf{B} + \lambda^2 \mathbf{B} = 0, \tag{37.60}$$

and the z-component is

$$\nabla^2 B_z + \lambda^2 B_z = 0. \tag{37.61}$$

In cylindrical geometry, this becomes Bessel's equation, with solutions of the form

$$B_z = \sum_{m,k} a_{m,k} J_m(\alpha r) e^{i(m\theta + kz)}, \tag{37.62}$$

with

$$\alpha^2 = \lambda^2 - k^2. \tag{37.63}$$

Equations for the other components are similarly found. It can be shown (but not here!) that only two of these solutions can have minimum energy: azimuthally symmetric solutions with $m = 0$,

$$\frac{B_z}{B_0} = J_0(\lambda r), \tag{37.64}$$

$$\frac{B_\theta}{B_0} = J_1(\lambda r), \tag{37.65}$$

and

$$\frac{B_r}{B_0} = 0; \tag{37.66}$$

and helical solutions with $m = 0$ and $m = 1$ components,

$$\frac{B_z}{B_0} = J_0(\lambda r) + a_{1,k} J_1(\alpha r) \cos(\theta + kz), \tag{37.67}$$

$$\frac{B_\theta}{B_0} = J_1(\lambda r) + \frac{a_{1,k}}{\alpha} \left[\lambda J_1'(\alpha r) + \frac{k}{\alpha r} J_1(\alpha r) \right] \cos(\theta + kz), \tag{37.68}$$

and

$$\frac{B_r}{B_0} = -\frac{a_{1,k}}{\alpha} \left[k J_1'(\alpha r) + \frac{\lambda}{\alpha r} J_1(\alpha r) \right] \sin(\theta + kz). \tag{37.69}$$

In all cases,

$$B_0 = \frac{\lambda a}{2 J_1(\lambda a)} \frac{\Phi}{2\pi}, \tag{37.70}$$

where

$$\Phi = 2\pi \int_0^a B_z r \, dr \tag{37.71}$$

is the total axial (or toroidal) flux. The helical distortions make no contribution to the toroidal flux, which is all carried by the azimuthally symmetric solution.

The azimuthally symmetric states, Eqs. (37.64, 37.65, 37.66), are called the *Bessel function model* (BFM). They are shown in the Fig. 37.1.

For these states, the total helicity and the toroidal flux are related to λa through

$$\frac{K_0}{\Phi^2} = \frac{L}{2\pi a} \left\{ \frac{\lambda a \left[J_0^2(\lambda a) + J_1^2(\lambda a) \right] - 2 J_0(\lambda a) J_1(\lambda a)}{J_1^2(\lambda a)} \right\}. \tag{37.72}$$

The details of the relaxed state are therefore completely determined by the two invariants K_0 and Φ: K_0/Φ^2 determines λa, and hence the field profiles, through Eq. (37.72); then Φ and λa determine the field amplitude through Eq. (37.70). The quantity K_0/Φ is related to the total volt-seconds available to sustain the discharge.

We now define two useful parameters:

$$F \equiv \frac{B_z(a)}{\langle B_z \rangle} = \frac{\pi a^2 B_z(a)}{\Phi} \tag{37.73}$$

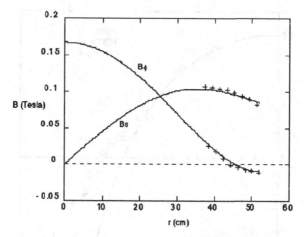

Fig. 37.1 Magnetic field profiles for the BFM. Some experimental measurements are also shown

where $\langle \ldots \rangle$ denotes the volume average, which is called the *field reversal parameter*; and

$$\Theta \equiv \frac{B_\theta(a)}{\langle B_z \rangle} = \frac{\pi a}{\mu_0} \frac{I}{\Phi}, \tag{37.74}$$

which is called the *pinch parameter*. The latter is related to the ratio of the total toroidal current to the total toroidal flux. For the BFM, it is easy to show that

$$F = \frac{\lambda a J_0(\lambda a)}{2 J_1(\lambda a)} \tag{37.75}$$

and

$$\Theta = \frac{\lambda a}{2}, \tag{37.76}$$

so that F and Θ are related by

$$F = \frac{\Theta J_0(2\Theta)}{J_1(2\Theta)}. \tag{37.77}$$

A plot of F versus Θ for the azimuthally symmetric states is shown as the solid line in Fig. 37.2. This is an example of an *F–Θ diagram*.

The theory predicts that the toroidal field at the wall will reverse sign with respect to its value on axis when $\lambda a > 2.4$ or $\Theta > 1.2$.

Thus, setting Θ by adjusting the current and flux predetermines the shape of the magnetic field profiles and the value of the toroidal field at the outer boundary. The F–Θ diagram defines a continuum of relaxed states, which could be "dialed in" by the operator of an experiment. Two regimes are of particular interest. The first corresponds to $\Theta \ll 1$. It is called the *tokamak regime*. In this case the fields are given by the small argument limits of the Bessel functions J_0 and J_1, so that

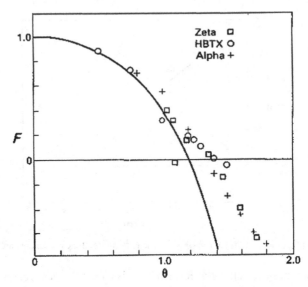

Fig. 37.2 An F–Θ diagram. The solid curve is the prediction of Taylor's theory. The points indicate experimental measurements

$B_z \approx B_0$, $B_\theta \approx (B_0 \Theta / 2)(r/a)$, and $B_z / B_\theta \approx 1/\Theta \gg 1$. These fields are sketched in the Fig. 37.3.

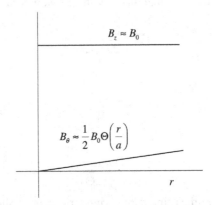

Fig. 37.3 Magnetic field profiles for the relaxed state when $\Theta \ll 1$. This is called a tokamak

The second regime corresponds to $\Theta > 1.2$ and is called the RFP *regime*. The fields are as sketched as the solid lines in Fig. 37.1.

For the helical states given by Eqs. (37.67, 37.68, 37.69), λa is now determined by the boundary condition $B_r(a) = 0$, i.e.,

$$k J_1'(\alpha a) + \frac{\lambda}{\alpha a} J_1(\alpha a) = 0, \tag{37.78}$$

where $\alpha = \sqrt{\lambda^2 - k^2}$ [see Eq. (37.69)], and solutions exist only for discrete values of λa. There are no solutions when $\lambda a < 3.11$ or $\Theta < 1.56$. Only the azimuthally symmetric states exist for lower Θ. However, when $\Theta > 1.56$, the helical state has the lowest energy. The minimum energy corresponds to $ka \approx 1.25 > 0$. The amplitude of the helical distortion, $a_{1,k}$, is then determined by K_0/Φ^2 [see Eq. (37.72)] with $\lambda a = 3.11$.

The predictions of the theory can be summarized as follows:

1. As the volt-seconds (expressed as K_0/Φ^2) increase, Θ will increase.
2. As Θ increases, $B_z(a)$ will decrease.
3. For $\Theta > 1.2$, $B_z(a)/B_z(0) < 0$.
4. At $\Theta = 1.56$ there will be the onset of a helical distortion.
5. As K_0/Φ^2 is further increased, the amplitude of the helical distortion will increase, but Θ will remain fixed at 1.56. The increase in volt-seconds does not drive more current; it is absorbed by the increased inductance due to the helical distortion of the plasma.

So, how does the relaxation theory compare with experiment? Very well in one case, pretty well in others, and not well in another.

Multi-Pinch

The multi-pinch is an axisymmetric toroidal plasma with a non-circular cross-section; the poloidal cross-section of the plasma exhibits equatorial, or up–down, symmetry, as shown in Fig. 37.4.

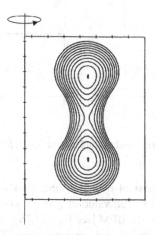

Fig. 37.4 Flux surfaces for the dipole state of the multi-pinch experiment

For such systems the periodic cylindrical approximation used in our previous discussions of this lecture does not apply, and toroidal effects must be included.

The calculation of the relaxed states goes through in much the same way as give previously, except that Eq. (37.60) is now expressed as a partial differential equation in the poloidal plane; the details will not be given here (see Taylor). It turns out that physically interesting axisymmetric ($n = 0$) solutions can be found. The lowest energy state possesses up–down symmetry; in analogy with our previous discussion, this is the only solution for low values of K_0/Φ^2. The field profiles are again parameterized by λa; K_0/Φ^2 determines λa, and then either K_0 or Φ determines the amplitude. There are also solutions that are not up–down symmetric (but still axisymmetric). These are analogous to the helically distorted states in the cylinder. These solutions do not exist for $\lambda a < 2.21$; when $\lambda a = 2.21$ these states have the lowest energy. As the volt-seconds are increased from a low value, λa increases until $\lambda a = 2.21$. As more volt-seconds are applied, λa (and hence the current) remains fixed while the amplitude of the up–down asymmetry increases.

These predictions are borne out well by experiment. The current saturation at $\lambda a = 2.21$ and the increase in up–down asymmetry are all observed. Details such as the dependence of the saturation level on toroidal flux are also predicted by the theory with quantitative accuracy (again, see Taylor).

Reversed-field Pinch

For the case of the RFP there is qualitative agreement between theory and experiment. Toroidal field reversal is observed, but it occurs at a larger value of Θ than predicted. [See the data points in the figure following Eq. (37.77)]. The pressure is not zero, and the parallel current (i.e., λa) is not constant throughout the plasma. However, most of the discrepancy occurs in the outer regions of the cylinder. These are sketched in Fig. 37.5.

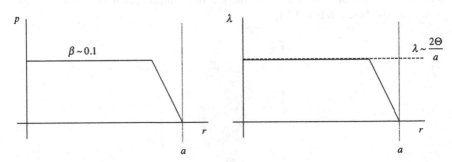

Fig. 37.5 Pressure (*left*) and normalized parallel current (*right*) for a partially relaxed RFP

Both profiles are nearly constant over the inner part of the discharge. The pressure corresponds to $\beta \sim 0.1$. The value of λ in the core of the plasma is in good agreement with the value for the BFM [see Eq. (37.76)]. Experimentally determined magnetic field profiles are shown as the points in Fig. 37.1. Again, deviation from the predictions of the theory occurs primarily near the outer boundary.

In the RFP, relaxation seems to be inhibited near the wall. This is because the fully relaxed condition $\lambda = $ constant is inconsistent with the boundary conditions for a resistive plasma at a perfectly conducting boundary. For a resistive plasma, the tangential electric field at the wall is

$$\hat{n} \times E = (\hat{n} \cdot B) V - (\hat{n} \cdot V) B + \eta \hat{n} \times J. \qquad (37.79)$$

Since $\hat{n} \times \mathbf{E} = \hat{n} \cdot \mathbf{B} = \hat{n} \cdot \mathbf{V} = 0$ at a perfectly conducting boundary, we must also require $\hat{n} \times \mathbf{J} = 0$, i.e., the tangential component of the current density must vanish (see, for example, the example of Hartmann flow in Lecture 10). This is inconsistent with non-vanishing λ; λ must decrease to zero at the wall. The RFP is said to exhibit *incomplete relaxation*. And, if the magnetic relaxation is necessarily incomplete, so is the pressure relaxation.

Further, the current saturation predicted to occur at $\Theta = 1.56$ is not observed. Recall that Taylor's helical state has $ka = 1.25$. However, in a cylinder of finite period length (as would result from straightening a torus), the wave number k is quantized as $k = n/R$ (with R the major radius). The predicted helical states thus require $na/R = 1.25$ or an aspect ratio $R/a = n/1.25$, where n is an integer. If the experiment is not so constructed, the helical states are impossible. (Helical fluctuations with $m = 1$ are observed in RFP experiments, but they have $ka < 0$ so they are not related to the helical relaxed state. They do, however, play an important role in the *dynamics* or relaxation.) In contrast, in the multi-pinch the distinction is between states of different up–down, rather than toroidal (or axial), symmetry. Quantization is not required, and there is good agreement with experiment.

Tokamak

For the tokamak, relaxation theory predicts $p = $ constant, $B_z = $ constant, and $B_\theta = $ constant $\times r$ (i.e., $J_z = $ constant). This is clearly not what is observed in experiments – *except* possibly after a "major disruption," a sudden event in which confinement is lost and the current is quenched. That state is consistent with the predictions of Taylor's theory ($p = J = 0$, $B_z = $ constant). Perhaps the major disruption is the manifestation of a "relaxation event" in the tokamak. But if so, why occur all the time? Why are tokamaks the leading candidate for a controlled fusion reactor?

This last discussion leads us to enquire into the dynamics responsible for relaxation. As mentioned, Taylor envisioned relaxation to result from small-scale turbulence in a resistive plasma. If this were true, then it should not know about the global geometry in which it is acting; it should not care if it is in a tokamak, an RFP, or a multi-pinch. It should apply equally well to all systems describable by resistive MHD.

However, we have seen that this is *not* what is found in experiments. There are large differences between relaxation (or lack thereof) in the multi-pinch, the RFP (and spheromak), and the tokamak. Perhaps the fundamental relaxation dynamics operates differently in each of these devices. Perhaps it knows about the geometry and the overall magnetic configuration. Recall that long wavelength motions can preferentially preserve the global helicity K_0 [see Eq. (37.59) and the preceding discussion]. These modes occur differently in the tokamak and the RFP. This can be seen in their q (safety factor) profiles, as sketched in Fig. 37.6.

The RFP has a decreasing q-profile. There are many long wavelength [low (m,n)] singular surfaces, and they become closely spaced near the location of the field reversal (the *reversal surface*). It is easy for them to interact nonlinearly, and provide quasi-continuous relaxation. Large, quasi-periodic oscillations are common in RFP plasmas, as shown in Fig. 37.7.

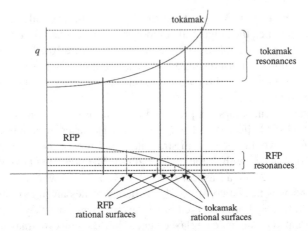

Fig. 37.6 Safety factor profiles for the tokamak (*top*) and RFP (*bottom*) configurations

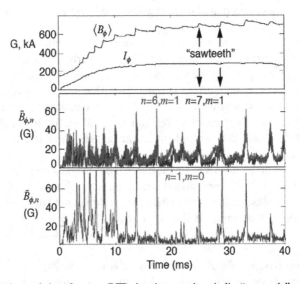

Fig. 37.7 Experimental data from an RFP showing quasi-periodic "sawteeth," which are interpreted as relaxation events[4]

They are called *sawtooth oscillations* (for historical reasons). They are characterized by (among other things) large increases in the amplitude of low (m,n) magnetic fluctuations with helical pitch corresponding to resonant perturbations. It is well established both experimentally and theoretically as a result of numerical simulation that MHD relaxation is associated with these events.

---

[4] S. C. Prager, et al., Nuclear Fusion **45**, S276 (2005).

The cyclic relaxation process that occurs in an RFP is indicated schematically in Fig. 37.8.

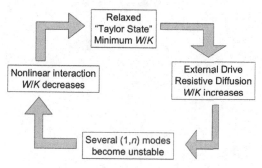

Fig. 37.8 Schematic diagram of the cyclic process associated with MHD relaxation

In contrast, the tokamak has an increasing q-profile and only a handful of low (m,n) rational surfaces. These are widely separated so it is difficult for them to interact with each other. The magnetic configuration prevents relaxation from occurring, even though low-level turbulence is always present. On the isolated occasion when these modes can seriously interact nonlinearly, disruptive-like events are predicted. Perhaps this disruptive behavior is just the tokamak seeking its preferred state.

Lecture 38
Dynamos: Magnetic Field Generation and Maintenance[1]

> *I saw the best minds of my generation destroyed by madness,*
> *... angelheaded hipsters burning for the ancient heavenly*
> *connection to the starry dynamo in the machinery of the night.*
> Allen Ginsburg, *Howl*

The study of dynamos is motivated by one of the most fundamental problems of physics: How to explain the structure, dynamics, and maintenance of magnetic fields in the universe. "Dynamo action" will be defined precisely later in this lecture. For now, we say that a dynamo is a process that can generate and amplify magnetic fields.

Although it is generally accepted that magnetic fields were not produced during the Big Bang, magnetic fields on the order of 10^{-24} T were likely created during the very early stages of the universe, before the onset of galaxy formation. The best current estimate of the magnetic field strength in existing galaxies is 10^{-10} T; how did the primordial field get amplified by 14 orders of magnitude? Further, galactic magnetic fields exhibit coherent structure on length scales that are much larger than the observed dynamical fluctuations. If the field is tied to the plasma, it should be tangled on relatively short length scales. How did this large-scale structure come about, and how is it maintained?

The resistive diffusion time (based on Coulomb collisions) for galaxies is estimated to be longer than the age of the universe, so there is no need to explain the lifetime of galactic fields. However, this is not the case for smaller astronomical bodies, such as the earth. The age of the earth is estimated to be about 4.5×10^9 years (the age of the solar system), but the resistive diffusion time is only 1.5×10^4 years, a difference of over 5 orders of magnitude. Any primordial field trapped during the earth's formation would have decayed by now, and yet the earth clearly remains magnetized. The terrestrial magnetic field is also dynamical; the geological record indicates that the polarity of the dipole field has reversed many times over the past

---

[1] Much of this lecture follows H. K. Moffatt, *Magnetic Field Generation in Electrically Conducting Fluids*, Cambridge University Press, Cambridge, UK (1978). We also acknowledge Ellen Zweibel and Fausto Cattaneo for many constructive suggestions concerning the content and presentation.

Schnack, D.D.: *Dynamos: Magnetic Field Generation and Maintenance*. Lect. Notes Phys. **780**, 261–281 (2009)
DOI 10.1007/978-3-642-00688-3_38

several million years, with a mean period of about 4×10^5 years. This record is indicated in the Fig. 38.1.

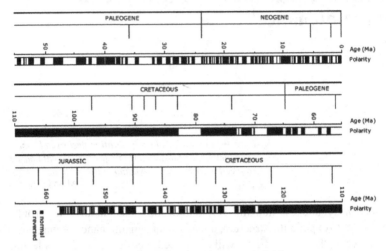

Fig. 38.1 Geological record of reversals of the earth's magnetic field

The age of the sun is also about 4.5×10^9 years, and the resistive diffusion time is on the order of 10^9 years, so, again, there is no clear need to explain the lifetime of the solar magnetic field. However, the solar field is incredibly dynamic and regular. Observations of the number of sunspots versus time show an almost regular period of 11 years, as shown in Fig. 38.2.

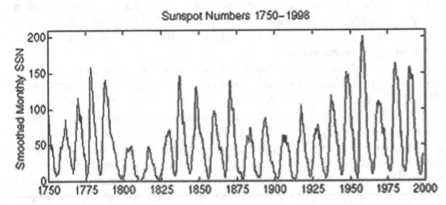

Fig. 38.2 The cyclic behavior of the number of sunspots (the "sunspot cycle") for a period of almost 250 years

The peaks of the curve are called "solar maxima," and the valleys are called "solar minima." Solar maxima are associated with dynamical activity such as solar flares and coronal mass ejections that can cause disruptions to terrestrial communications and damage satellites. The oscillations in the sunspot number are strongly correlated with reversals of the dipole field with a 22-year period. Further, during the beginning of a solar cycle (the end of solar minimum), the sunspots first appear at high latitudes (near the poles of the sun) and then migrate toward the equator as solar maximum is approached. During the next cycle the polarity of the spots is reversed, correlating with the reversal of the large-scale dipole field. This behavior is captured in the "butterfly diagram," which indicates the observed latitude of sunspots as a function of time. This is shown over several solar cycles in Fig. 38.3.

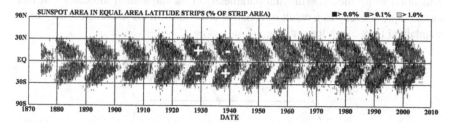

Fig. 38.3 Latitude of the appearance of new sunspots over several sunspot cycles. The plus (+) and minus (−) signs indicate the polarity of the sunspots. This is called the "butterfly diagram"

What is the origin of this dynamical behavior?

The solar magnetic field also exhibits structure on scale lengths much longer than the observed velocity fluctuations. Figure 38.4 (left) shows what is called the "granulation" of the photosphere or visible surface of the sun. These structures are thought to be the tops of thermal plumes generated by convection below the photosphere. Their characteristic size is about 10^6 m. Presumably this is the scale on which the coronal magnetic field is driven. Figure 38.4 (right) shows bright loops that are the characteristic structures in the solar corona. They are believed to highlight the

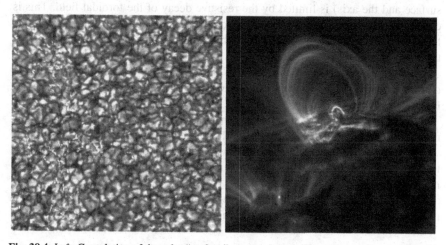

Fig. 38.4 *Left:* Granulation of the solar "surface," or photosphere. The cells are thought to be the tops of plumes of rising gas. *Right:* Bright loops in the solar corona. They are believed to outline the magnetic field structure

structure of the magnetic field. Their characteristic scale is around $10^7 - 10^8$ m. As is the case in galaxies, how can the short-wavelength driving mechanism result in long-wavelength coherent magnetic field structure?[2]

A mechanism for field generation and maintenance is also needed to explain the behavior of laboratory plasmas, in particular the RFP. In Fig. 38.5 we sketch the toroidal flux as a function of time for an RFP discharge. The solid line represents experimental results. The dashed line indicates the results of a transport simulation using the experimental parameters. This calculation indicates that flux should rapidly decay as a result of resistive diffusion.[3] In the experiment the flux is maintained for a much longer time.

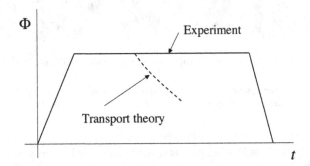

Fig. 38.5 A sketch of the toroidal flux in an RFP plasma as a function of time. The dashed line indicates the best results of "transport theory" that does not include a dynamo mechanism

After toroidal field reversal is achieved in the RFP, only negative toroidal flux can be supplied by the external circuit. Without some field generation mechanism, the lifetime of the *positive* toroidal flux (due to the toroidal field between the reversal surface and the axis) is limited by the resistive decay of the toroidal field. This is especially apparent in the case where the toroidal field is held fixed (and negative) at the outer boundary. The toroidal flux will become negative on the resistive diffusion time, as indicated in Fig. 38.6.

Instead, we have seen in Lecture 37 that toroidal flux is generated during quasi-periodic bursts of activity, called sawteeth. The relevant figure is repeated as Fig. 38.7.

What is the relationship, if any, between "relaxation" and "dynamo"? What is the relationship between the behavior of the RFP plasma and astrophysical plasmas?

For all the cases discussed above, the magnetic Prandtl number is very small, so that resistivity is the dominant dissipation mechanism. It is usual to ignore viscosity in the theory.

Consider a fluid that occupies a volume V with surface S. It is characterized by a resistive diffusivity $\lambda = \eta/\mu_0$. It is surrounded by a vacuum of volume $\hat{V}$; we take

---

[2] For more details about the sun and the solar corona, see Eric R. Priest, *Solar Magnetohydrodynamics*, D. Reidel, Dortrecht (1982).

[3] E. J. Caramana and D. A. Baker, Nuclear Fusion **24**, 423 (1984).

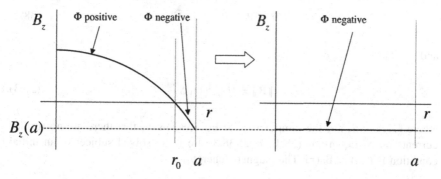

Fig. 38.6 Initial (*left*) and final (*right*) axial magnetic field profiles of an RFP when the magnetic field is fixed at the outer boundary

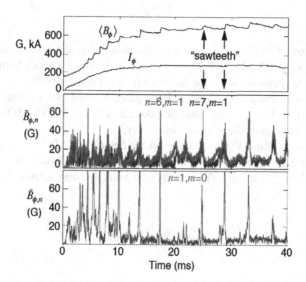

Fig. 38.7 Experimental data from an RFP showing quasi-periodic "sawteeth," which are interpreted as relaxation events [4]

the volume $V_\infty = V + \hat{V}$ to comprise the entire universe! All flows $\mathbf{V}$ and currents $\mathbf{J}$ flow within V. The flow satisfies $\nabla \cdot \mathbf{V} = 0$ within V, and $\mathbf{V} \cdot \hat{\mathbf{n}} = 0$ on S. The fluid has a characteristic length scale $L \sim V^{1/3}$. The magnetic field $\mathbf{B}$, which occupies V_∞, is produced entirely by $\mathbf{J}$. Then at large distance from the fluid $\mathbf{B}$ must be a dipole field, so $B \sim 1/r^3$ as $|r| \to \infty$. The field $\mathbf{B}$ evolves according to

$$\frac{\partial \mathbf{B}}{\partial t} = \nabla \times (\mathbf{V} \times \mathbf{B}) + \lambda \nabla^2 \mathbf{B} \qquad \text{in } V, \tag{38.1}$$

---

[4] S. C. Prager, et al., Nuclear Fusion **45**, S276 (2005).

$$\nabla \times \mathbf{B} = 0 \qquad \text{in } \hat{V}, \tag{38.2}$$

and

$$[\![\mathbf{B}]\!] = 0 \qquad \text{on } S, \tag{38.3}$$

where $[\![\ldots]\!]$ indicates the jump across S; Eq. (38.3) says that there are no surface currents on S. Equations (38.1, 38.2, 38.3) are to be solved subject to an initial condition $\mathbf{B}(\mathbf{r}, 0) = \mathbf{B}_0(\mathbf{r})$. The magnetic energy is

$$M(t) = \frac{1}{2\mu_0} \int_{V_\infty} B^2 dV. \tag{38.4}$$

It is finite since $B^2 dV \sim 1/r^3$ is bounded at infinity. We assume that initially $M(0) = M_0 > 0$. Clearly, if $\mathbf{V} = 0$ then $M(t) \to 0$ as $t \to \infty$. Some non-zero velocity field is necessary to counteract the effects of resistive diffusion.

We can now define *dynamo action*: For a given $\mathbf{V}$ and λ, we say that $\mathbf{V}$ *acts as a dynamo* if $M(t) \neq 0$ as $t \to \infty$. This includes cases where $M(t)$ tends to a constant, where it fluctuates, and where it goes to infinity.

What about the flow $\mathbf{V}(\mathbf{r}, t)$? Where did *it* come from? In "classical" dynamo theory, which we will review here, $\mathbf{V}$ is taken to be any *kinematically possible* flow field. What does *this* mean? The density and velocity must satisfy the continuity equation

$$\frac{\partial \rho}{\partial t} + \nabla \cdot \rho \mathbf{V} = 0, \tag{38.5}$$

with

$$\mathbf{V} \cdot \hat{\mathbf{n}} = 0 \tag{38.6}$$

on S. The *joint field* $[\mathbf{V}(\mathbf{r}, t), \rho(\mathbf{r}, t)]$ is said to be kinematically possible if Eqs. (38.5) and (38.6) are satisfied. We remark that *only a small subset of these fields are also dynamically possible*, i.e., satisfy the equation of motion; the allowable flows in classical dynamo theory are not so constrained. For this reason, classical dynamo theory is also called *kinematic dynamo theory*; the *back reaction* of the magnetic field on the flow, through the $\mathbf{J} \times \mathbf{B}$ force, is not considered. This may be valid if the kinetic energy is much greater than the magnetic energy. We also note that Eq. (38.5) can be re-written as

$$\frac{d\rho}{dt} = -\rho \nabla \cdot \mathbf{V}, \tag{38.7}$$

where $d\rho/dt = \partial\rho/\partial t + \mathbf{V} \cdot \nabla\rho$ is the Lagrangian derivative. If the fluid is incompressible, so that $d\rho/dt = 0$, then kinematically possible flows are defined by the conditions $\nabla \cdot \mathbf{V} = 0$ and $\mathbf{V} \cdot \hat{\mathbf{n}} = 0$ on S. This will be the case in what follows.

Why treat the velocity field this way? There are two reasons. First, the kinetic energy is much larger than the magnetic energy in many astrophysical settings, so it may be a good physical approximation. Second, and perhaps more important, the resulting theory is *linear*, and therefore amenable to analytic treatment; if $\mathbf{V}(\mathbf{r}, t)$ is a given function, then Eq. (38.1) is linear in $\mathbf{B}$. Formally, this treatment is justified if the initial ($t = 0$) magnetic energy is much smaller than the kinetic energy. If successful, the theory will then describe the initial stage of dynamo action, preceding from a state of weak magnetization. The questions classical dynamo theory asks are as follows: (a) what kind of flows $\mathbf{V}(\mathbf{r}, t)$ (if any) can produce a dynamo? and (b) how can we test whether a given $\mathbf{V}(\mathbf{r}, t)$ produces a dynamo?

One constraint on the parameters of the problem can be obtained by considering the time evolution of the magnetic energy, given by Eq. (38.4). The magnetic energy changes according to

$$\frac{dM}{dt} = \int_V \mathbf{V} \cdot \mathbf{J} \times \mathbf{B} dV - \eta \int_V J^2 dV, \qquad (38.8)$$

where the integrals are taken over V because both the velocity and the current density vanish in $\hat{V}$. The first term is the rate of production of magnetic energy, and the second term is the rate of Ohmic dissipation. If the flow is laminar (i.e., not dominated by small-scale turbulence), their magnitudes can be estimated as

$$\int_V \mathbf{V} \cdot \mathbf{J} \times \mathbf{B} dV \sim (VJB) L^3 \sim V \left(\frac{B}{\mu_0 L}\right) BL^3 = \frac{V}{L} \left(\frac{B^2 L^3}{\mu_0}\right) \approx \frac{V}{L} M \quad (38.9)$$

and

$$\eta \int_V J^2 dV \sim \eta J^2 L^3 \sim \left(\frac{\eta}{\mu_0}\right) \left(\frac{B^2 L^3}{\mu_0}\right) \frac{1}{L^2} \approx \frac{\lambda}{L^2} M. \qquad (38.10)$$

For dynamo action we require $dM/dt > 0$, which yields $VL/\lambda \equiv R_m > 1$, where R_m is the global magnetic Reynolds' number. A more rigorous estimate is $R_m > \pi^2 \sim 10$. This result is not very insightful, but true nonetheless; we could have anticipated that the dynamics should dominate dissipation for magnetic field generation to occur. In practice, we expect that $R_m \gg 1$ will be required.

Now consider the important case where the system has some preferred axis, $O–A$, say, as in Fig. 38.8.

This may be because the system is rotating, as is the case for planets, stars, and galaxies, or because of inherent geometry, as in a magnetic fusion device. Such a configuration has two types of magnetic fields: a poloidal field, $\mathbf{B}_P$, which lies in

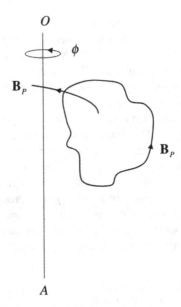

Fig. 38.8 Toroidal and poloidal magnetic fields in a system with a preferred axis

the (R,Z) plane everywhere parallel to the axis O–A; and, a toroidal field $\mathbf{B}_T$ in the $\hat{\mathbf{e}}_\phi$ direction, which encircles the axis O–A. How could a dynamo work in this situation? In order to sustain both fields, we need to devise a steady loop, denoted as $\mathbf{B}_P \rightleftarrows \mathbf{B}_T$, in which poloidal field is converted into toroidal field, and toroidal field is in turn converted into poloidal field; they sustain each other. The first part of the loop, $\mathbf{B}_P \rightarrow \mathbf{B}_T$, can be easily accomplished with differential rotation about the axis O–A, as shown in Fig. 38.9.

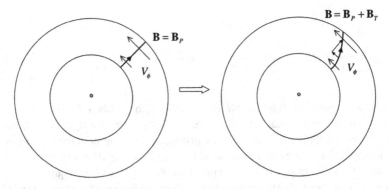

Fig. 38.9 Generation of toroidal field from poloidal field by differential rotation

If the velocity $\mathbf{V}_T(\mathbf{r}) = V_\phi \hat{\mathbf{e}}_\phi$ does not correspond to rigid body rotation, then some of the poloidal field that was originally parallel to the axis will be "bent" into the toroidal direction, thus producing toroidal field. Now we just need to get $\mathbf{B}_P$ from $\mathbf{B}_T$. It turns out that this is not simple.

Now let the system have axial symmetry, so that everything is independent of the toroidal angle ϕ. With $\mathbf{B} = \nabla \times \mathbf{A}$, the induction equation is

$$\frac{\partial \mathbf{A}}{\partial t} = -\mathbf{E} = \mathbf{V} \times \mathbf{B} - \eta \mathbf{J}. \tag{38.11}$$

We know from our previous studies of toroidal equilibria that in an axisymmetric system we can write $\psi = RA_\phi$, so that

$$RB_R = -\frac{\partial \psi}{\partial z}, \tag{38.12}$$

$$RB_z = \frac{\partial \psi}{\partial R}, \tag{38.13}$$

and

$$\mu_0 J_\phi = -\Delta^* \psi \equiv -\nabla \cdot \left(\frac{1}{R} \nabla \psi \right). \tag{38.14}$$

If we also decompose the velocity as $\mathbf{V} = \mathbf{V}_P + \mathbf{V}_T$, then $\mathbf{V} \times \mathbf{B} = -\mathbf{V}_P \cdot \nabla \psi$, and Eq. (38.11) becomes

$$\frac{\partial \psi}{\partial t} = -\mathbf{V}_P \cdot \nabla \psi + \lambda \Delta^* \psi. \tag{38.15}$$

We now assume that

$$\nabla \cdot \mathbf{V} = \nabla \cdot \mathbf{V}_P = 0, \tag{38.16}$$

(so that $\mathbf{V}$ is kinematically possible) and

$$\mathbf{V}_P \cdot \nabla \lambda = 0, \tag{38.17}$$

(so that λ is advected with the fluid). We multiply Eq. (38.15) by ψ/λ and integrate over all space to obtain

$$\frac{d}{dt} \int_{V_\infty} \frac{1}{2\lambda} \psi^2 dV = - \int_{V_\infty} \frac{1}{\lambda} \psi \mathbf{V}_P \cdot \nabla \psi dV + \int_{V_\infty} \psi \Delta^* \psi dV. \tag{38.18}$$

Using Eqs. (38.16) and (38.17), the integrand in the first term on the right-hand side can be written as $(\psi/\lambda) \mathbf{V}_P \cdot \nabla \psi = \nabla \cdot (\mathbf{V}_P \psi^2 / 2\lambda)$. Then, integrating the divergence,

$$\frac{d}{dt} \int_{V_\infty} \frac{1}{2\lambda} \psi^2 dV = -\int_{S_\infty} \frac{1}{2\lambda} \psi^2 \mathbf{V}_P \cdot \hat{\mathbf{n}} dS + \int_{V_\infty} \psi \Delta^* \psi dV. \tag{38.19}$$

As $r \to \infty$, $\psi \sim 1/r$, so the surface integral scales like $\mathbf{V}_P \cdot \hat{\mathbf{n}}/\lambda$, so the surface integral vanishes if $\mathbf{V}_P \cdot \hat{\mathbf{n}}/\lambda \to 0$ as $r \to \infty$. This is trivially satisfied if $\mathbf{V}_P \cdot \hat{\mathbf{n}} = 0$ on S_∞, or $\mathbf{V}_P = 0$ in $\hat{V}$. Thus

$$\frac{d}{dt} \int_{V_\infty} \frac{1}{2\lambda} \psi^2 dV = \int_{V_\infty} \psi \Delta^* \psi dV. \tag{38.20}$$

From Eq. (38.14), we can write $\Delta^* \psi = \nabla \cdot \mathbf{f}$, where $\mathbf{f} = \nabla \psi / R$. Then using the identity $\psi \nabla \cdot \mathbf{f} = \nabla \cdot (\psi \mathbf{f}) - \mathbf{f} \cdot \nabla \psi$ and integrating the divergence, we find

$$\frac{d}{dt} \int_{V_\infty} \frac{1}{2\lambda} \psi^2 dV = \int_{S_\infty} \frac{1}{R} \psi \hat{\mathbf{n}} \cdot \nabla \psi dS - \int_{V_\infty} \frac{1}{R} |\nabla \psi|^2 dV. \tag{38.21}$$

The surface integral scales as $1/r^2$ as $r \to \infty$, so, finally,

$$\frac{d}{dt} \int_{V_\infty} \frac{1}{2\lambda} \psi^2 dV = -\int_{V_\infty} \frac{1}{R} |\nabla \psi|^2 dV. \tag{38.22}$$

The integral on the right-hand side is positive definite, so that the left-hand side must continually decrease with time. Therefore ψ^2 (and hence $\mathbf{B}_P$ and $M(t)$) must vanish as $t \to \infty$. Thus, *dynamo action is impossible in axisymmetric systems*. This famous result is known as *Cowling's Theorem*.

Cowling's Theorem is one of many *antidynamo theorems*. Others state that $\mathbf{B}_T$ cannot be maintained in axisymmetry, that dynamo action is impossible from purely toroidal flows, and that dynamo action is impossible from plane two-dimensional motions. Together, they imply that dynamos require some ingredient that is symmetry breaking, like three-dimensionality. The step $\mathbf{B}_T \to \mathbf{B}_P$ must come from complex motions. So, dynamos may be possible, but they cannot be simple!

A relatively simple three-dimensional flow that can amplify magnetic field is called the "stretch, twist, fold" flow.[5] It is illustrated in Fig. 38.10.

The initial condition consists of a circular flux tube. It is then stretched to larger diameter. Since both the volume of, and the flux in, the tube must be conserved (the flow is incompressible), the stretching increases the magnitude of the magnetic field in the tube. The twist and fold stages return the tube to its original diameter, but with

---

[5] The "stretch-fold-twist" mechanism was first introduced by S. I. Vainshtain and Ya. B. Zeldovich, Sov. Phys. Usp. **15**, 159 (1972); it is discussed in detail by H. K. Moffatt and M. R. E. Proctor, J. Fluid. Mech. **154**, 493 (1985); who give a specific velocity field that will accomplish the desired deformation.

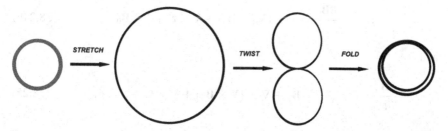

Fig. 38.10 "Stretch, twist, fold" flow leading to amplification of the magnetic field

a larger magnetic field. The field will increase each time the process is repeated. This leads to field amplification, albeit on a small length scale. We remark that this flow is quite contrived and unlikely to occur in nature exactly as diagrammed above. But it is kinematically possible, and that is all that matters in classical dynamo theory. One could conceive of flows like this occurring irregularly as a result of turbulence. Then, on the average, field amplification might occur.

The example of stretch, twist, fold flow implies that the axisymmetry required by Cowling's Theorem might be broken at short length scales if there is a small random (and therefore non-axisymmetric) component of the velocity field. As we did with turbulence (Lecture 36), we write the flow and magnetic field as

$$\mathbf{V}(\mathbf{r}, t) = \mathbf{V}_0(\mathbf{r}, t) + \tilde{\mathbf{V}}(\mathbf{r}, t) \tag{38.23}$$

and

$$\mathbf{B}(\mathbf{r}, t) = \mathbf{B}_0(\mathbf{r}, t) + \tilde{\mathbf{B}}(\mathbf{r}, t), \tag{38.24a}$$

where $f_0 = \langle f \rangle$ is an axisymmetric mean component and $\tilde{f}$ is a small, three-dimensional random component (i.e., $|\tilde{f}| << |f_0|$). Of course, $\langle \tilde{f} \rangle = 0$. The random component may arise from either turbulence or a superposition of waves. For example, in the case of a thermally stratified fluid in the presence of gravity, if $\mathbf{g} \cdot \nabla T > 0$ (strongly heated from below) the fluid is unstable (because of thermal expansion and buoyancy), and the random component will be turbulent; if $\mathbf{g} \cdot \nabla T > 0$ (so that the hot fluid is above the cold fluid), the configuration is stable and the random component will be the result of waves. If L is the scale of spatial variation for the mean quantities and l_0 is the scale of variation for the random component, then, for the case of turbulence, l_0 is the size of the energy containing eddies, and for the case of waves, l_0 is the wavelength.

The *ansatz* represented by Eqs. (38.23) and (38.24) implies a separation of scales in both space and time; that is, $l_0 << L$ and $l_0/\tilde{V} << L/V_0$. The resulting theory is called *mean-field dynamo theory*.

From our turbulence lectures (Lecture 36), we know that the mean and random components of the magnetic field evolve according to

$$\frac{\partial \mathbf{B}_0}{\partial t} = \nabla \times (\mathbf{V}_0 \times \mathbf{B}_0) + \nabla \times \mathcal{E} + \lambda \nabla^2 \mathbf{B}_0 \qquad (38.24\text{b})$$

and

$$\frac{\partial \tilde{\mathbf{B}}}{\partial t} = \nabla \times \left(\mathbf{V}_0 \times \tilde{\mathbf{B}}\right) + \nabla \times \left(\tilde{\mathbf{V}} \times \mathbf{B}_0\right) + \nabla \times \mathbf{G} + \lambda \nabla^2 \tilde{\mathbf{B}}, \qquad (38.25)$$

where

$$\mathcal{E} = \langle \tilde{\mathbf{V}} \times \tilde{\mathbf{B}} \rangle \qquad (38.26)$$

and

$$\mathbf{G} = \tilde{\mathbf{V}} \times \tilde{\mathbf{B}} - \langle \tilde{\mathbf{V}} \times \tilde{\mathbf{B}} \rangle. \qquad (38.27)$$

The goal of the theory will be to express $\mathcal{E}$ in terms of $\mathbf{V}_0$ and $\mathbf{B}_0$, i.e., we will look for a closure relation.

Suppose that $\tilde{\mathbf{B}} = 0$ at $t = 0$. Then, at $t = 0$,

$$\frac{\partial \tilde{\mathbf{B}}}{\partial t} = \nabla \times \left(\tilde{\mathbf{V}} \times \mathbf{B}_0\right), \qquad (38.28)$$

so that $\tilde{\mathbf{B}}$ is *linear* in $\mathbf{B}_0$. It follows that $\mathcal{E} = \langle \tilde{\mathbf{V}} \times \tilde{\mathbf{B}} \rangle$ is also linear in $\mathbf{B}_0$ (since $\tilde{\mathbf{B}}$ is linear in $\mathbf{B}_0$. We therefore anticipate a closure expression of the form

$$\mathcal{E}_i = \alpha_{ij} B_{0j} + \beta_{ijk} \frac{\partial B_{0j}}{\partial x_k} + \gamma_{ijkl} \frac{\partial^2 B_{0j}}{\partial x_k \partial x_l} + \cdots. \qquad (38.29)$$

It will be important to note that α_{ij}, β_{ijk}, γ_{ijkl}, etc., are pseudo-tensors, since $\mathcal{E}$ is a vector and $\mathbf{B}_0$ is a pseudo-vector. Since $\tilde{\mathbf{B}}$ depends on $\tilde{\mathbf{V}}$, $\mathbf{V}_0$, and λ (in addition to the linear dependence on $\mathbf{B}_0$), we expect that the α_{ij}, β_{ijk}, γ_{ijkl}, etc., will be completely determined by $\mathbf{V}_0$, λ, and the statistical properties of $\tilde{\mathbf{V}}$.

Now consider the case $\mathbf{V}_0 = $ constant. If we transform to a frame of reference moving with $\mathbf{V}_0$, then $\mathbf{V}_0 \times \mathbf{B}_0 \rightarrow 0$, and $\mathbf{B}_0$ evolves according to

$$\frac{\partial B_{0i}}{\partial t} = \varepsilon_{ijk} \frac{\partial}{\partial x_j} \left(\alpha_{kl} B_{0l} + \beta_{jlm} \frac{\partial B_{0l}}{\partial x_m} + \ldots \right) + \lambda \nabla^2 B_{0i}. \qquad (38.30)$$

If $\mathbf{B}_0$ is *weakly non-uniform*, then $\partial B_{0l}/\partial x_j >> \partial^2 B_{0l}/\partial x_j \partial x_m$ and the first term dominates. (However, β_{jlm} will still contribute to an "eddy resistivity" that will enhance λ, i.e., $\lambda_e = \lambda + \beta$. We will ignore this effect from now on, and concentrate on the effect of α.) The we can write

$$\mathcal{E}_i = \alpha_{ij} B_{0j}. \qquad (38.31)$$

We can decompose α_{ij} into symmetric and antisymmetric parts as

$$\alpha_{ij}^{(s)} = \frac{1}{2}\left(\alpha_{ij} + \alpha_{ji}\right) \tag{38.32}$$

and

$$\alpha_{ij}^{(A)} = \frac{1}{2}\left(\alpha_{ij} - \alpha_{ji}\right). \tag{38.33}$$

The antisymmetric part, Eq. (38.33), can be written as

$$\alpha_{ij}^{(A)} = -\varepsilon_{ijk}a_k, \tag{38.34}$$

where

$$a_k = -\frac{1}{2}\varepsilon_{lmk}\alpha_{lm}, \tag{38.35}$$

(you can work it out!) so that $\alpha_{ij} = \alpha_{ij}^{(s)} - \varepsilon_{ijk}a_k$. Then Eq. (38.31) becomes

$$
\begin{aligned}
\mathcal{E}_i &= \left(\alpha_{ij}^{(s)} - \varepsilon_{ijk}a_k\right)B_{0j}\\
&= \alpha_{ij}^{(s)}B_{0j} - \varepsilon_{ijk}B_{0j}a_k\\
&= \alpha_{ij}^{(s)}B_{0j} + (\mathbf{a} \times \mathbf{B}_0)_i,
\end{aligned} \tag{38.36}
$$

so that the antisymmetric part of α_{ij} merely contributes an addition to the mean velocity, i.e., $\mathbf{V}_0 \to \mathbf{V}_0 + \mathbf{a}$. Only the symmetric part contributes to the dynamo. We will therefore concentrate on the symmetric part of α_{ij}; from now on the notation α_{ij} will refer to its symmetric part.

When $\tilde{\mathbf{V}}$ is statistically *isotropic* and *homogeneous*, the statistical properties of $\tilde{\mathbf{V}}$ are invariant under rotation. In this case α_{ij} must also be isotropic, so that

$$\alpha_{ij} = \alpha\delta_{ij} \tag{38.37}$$

and $\mathbf{a} = 0$. Now, α must be a pseudo-scalar. That means that it must change sign under coordinate inversion $\mathbf{r}' = -\mathbf{r}$. Since α depends on the statistical properties of $\tilde{\mathbf{V}}$, this implies that $\tilde{\mathbf{V}}$ cannot be statistically invariant under inversions, i.e., α can be non-zero only if $\tilde{\mathbf{V}}$ lacks reflectional symmetry. This imposes another constraint on flows that can produce dynamo action.

If $\alpha \neq 0$, then $\mathcal{E} = \alpha\mathbf{B}_0$ and

$$\mathbf{J} = \sigma\mathcal{E} = \sigma\alpha\mathbf{B}_0, \tag{38.38a}$$

where $\sigma = 1/\eta$ is the electrical conductivity. The field $\mathcal{E} = \left\langle \tilde{\mathbf{V}} \times \tilde{\mathbf{B}} \right\rangle$ therefore produces a mean current that is *parallel* to $\mathbf{B}_0$. Without fluctuations $\mathbf{J} = \sigma\mathbf{V}_0 \times \mathbf{B}_0$, which is *perpendicular* to $\mathbf{B}_0$. The appearance of this parallel mean current is called the α-*effect*. Since mean toroidal current due to the fluctuations is $\mathbf{J}_T = \sigma\alpha\mathbf{B}_T$, and since $\nabla \times \mathbf{B}_P = \mu_0\mathbf{J}_T = \mu_0\sigma\alpha\mathbf{B}_T$, the mean toroidal field acts as a source for the poloidal field, and we have a way of closing the loop $\mathbf{B}_P \rightleftarrows \mathbf{B}_T$ and producing a dynamo. Now, all we need to do is to calculate α!

We have seen that $\tilde{\mathbf{V}}$ must lack reflectional symmetry in order to produce dynamo action. It turns out that this means that $\langle\tilde{\mathbf{V}} \cdot \tilde{\boldsymbol{\omega}}\rangle = 0$, where $\tilde{\boldsymbol{\omega}} = \nabla \times \tilde{\mathbf{V}}$ is the random vorticity. The quantity $\mathbf{V} \cdot \boldsymbol{\omega}$ is called the *kinetic helicity* (as opposed to the magnetic helicity). The physical picture is illustrated in Fig. 38.11.

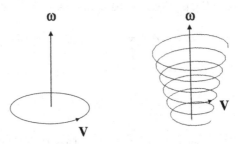

Fig. 38.11 *Left:* Rotating flow with no kinetic helicity. *Right:* Rotating flow with positive kinetic helicity

The flow on the left has vorticity but no kinetic helicity. The flow on the right has finite kinetic helicity. Therefore, the random flow field must have some twist, or handedness, in order to produce dynamo action.

We now proceed to calculate α. In order to compute $\mathcal{E} = \langle\tilde{\mathbf{V}} \times \tilde{\mathbf{B}}\rangle$, we must express $\tilde{\mathbf{B}}$ in terms of $\mathbf{B}_0$, λ, and the properties of $\tilde{\mathbf{V}}$. The fluctuating field satisfied Eq. (38.25). For the case $\mathbf{V}_0 = 0$, which is all we will consider here, this is

$$\frac{\partial\tilde{\mathbf{B}}}{\partial t} = \nabla \times (\tilde{\mathbf{V}} \times \mathbf{B}_0) + \nabla \times \mathbf{G} + \lambda\nabla^2\tilde{\mathbf{B}}. \qquad (38.38b)$$

The term containing $\mathbf{G} = \tilde{\mathbf{V}} \times \tilde{\mathbf{B}} - \langle\tilde{\mathbf{V}} \times \tilde{\mathbf{B}}\rangle$ will cause difficulty in solving this equation, and we look for circumstances under which it can be ignored. The magnitudes of the terms in this equation are, approximately,

$$\underbrace{\frac{\partial\tilde{\mathbf{B}}}{\partial t}}_{\tilde{B}/t_0} = \underbrace{\nabla \times (\tilde{\mathbf{V}} \times \mathbf{B}_0)}_{B_0\tilde{V}/l_0} + \underbrace{\nabla \times \mathbf{G}}_{\tilde{V}\tilde{B}/l_0} + \underbrace{\lambda\nabla^2\tilde{\mathbf{B}}}_{\lambda\tilde{B}/l_0^2}, \qquad (38.39)$$

where $\tilde{B} = \langle\tilde{\mathbf{B}}^2\rangle^{1/2}$, $\tilde{V} = \langle\tilde{\mathbf{V}}^2\rangle^{1/2}$, and l_0 and t_0 are, respectively, the length and time scales associated with the fluctuations. There are two cases of interest: (1) random waves, for which $\tilde{V}t_0/l_0 \approx \tilde{V}k_0/\omega_0 << 1$ (or $\tilde{V} << \omega_0/k_0$, the phase velocity of the waves); and, (2) turbulence, for which $\tilde{V}t_0/l_0 \approx 1$.

For the case of random waves, we have

$$\frac{|\nabla \times \mathbf{G}|}{|\partial \tilde{\mathbf{B}}/\partial t|} \approx \frac{\tilde{V}\tilde{B}}{l_0} \frac{t_0}{\tilde{B}} = \frac{\tilde{V}t_0}{l_0} << 1, \tag{38.40}$$

and the term $\nabla \times \mathbf{G}$ can be ignored, so that

$$\frac{\partial \tilde{\mathbf{B}}}{\partial t} = \nabla \times \left(\tilde{\mathbf{V}} \times \mathbf{B}_0 \right) + \lambda \nabla^2 \tilde{\mathbf{B}}. \tag{38.41}$$

For the case of turbulence, $\tilde{V}t_0/l_0 \approx 1$ so that formally the term $\nabla \times \mathbf{G}$ must be retained. However, both $\nabla \times \mathbf{G}$ and $\partial \tilde{\mathbf{B}}/\partial t$ can be neglected in comparison with $\lambda \nabla^2 \tilde{\mathbf{B}}$ if

$$\frac{|\partial \tilde{\mathbf{B}}/\partial t|}{|\lambda \nabla^2 \tilde{\mathbf{B}}|} \approx \frac{\tilde{B}}{t_0} \frac{l_0^2}{\lambda \tilde{B}} = \frac{\tilde{V}l_0}{\lambda} << 1, \tag{38.42}$$

where $\tilde{V} = l_0/t_0$. This is equivalent to $R_{m0} << 1$, where R_{m0} is the magnetic Reynolds' number at the l_0 length scale. In this case,

$$0 = \nabla \times \left(\tilde{\mathbf{V}} \times \mathbf{B}_0 \right) + \lambda \nabla^2 \tilde{\mathbf{B}}. \tag{38.43}$$

In both cases, $\tilde{\mathbf{B}}$ is generated from $\tilde{\mathbf{V}}$. For the case of turbulence (38.43), the process is instantaneous because of the dominance of diffusion. However, the solutions of Eq. (38.41) for random waves should approach the solutions of Eq. (38.43) in the limit $R_{m0} << 1$. Therefore, we can study the case of random waves and obtain the turbulent results in the limit $\lambda \to \infty$. These approximations are collectively called *first-order smoothing*.

An Aside on Some Statistical Definitions for Random Fields

The purpose here is to introduce the kinetic helicity spectrum, Eq. (38.53), which will appear in what follows. The details can be skipped without loss of continuity in the presentation, but are included here for completeness. For many more details, see G. K. Batchelor, *Theory of Homogeneous Turbulence*, Cambridge, 1953.

The Fourier transform pair for the velocity $\tilde{\mathbf{V}}$ is

$$\tilde{\mathbf{V}}(\mathbf{k}, \omega) = \frac{1}{(2\pi)^4} \iint \tilde{\mathbf{V}}(\mathbf{r}, t) e^{-i(\mathbf{k}\cdot\mathbf{r} - \omega t)} d\mathbf{r} dt \tag{38.44}$$

and

$$\tilde{\mathbf{V}}(\mathbf{r}, t) = \iint \tilde{\mathbf{V}}(\mathbf{k}, \omega) e^{i(\mathbf{k}\cdot\mathbf{r} - \omega t)} d\mathbf{k} d\omega. \tag{38.45}$$

Since $\tilde{\mathbf{V}}(\mathbf{r}, t)$ is real, $\tilde{\mathbf{V}}(-\mathbf{k}, -\omega) = \tilde{\mathbf{V}}^*(\mathbf{k}, \omega)$, and since $\nabla \cdot \tilde{\mathbf{V}}(\mathbf{r}, t) = 0$, $\mathbf{k} \cdot \tilde{\mathbf{V}}(\mathbf{k}, \omega) = 0$. The *correlation tensor* is defined as

$$R_{ij}(\boldsymbol{\xi}, \tau) = \langle V_i(\mathbf{r}, t) V_j(\mathbf{r} + \boldsymbol{\xi}, t + \tau) \rangle, \tag{38.46}$$

where $\langle F \rangle = \int F P(\mathbf{u}_1, \mathbf{u}_2, \ldots) d\mathbf{u}_1 d\mathbf{u}_2 \ldots$ is the probability average of F, and P is the joint probability distribution function (i.e., it is the probability of finding the variable $\mathbf{u}_1$ between $\mathbf{u}_1$ and $\mathbf{u}_1 + d\mathbf{u}_1$, *and* $\mathbf{u}_2$ between $\mathbf{u}_2 + d\mathbf{u}_2$, etc.).

Now consider the mean quantity

$$\langle \tilde{V}_i(\mathbf{k}, \omega)_i \, \tilde{V}_j^*(\mathbf{k}, \omega) \rangle =$$
$$\frac{1}{(2\pi)^8} \iint \iint \langle \tilde{V}_i(\mathbf{r}, t)_i \tilde{V}_j^*(\mathbf{r}, t) \rangle e^{-i(\mathbf{k}\cdot\mathbf{r} - \mathbf{k}'\cdot\mathbf{r}' - \omega t - \omega' t')} d\mathbf{r} d\mathbf{r}' dt dt'.$$
$$\tag{38.47}$$

Since

$$\iint e^{i(\mathbf{k}-\mathbf{k}')\cdot\mathbf{x}} e^{i(\omega-\omega')t} d\mathbf{x} dt = (2\pi)^4 \delta(\mathbf{k} - \mathbf{k}') \delta(\omega - \omega'), \tag{38.48}$$

we can write Eq. (38.47) as

$$\langle \tilde{V}_i(\mathbf{k}, \omega)_i \, \tilde{V}_j^*(\mathbf{k}, \omega) \rangle = \Phi_{ij}(\mathbf{k}, \omega) \delta(\mathbf{k} - \mathbf{k}') \delta(\omega - \omega'), \tag{38.49}$$

where $\Phi_{ij}(\mathbf{k}, \omega)$, which is called the *spectrum tensor* of $\tilde{\mathbf{V}}(\mathbf{r}, t)$, is the Fourier transform of the correlation tensor, Eq. (38.46). The spectrum tensor has the Hermitian property $\Phi_{ij}(\mathbf{k}, \omega) = \Phi_{ij}(-\mathbf{k}, -\omega) = \Phi_{ji}^*(\mathbf{k}, \omega)$, and since $\nabla \cdot \tilde{\mathbf{V}}(\mathbf{r}, t) = 0$, $k_i \Phi_{ij}(\mathbf{k}, \omega) = k_j \Phi_{ij}(\mathbf{k}, \omega) = 0$.

The *energy spectrum* is defined as

$$E(k, \omega) = \frac{1}{2} \int_{S_k} \Phi_{ii}(\mathbf{k}, \omega) \, dS, \tag{38.50}$$

where S_k is a sphere of radius k in $\mathbf{k}$-space. Also,

$$\frac{1}{2} \langle \tilde{V}^2 \rangle = \frac{1}{2} R_{ii}(0, 0)$$
$$= \frac{1}{2} \iint \Phi_{ii}(\mathbf{k}, \omega) d\mathbf{k} d\omega$$
$$= \iint E(k, \omega) d\mathbf{k} d\omega. \tag{38.51}$$

Therefore, $\rho E(k, \omega) dk d\omega$ is to be interpreted as the kinetic energy in the range between k and $k + dk$, and ω and $\omega + d\omega$; $E(k, \omega) > 0$ for all k and ω.

The vorticity is defined as $\tilde{\boldsymbol{\omega}} = \nabla \times \tilde{\mathbf{V}}$, its transform is $\tilde{\boldsymbol{\omega}}(\mathbf{k}, \omega) = i\mathbf{k} \times \tilde{\mathbf{V}}(\mathbf{k}, \omega)$, and its spectrum tensor is $\Omega_{ij}(\mathbf{k}, \omega) = \varepsilon_{imn}\varepsilon_{jpq}k_m k_p \Phi_{nq}(\mathbf{k}, \omega)$. Using the properties of Φ_{ij}, one finds $\Omega_{ii}(\mathbf{k}, \omega) = k^2\Phi_{ii}(\mathbf{k}, \omega)$, so that

$$\frac{1}{2}\langle \tilde{\omega}^2 \rangle = \iint k^2 E(k, \omega) dk d\omega. \tag{38.52}$$

The *kinetic helicity spectrum* is defined as

$$F(k, \omega) = i \int_{S_k} \varepsilon_{ikl} k_k \Phi_{il}(\mathbf{k}, \omega) dS. \tag{38.53}$$

Then

$$\langle \tilde{\mathbf{V}} \cdot \tilde{\boldsymbol{\omega}} \rangle = i\varepsilon_{ikl} \iint k_k \Phi_{il}(\mathbf{k}, \omega) d\mathbf{k} d\omega \tag{38.54}$$

$$= \iint F(k, \omega) dk d\omega. \tag{38.55}$$

From the above, it can be shown that $\langle \tilde{\mathbf{V}} \cdot \tilde{\boldsymbol{\omega}} \rangle = 0$ if $\tilde{\mathbf{V}}$ is reflectionally symmetric. We have seen that lack of reflectional symmetry is required for the α-effect, so $F(k, \omega) \neq 0$ is an important property of flow fields that can produce dynamo action.

Now let $\mathbf{B}_0$ be uniform and steady, and $\nabla \cdot \tilde{\mathbf{V}} = 0$. Then, with first-order smoothing, the induction equation is

$$\frac{\partial \tilde{\mathbf{B}}}{\partial t} - \lambda \nabla^2 \tilde{\mathbf{B}} = (\mathbf{B}_0 \cdot \nabla) \tilde{\mathbf{V}}, \tag{38.56}$$

which is an inhomogeneous equation for $\tilde{\mathbf{B}}$. Consider the case of a single wave

$$\tilde{\mathbf{V}}(\mathbf{r}, t) = \tilde{V}_0 \left[\sin(kz - \omega t)\hat{\mathbf{e}}_x + \cos(kz - \omega t)\hat{\mathbf{e}}_y \right], \tag{38.57}$$

with $\mathbf{k} = k\hat{\mathbf{e}}_z$ and $k > 0$, $\omega > 0$, and assume a solution of the form $\tilde{\mathbf{B}}(\mathbf{r}, t) = \mathrm{Re}\tilde{\mathbf{B}}_0 \exp[i(\mathbf{k} \cdot \mathbf{r} - \omega t)]$. Substituting into Eq. (38.56), we find

$$\tilde{\mathbf{B}}_0 = \frac{i\mathbf{k} \cdot \mathbf{B}_0}{-i\omega + \lambda k^2} \tilde{\mathbf{V}}_0, \tag{38.58}$$

so that

$$\tilde{\mathbf{B}}(\mathbf{r}, t) = \text{Re} \frac{i\mathbf{k} \cdot \mathbf{B}_0}{-i\omega + \lambda k^2} \tilde{\mathbf{V}}(\mathbf{r}, t) \tag{38.59}$$

or

$$\tilde{\mathbf{B}}(\mathbf{r}, t) = \frac{\mathbf{k} \cdot \mathbf{B}_0}{\omega^2 + \lambda^2 k^4} \left(-\omega \tilde{\mathbf{V}} + \lambda k \tilde{\mathbf{u}} \right), \tag{38.60}$$

where

$$\tilde{\mathbf{u}}(\mathbf{r}, t) = \tilde{V}_0 \left[\cos (kz - \omega t) \hat{\mathbf{e}}_x - \sin (kz - \omega t) \hat{\mathbf{e}}_y \right]. \tag{38.61}$$

Notice that, in Eq. (38.60), the diffusivity λ introduces a phase shift between $\tilde{\mathbf{B}}$ and $\tilde{\mathbf{V}}$.

The computation of $\mathcal{E} = \langle \tilde{\mathbf{V}} \times \tilde{\mathbf{B}} \rangle$ is now straightforward. The result for α_{ij} is $\alpha_{ij} = \alpha^{(3)} \delta_{i3} \delta_{j3}$, where

$$\alpha^{(3)} = -\frac{\lambda \tilde{V}_0^2 k^3}{\omega^2 + \lambda^2 k^4}. \tag{38.62}$$

Notice the following: (1) The tensor α_{ij} is anisotropic, i.e., the only non-zero component is α_{33}; it knows about the direction of the wave. (2) $\alpha_{ij} \to 0$ as $\lambda \to 0$, so that some diffusion is necessary for the α-effect. This is related to the phase shift noted above. Without this phase shift, we would have $\tilde{\mathbf{V}} \times \tilde{\mathbf{B}} = 0$, and no possibility of a dynamo. (3) $\mathcal{E} = \langle \tilde{\mathbf{V}} \times \tilde{\mathbf{B}} \rangle$ is uniform in space, so that $\mathbf{G} = \tilde{\mathbf{V}} \times \tilde{\mathbf{B}} - \langle \tilde{\mathbf{V}} \times \tilde{\mathbf{B}} \rangle = 0$, i.e., first-order smoothing is exact in this case. This will not be true in general.

The case of a superposition of random wave is similar, but requires more calculation. One must now take the Fourier transform of Eq. (38.56), solve, and then invert the transform and compute $\mathcal{E} = \langle \tilde{\mathbf{V}} \times \tilde{\mathbf{B}} \rangle$. The result, after a long calculation, is $E_i = \alpha B_{0i}$ with

$$\alpha = -\frac{1}{3} \lambda \iint \frac{k^2 F(k, \omega)}{\omega^2 + \lambda^2 k^4} dk d\omega, \tag{38.63}$$

where $F(k, \omega)$ is the kinetic helicity spectrum defined in Eq. (38.55). As anticipated, the α-effect requires a flow field with $\langle \tilde{\mathbf{V}} \cdot \tilde{\boldsymbol{\omega}} \rangle \neq 0$.

We remark on the role played by dissipation (i.e., λ) in dynamo action. It seems intuitive that too much dissipation will simply make everything diffuse away, so that the dynamo will cease to operate. On the other hand, we have just seen [see Eqs. (38.62, 38.63)] that *some* diffusion is necessary for the α-effect; without it, dynamo action is impossible. Of further interest is the rate γ at which the magnetic field is generated by the dynamo. Heuristically, we expect $\gamma \sim \lambda^s$, where $0 \le s \le 1$. Dynamos for which $0 < s \le 1$ are called *slow dynamos*, since their rate depends on the small dissipation λ. An example is the α-effect. Most of the known dynamo mechanisms are of this type. A dynamo for which γ is independent of λ (i.e., $s = 0$) is called a *fast dynamo*. These dynamos are of special interest because they may be required to account for magnetic field generation in astrophysical settings where λ is extremely small. For example, the 22-year reversal cycle of the solar magnetic field occurs much faster that resistive diffusion. Unfortunately, except for some contrived examples, not many fast dynamos have been found theoretically. An example is the "stretch, twist, fold" flow shown in Fig. 38.10. However, continuation of the process for a large number of steps will lead to magnetic field structure on a very fine spatial scale. If the system possesses any dissipation at all (as all real systems do), its presence will eventually be felt and may affect the final rate of field generation.[6] The *quantitative* role of diffusion in allowing or inhibiting dynamos remains an unresolved issue.

We now consider the evolution of the axisymmetric mean fields in cylindrical coordinates. We write $\mathbf{V} = R\omega(R, z)\hat{\mathbf{e}}_\phi + \mathbf{V}_P$ and $\mathbf{B} = B(R, z)\hat{\mathbf{e}}_\phi + \mathbf{B}_P$, where $\mathbf{B}_P = \nabla \times [A(R, z)\hat{\mathbf{e}}_\phi]$. With $\mathbf{E} = \alpha\mathbf{B}$, Eq. (38.24) becomes the pair of equations

$$\frac{\partial B}{\partial t} + R\mathbf{V}_P \cdot \nabla \left(\frac{B}{R}\right) = R(\mathbf{B}_P \cdot \nabla\omega) + \hat{\mathbf{e}}_\phi \cdot \nabla \times (\alpha\mathbf{B}_P)$$
$$+ \lambda\left(\nabla^2 - \frac{1}{R^2}\right)B \qquad (38.64)$$

and

$$\frac{\partial A}{\partial t} + \frac{1}{R}\mathbf{V}_P \cdot \nabla(RA) = \alpha B + \lambda\left(\nabla^2 - \frac{1}{R^2}\right)A. \qquad (38.65)$$

The toroidal field B has two sources, given by the first two terms on the right-hand side of Eq. (38.64). The first term, $R(\mathbf{B}_P \cdot \nabla\omega)$, is due to differential rotation in the mean flow. The second term, $\hat{\mathbf{e}}_\phi \cdot \nabla \times (\alpha\mathbf{B}_P)$, represents the generation of toroidal field directly from the poloidal field as a result of the α-effect. The poloidal field, Eq. (38.65), has a single source term, αB, which comes directly from the toroidal field and the α-effect. This reconfirms Cowling's theorem that, without the symmetry breaking due to three-dimensional motions (as encapsulated in α), dynamo action is impossible.

---

[6] The magnetic helicity $K = \int \mathbf{A} \cdot \mathbf{B}dV$ is conserved in ideal MHD. This implies that the magnetic field cannot be amplified in a fluid that has net magnetic helicity, i.e., $K \neq 0$. Breaking the constraint of helicity conservation requires dissipation. It can be shown that this is true even if $K = 0$. Therefore, even fast dynamos must be dependent on dissipation, even if the overall rate of magnetic energy production is independent of it. These are called *diffusive fast dynamos*. An example is the stretch–fold–twist flow. For more discussion, see H. K. Moffatt and M. R. E. Proctor, J. Fluid. Mech. **154**, 493 (1985).

The ratio of the two sources of the toroidal field is

$$\frac{|R\,(\mathbf{B_P}\cdot\nabla\omega)|}{|\hat{\mathbf{e}}_\phi\cdot\nabla\times(\alpha\mathbf{B_P})|} \approx \frac{L B_P\,|\nabla\omega|}{|\alpha_0|\,B_P/L} = \frac{L^2\,|\nabla\omega|}{|\alpha_0|}, \tag{38.66}$$

where α_0 is some characteristic value of α. If $|\alpha_0| \gg L^2\,|\nabla\omega|$, then the effect of differential rotation is negligible and the only source of toroidal field is α. Since α is also the source of poloidal field, these dynamos are called α^2-dynamos. Conversely, if $|\alpha_0| \ll L^2\,|\nabla\omega|$, then differential rotation acts as a source of the toroidal field, and the poloidal field is generated by α. These dynamos are called $\alpha - \omega$ dynamos. They may be important in astrophysical systems that are dominated by rotation.

Now, almost all of the preceding theory has depended on the assumption of scale separation, as embodied in Eqs. (38.23) and (38.24). Similar assumptions were required to make progress in theories of turbulence; see Lecture 36. However, in that lecture we saw that MHD turbulence may not be self-similar, with the implication that the system may evolve toward an ensemble of small-scale, spiky, patchy structures. In essence, the characteristic scale length of the system, L, tends to become smaller and smaller as the turbulence develops. This throws into doubt the assumption of separation of scales required to make theoretical progress in the first place. This remains an unresolved issue. The development of classical dynamo theory should be viewed with this situation in mind.

So, what *really* happens at small scales? Can small-scale turbulence *really* generate and sustain large-scale fields? In recent years computers have become just barely powerful enough to begin to answer this question through direct numerical simulation (DNS) of the MHD equations. One starts with a small, random $\tilde{\mathbf{V}}$ and $\tilde{\mathbf{B}}$ fields with $\langle\tilde{\mathbf{V}}\cdot\tilde{\omega}\rangle \neq 0$, integrates the resistive MHD equations forward in time with $R_m \gg 1$ ($\sim 10^3$), including the back-reaction, and looks for field amplification. The result is that there *is* a dynamo due to the α-effect, i.e., the initial magnetic energy is amplified. However, this energy is found to inevitably saturate, or stop growing, at a low amplitude. Further, it seems that the process generates more $\langle B^2\rangle$ than it does $\langle B\rangle^2$, i.e., it can generate energy but not mean flux.[7] It appears that the way to generate and sustain mean flux is to have some around in the initial state, which begs the question of how it got there in the first place. At the present time it does not appear that small-scale turbulence alone is sufficient to explain the sustainment and dynamics of solar and other magnetic fields.

Nonetheless, DNS *does* seem to do a pretty good job of reproducing the global properties of the geo-dynamo.[8] The large-scale field is sustained against resistive diffusion and demonstrates reversals of polarity. People have also applied DNS to the global sun (as opposed to a small slice of it). They get a dynamical field, but

[7] S. I. Vainshtein and F. Cattaneo, Ap. J. **393**, 165 (1992); F. Cattaneo and D. W. Hughes, Phys. Rev. E **54**, 4232 (1996); S. Boldyrev, F. Cattaneo, and R. Rosner, Phys. Rev. Lett. **95**, 255001 (2005); F. Cattaneo and D. F. Hughes, Fluid Mech. **553**, 401 (2006).

[8] G. A. Glatzmaier and P. H. Roberts, Nature **377**, 203 (1995).

not detailed agreement with observations.[9] For example, they cannot reproduce the butterfly diagram.

What about laboratory plasmas? Historically, some sort of dynamo mechanism has been invoked to explain the generation and sustainment of the toroidal flux in the RFP. But what really happens? The RFP is a driven system. There is an applied toroidal voltage that drives the current. This voltage (or toroidal electric field) constantly supplies poloidal flux to the system (i.e., it drives current). As we saw in Lecture 37, this tends to drive the system away from its preferred, relaxed condition, resulting in the destabilization of long-wavelength, low-frequency (i.e., $m = 1$) MHD modes. The nonlinear interaction of these modes produces a mean parallel electric field $\mathcal{E} = \langle \tilde{\mathbf{V}} \times \tilde{\mathbf{B}} \rangle$ that tends to suppress parallel current in the core and drive parallel current in the edge. This is precisely what is needed to produce toroidal flux. This is an α-effect that accounts the first part of the dynamo loop, $\mathbf{B}_P \rightarrow \mathbf{B}_T$. Results from numerical simulation of this process are shown in Fig. 38.12. (E_f is the electric field due to the α-effect).

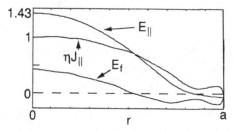

Fig. 38.12 Parallel components of the total, Ohmic, and fluctuating electric fields in an RFP [10]

But, as far as I know, there is no evidence of the second half of the loop, $\mathbf{B}_T \rightarrow \mathbf{B}_P$. Instead, $\mathbf{B}_T$ is simply dissipated resistively.

Magnetic energy comes from the external circuitry into the discharge as poloidal flux. It is converted into toroidal flux by the α-effect produced by the nonlinear interaction of long-wavelength MHD instabilities. Small-scale turbulence seems to play little or no role in this process. The toroidal flux is then dissipated by resistivity. If one turns off the driving force (the toroidal voltage), the discharge terminates immediately. In the end, the RFP seems to exhibit flux transport and conversion as an aspect of the relaxation process, but it does not appear to fit the definition of a classical dynamo.

Of course, both the geo-dynamo and the solar dynamo also occur in driven systems; in both cases the drive is due to thermal convection as a result of strong heating from below. Perhaps similar processes as are observed in the RFP also occur throughout the universe.

---

[9] See, for example, M. S. Miesch, Journal of Physics: Conf. Series **118**, 012031 (2008), and references therein.

[10] Y. L. Ho, S. C. Prager, and D. D. Schnack, Phys. Rev. Lett. **62**, 1504 (1989).

Appendix

In April 2006, I gave three lectures at UW Madison on extended Magnetohydrodynamics (MHD). These were the last three lectures of the semester for the course in MHD. In doing so, I prepared a set of notes for the students, which are reproduced here as an Appendix. The goal was to show where MHD comes from (what came before) and how to extend the MHD model to incorporate additional physical effects (what comes after), hence *extended MHD*. They are meant to be self-contained and do not refer directly to any of the lectures in the body of this volume. I hope you find them interesting and not too confusing.

Schnack, D.D.: *Appendix*. Lect. Notes Phys. **780**, 283–317 (2009)
DOI 10.1007/978-3-642-00688-3_BM2   © Springer-Verlag Berlin Heidelberg 2009

Fluid Models of Magnetized Plasmas

1 Introduction

You are now all experts in MHD. As you know, ideal MHD describes the dynamics of a perfectly conducting fluid in the presence of electric and magnetic fields. The fluid obeys the usual laws of hydrodynamics, with the addition of the Lorentz body force. The electromagnetic fields obey the "pre-Maxwell" equations (i.e., the laws of electrodynamics as they were known before the work of James Clerk Maxwell). These two systems are coupled by "Ohm's law," which states that the electric field vanishes in a frame of reference moving with the fluid. The electromagnetic fields affect the motion of the fluid, and the fluid in turn modifies the electromagnetic fields. The ideal MHD equations are summarized in Table 1.

Table 1 Ideal MHD Equations

Fluid	Ohm	Field
$\partial n/\partial t = -\nabla \cdot (n\mathbf{V})$	$\mathbf{E} = -\mathbf{V} \times \mathbf{B}$	$\partial \mathbf{B}/\partial t = -\nabla \times \mathbf{E}$
$Mnd\mathbf{V}/dt = -\nabla P + \mathbf{J} \times \mathbf{B}$		$\mu_0 \mathbf{J} = \nabla \times \mathbf{B}$
$dP/dt = -\gamma P \nabla \cdot \mathbf{V}$		

The fluid equations deal with the macroscopic properties of the fluid. That is, all physical quantities are "averaged over elements of volume which are 'physically infinitesimal', ignoring the microscopic variations of the quantities which result from the molecular structure of matter." This is generally a good approximation for fluids such as water or gasses such as air. Interparticle collisions, which in these cases occur on length scales (the "mean-free path") that are much smaller than any "physically infinitesimal" volume element, tend to "average out" the effects of individual particle motion. However, in low-density, strongly magnetized plasmas, the mean-free path between particle collisions can be comparable to the macroscopic length scales, and we all know that these plasmas often exhibit macroscopic properties that reflect the fact that the medium is made up of individual charged particles (ions and electrons). Examples are the cyclotron gyration of charged particles in a magnetic field and the average drift of these particle orbits relative to the field.

These issues raise the following questions:

1. In what sense can a low-density magnetized plasma be modeled as a fluid?
2. Are there other fluid descriptions besides ideal MHD?
3. What happened to the electric charge density?
4. What happened to the displacement current?
5. What is "Ohm's law," and where does it come from?
6. Where does MHD fit into this picture?

In what follows I will attempt to address these issues.

2 Models with Reduced Degrees of Freedom

A plasma consists of N individual particles. Each particle has a position and a velocity, and each obeys the laws of Hamiltonian mechanics, i.e.,

$$\dot{\mathbf{q}}_k = \frac{\partial H}{\partial \mathbf{p}_k}, \tag{1}$$

$$\dot{\mathbf{p}}_k = -\frac{\partial H}{\partial \mathbf{q}_k}, \tag{2}$$

$$\dot{E}_k = \frac{\partial H}{\partial t}, \tag{3}$$

for $k = 1, 2, \ldots, N$, where $\mathbf{q}_k$ are the generalized coordinates of the particles, $\mathbf{p}_k$ are the conjugate momenta, E_k is the energy, and $H(\mathbf{q}_1, \mathbf{q}_2, \ldots, \mathbf{q}_N, \mathbf{p}_1, \mathbf{p}_2, \ldots, \mathbf{p}_N, t)$ is the *Hamiltonian* for the system. If the Hamiltonian is independent of time, then the energy is constant. For point particles, there are $6N$ degrees of freedom (dependent variables).

The system, as defined by a point in the $6N$-dimensional phase space $(\mathbf{q}_k, \mathbf{p}_k)$, evolves in time along a precise trajectory according to Eqs. (1), (2), and (3). In principle, once the Hamiltonian is known, the state of the system at any time t can be determined by integrating Eqs. (1), (2), and (3) forward in time from its state at $t = 0$. The problem with this approach is twofold. First, there is a tremendous amount of information required to describe the system. Second, in any practical sense, it is not possible to know the initial conditions with sufficient accuracy to make this a useful procedure. It has proven useful to seek a statistical approach.

Instead of describing the system by its precise coordinates in the $6N$-dimensional phase space defined by the variables $(\mathbf{q}_k, \mathbf{p}_k)$, we consider an *ensemble* of all possible realizations of the system in phase space (corresponding to all possible initial conditions of the system), and we make the a priori assumption that, in its motion through phase space in time, an individual system will pass arbitrarily close to all points in phase space consistent with any constraints on the system (such as constant energy, for example). Then we might expect that the average properties of a realization of a single system over time will be equivalent to the average of an ensemble of such systems over phase space. This is called the *ergodic hypothesis* and is the basis for statistical mechanics. (Another related hypothesis is sometimes called equal a proiri probability for different regions in phase space; the system is just as likely to be found at one point as another.) We can then define the function

$$F_N(\mathbf{q}_1, \mathbf{q}_2, \ldots, \mathbf{q}_N, \mathbf{p}_1, \mathbf{p}_2, \ldots, \mathbf{p}_N, t)d\mathbf{q}_1 d\mathbf{q}_2 \ldots d\mathbf{q}_N d\mathbf{p}_1 d\mathbf{p}_2 \ldots d\mathbf{p}_N dt \tag{4}$$

as the probability of finding particle 1 with position between $\mathbf{q}_1$ and $\mathbf{q}_1 + d\mathbf{q}_1$ and momentum between $\mathbf{p}_1$ and $\mathbf{p}_1 + d\mathbf{p}_1$, particle 2 with position between $\mathbf{q}_2$ and $\mathbf{q}_2 + d\mathbf{q}_1$ and momentum between $\mathbf{p}_2$ and $\mathbf{p}_2 + d\mathbf{p}_2$, and ... particle N with position between $\mathbf{q}_N$ and $\mathbf{q}_N + d\mathbf{q}_N$ and momentum between $\mathbf{p}_N$ and $\mathbf{p}_N + d\mathbf{p}_N$, at a time between t and $t + dt$. Clearly, $\int F_N(\mathbf{q}, \mathbf{p})d^N\mathbf{q}d^N\mathbf{p} = 1$. This description does not reduce the number of degrees of freedom, but it does eliminate the need for precise knowledge of the initial conditions. The time evolution of F_N is given by *Liouville's theorem*,

$$\frac{dF_N}{dt} = \frac{\partial F_N}{\partial t} + \sum_{k=1}^{N}\left[\dot{\mathbf{q}}_k \cdot \frac{\partial F_N}{\partial \mathbf{q}_k} + \dot{\mathbf{p}}_k \cdot \frac{\partial F_N}{\partial \mathbf{p}_k}\right] = 0, \tag{5}$$

so that F_N remains constant as a volume element moves about in phase space. The validity of Liouville's theorem only depends on the following:

1. The ergodic hypothesis is true.
2. The number of systems N remains constant.
3. The system obeys the laws of Hamiltonian dynamics (Eqs. 1, 2, 3).

While the phase space of Liouville's theorem still has $6N$ degrees of freedom, the fact that it encapsulates the dynamics of a volume element of phase space opens the door to formulations that greatly reduce the amount of information required to describe the system. For example, it is possible to define a single-particle distribution function by integrating F_N over the positions and coordinates of all the other particles:

$$F_1(\mathbf{q}_1, \mathbf{p}_1, t)d\mathbf{q}_1 d\mathbf{p}_1 dt = \int\int \cdots \int F_N(\mathbf{q}_1, \mathbf{q}_2, \cdots, \mathbf{q}_N, \mathbf{p}_1, \mathbf{p}_2, \ldots, \mathbf{p}_N, t)$$
$$\times d\mathbf{q}_2 \ldots d\mathbf{q}_N d\mathbf{p}_2 \ldots d\mathbf{p}_N dt. \tag{6}$$

This is the probability of finding particle 1 between $\mathbf{q}_1$ and $\mathbf{q}_1 + d\mathbf{q}_1$ with momentum between $\mathbf{p}_1$ and $\mathbf{p}_1 + d\mathbf{p}_1$, at a time between t and $t + dt$, for all possible configurations of the remaining particles. (There is an implicit assumption that the particles are identical, so that, in place of "particle," we can say "a single particle.")

It is now convenient, and consistent with convention, to define the *one-particle distribution function* for a species of type α ($=e, i$, for ions and electron), $f_\alpha^{(1)}$, as

$$n_\alpha f_\alpha^{(1)}(\mathbf{q}_1, \mathbf{p}_1, t) = N_\alpha F_1(\mathbf{q}_1, \mathbf{p}_1, t), \tag{7}$$

where $n_\alpha = N_\alpha/V$, N_α is the number of particles of species α, and $V = \int d^N q$ is the system volume. In what follows, we also make the association $\mathbf{q} \to \mathbf{x}$, and $\mathbf{p} \to M\mathbf{v}$, so that $\dot{\mathbf{q}} \to \mathbf{v}$ and $\dot{\mathbf{p}} \to \mathbf{F}$, where $\mathbf{F} = M\mathbf{a}$ is the force (due to both external application and interparticle interactions), and $\mathbf{a}$ is the acceleration. Then a dynamical equation for $f_\alpha^{(1)}$ is found by applying the averaging procedure defined in Eq. (6) directly to Liouville's theorem, Eq. (5). After some algebra, the result is

$$\frac{df_\alpha^{(1)}}{dt} = \frac{\partial f_\alpha^{(1)}}{\partial t} + \mathbf{v}_1 \cdot \frac{\partial f_\alpha^{(1)}}{\partial \mathbf{x}_1} + \mathbf{a}_1 \cdot \frac{\partial f_\alpha^{(1)}}{\partial \mathbf{v}_1} = \left(\frac{\partial f_\alpha^{(1)}}{\partial t}\right)_c, \qquad (8)$$

where

$$\left(\frac{\partial f_\alpha^{(1)}}{\partial t}\right)_c = -\sum_\beta \int \left(\mathbf{a}_{1\beta} - \langle \mathbf{a}_{1\beta}^{int} \rangle\right) \cdot \frac{\partial f_{\alpha\beta}^{(2)}}{\partial t} d\mathbf{x}_\alpha d\mathbf{x}_\beta. \qquad (9)$$

Here, $\mathbf{a}_1$ is the *total* acceleration felt by particle 1 due to *all* forces (both external and internal), $\mathbf{a}_{1\beta}$ is the *total* acceleration felt by particle 1 due to all other particles of species β, and $\langle \mathbf{a}_{1\beta}^{int} \rangle$ is the *average* acceleration felt by particle 1 due to all other particles of species β. The difference between the last two quantities is therefore the acceleration due to nearest neighbor interactions ("collisions"), and Eq. (9) is a measure of the *binary collision rate*. Equation (8) is called the *kinetic equation*.

Note that Eq. (9) depends on $f_{\alpha\beta}^{(2)}(\mathbf{x}_1, \mathbf{x}_\beta, \mathbf{v}_1, \mathbf{v}_\beta, t)$, the *two-body distribution function*, defined in a manner analogous to Eq. (6). This is unknown. An equation for it can be found by integrating Liouville's theorem this time over the coordinates and momenta of particles 3 through N. The result is an equation analogous to Eq. (8), but with the right-hand side depending on $f^{(3)}$, the three-body distribution function. In this way, one can generate a sequence of n equations for the first n distribution functions that will ultimately depend on the $(n + 1)$-body distribution function. There is always one more unknown than equations; the system is never *closed*. Obtaining a solvable system of equations requires obtaining a relationship that expresses the $(n + 1)$-body distribution function in terms of the previous n-body (or lower) distribution functions. Such an expression is called a *closure relation*. Finding suitable closures is one of the primary tasks of theoretical plasma physics.

For example, one approximation is to write the two-body distribution function as the product of one-body distribution functions, i.e., $f_{\alpha\beta} = f_\alpha f_\beta$. This ignores two-body correlations. Since the plasma state is dominated by long range forces, in many cases this is an excellent approximation. The *collision operator* is then defined as

$$\left(\frac{\partial f_\alpha^{(1)}}{\partial t}\right)_c \equiv \sum_\beta C_{\alpha\beta}(f_\alpha, f_\beta). \qquad (10)$$

Specific expressions for $C_{\alpha\beta}$ are derived by kinetic theory. These often appear as differential operators, and specific expressions are not of interest here. However, we note that, for a plasma in local thermodynamic equilibrium, the distribution function is Maxwellian and $df_M/dt = 0$, so that we require that $C(f_M) = 0$.

For a system of charged particles, such as plasma, the total force is just the sum of the electric and magnetic forces. The *plasma kinetic equation* (PKE) is therefore written as

$$\frac{\partial f_\alpha}{\partial t} + \mathbf{v} \cdot \frac{\partial f_\alpha}{\partial \mathbf{x}} + \left(\frac{q_\alpha}{M_\alpha}\right) \langle \mathbf{E} + \mathbf{v} \times \mathbf{B} \rangle \cdot \frac{\partial f_\alpha}{\partial \mathbf{v}} = \sum_\beta C_{\alpha\beta}(f_\alpha, f_\beta), \qquad (11)$$

where $\langle \mathbf{E} \rangle$ and $\langle \mathbf{B} \rangle$ are the sum of the external and average internal fields. They satisfy Maxwell's equations

$$\nabla \cdot \langle \mathbf{E} \rangle = \langle \rho_q \rangle / \epsilon_0, \qquad (12)$$

$$\frac{\partial \langle \mathbf{B} \rangle}{\partial t} = -\nabla \times \langle \mathbf{E} \rangle, \qquad (13)$$

and

$$\nabla \times \langle \mathbf{B} \rangle = \mu_0 \langle \mathbf{J} \rangle + \frac{1}{c^2} \frac{\partial \langle \mathbf{E} \rangle}{\partial t}, \qquad (14)$$

where $\langle \rho_q \rangle$ and $\langle \mathbf{J} \rangle$ are the average charge and current densities. These equations must be solved simultaneously with Eq. (11).

Going from the dynamic description of individual, interacting particles (1, 2, 3) to the statistical description (11) represents a tremendous reduction ($6N \to 6$) in the number of degrees of freedom needed to describe the evolution of the system. [In doing so we have also lost a tremendous amount of information about the system ($\sim 6(N-1)$). In particular, we have lost all information resulting from direct two-body (and higher) interactions. Even though the PKE is a very good approximation, it is still just an approximation.] Unfortunately, the coupled system consisting of Eqs. (11, 12, 13, 14) can still only be solved in the simplest circumstances, and the validity of these solutions is often restricted to very short times and quite small distances. To make progress, we require an even further reduction of the theoretical model.

3 Moment (Transport) Equations

The plasma kinetic equation provides a statistical description of plasma dynamics in phase space. Recall that, corresponding to any probability distribution $f_\alpha(\mathbf{x}, \mathbf{v}, t)$, the *expectation value*, or average, of any function of the velocity is $\langle g(\mathbf{x}, t) \rangle = \int g(\mathbf{x}, \mathbf{v}, t) f_\alpha(\mathbf{x}, \mathbf{v}, t) d^3 v$. It is therefore convenient to introduce the concept of *velocity moments* of the distribution function, which provide the local (in space) average value of familiar (or theoretically useful) physical quantities. In particular, we define the following:
The *density*:

$$n_\alpha(\mathbf{x}, t) \equiv \int f_\alpha(\mathbf{x}, \mathbf{v}, t) d^3 v. \qquad (15)$$

The *velocity*:

$$n_\alpha \mathbf{V}_\alpha(\mathbf{x}, t) \equiv \int \mathbf{v} f_\alpha(\mathbf{x}, \mathbf{v}, t) d^3 v. \qquad (16)$$

The *temperature* and *pressure*:

$$n_\alpha T = P_\alpha \equiv \int \frac{1}{2} M_\alpha v'^2 f_\alpha(\mathbf{x}, \mathbf{v}, t) d^3 v', \tag{17}$$

where $\mathbf{v}' = \mathbf{v} - \mathbf{V}_\alpha$ is the "random" component of the velocity.
The *heat flux*:

$$\mathbf{q}_\alpha \equiv \int \frac{1}{2} M_\alpha v'^2 \mathbf{v}' f_\alpha(\mathbf{x}, \mathbf{v}, t) d^3 v'. \tag{18}$$

The *stress tensor*:

$$\Pi_\alpha \equiv \int M_\alpha \left[\mathbf{v}'\mathbf{v}' - \frac{1}{3} v'^2 \mathbf{I} \right] f_\alpha(\mathbf{x}, \mathbf{v}, t) d^3 v'. \tag{19}$$

The *total pressure tensor*:

$$\mathbf{P}_\alpha \equiv \int M_\alpha \mathbf{v}'\mathbf{v}' f_\alpha(\mathbf{x}, \mathbf{v}, t) d^3 v' = P_\alpha \mathbf{I} + \Pi_\alpha. \tag{20}$$

The *total stress tensor*:

$$\mathsf{P}_\alpha \equiv \int M_\alpha \mathbf{v}'\mathbf{v}' f_\alpha(\mathbf{x}, \mathbf{v}, t) d^3 v' = M_\alpha n_\alpha \mathbf{V}_\alpha \mathbf{V}_\alpha + \mathbf{P}_\alpha. \tag{21}$$

The *total energy flux*:

$$\mathbf{Q}_\alpha \equiv \int \frac{1}{2} M_\alpha v^2 \mathbf{v} f_\alpha(\mathbf{x}, \mathbf{v}, t) d^3 v = \mathbf{q}_\alpha + \left(\frac{5}{2} P_\alpha + \frac{1}{2} M_\alpha n_\alpha V_\alpha^2 \right) \mathbf{V}_\alpha + \mathbf{V}_\alpha \cdot \Pi_\alpha. \tag{22}$$

The *energy-weighted total stress tensor*:

$$\mathsf{r}_\alpha \equiv \int \frac{1}{2} M_\alpha v^2 \mathbf{v}\mathbf{v} f_\alpha(\mathbf{x}, \mathbf{v}, t) d^3 v, \tag{23}$$

and so on.

It is also convenient to define the following velocity moments of the collision operator:
The *frictional force*:

$$\mathbf{F}_{1\alpha} \equiv \int M_\alpha \mathbf{v}' C_\alpha(f_\alpha) d^3 v'. \tag{24}$$

The *heat frictional force*:

$$\mathbf{F}_{2\alpha} \equiv \int M_\alpha \mathbf{v} \left[\frac{1}{2} M_\alpha v^2 - \frac{5}{2} \right] C_\alpha(f_\alpha) d^3 v. \tag{25}$$

The *heating*:

$$Q_\alpha = \int \frac{1}{2} M_\alpha v'^2 C_\alpha(f_\alpha) d^3 v. \tag{26}$$

Equations for the time evolution of these moments can be found by taking successive velocity moments of the PKE, Eq. (11). After considerable algebra, these are found to be as follows:

Density (or mass) conservation:

$$\frac{\partial n_\alpha}{\partial t} + \nabla \cdot (n_\alpha \mathbf{V}_\alpha) = 0. \tag{27}$$

Momentum conservation:

$$M_\alpha n_\alpha \frac{d\mathbf{V}_\alpha}{dt} = n_\alpha q_\alpha (\mathbf{E} + \mathbf{V}_\alpha \times \mathbf{B}) - \nabla P_\alpha - \nabla \cdot \Pi_\alpha + \mathbf{F}_{1\alpha}, \tag{28}$$

where $d\mathbf{V}/dt = \partial \mathbf{V}/\partial t + \mathbf{V} \cdot \nabla \mathbf{V}$.

Energy conservation:

$$\frac{\partial}{\partial t}\left(\frac{3}{2}P_\alpha + \frac{1}{2}M_\alpha n_\alpha V_\alpha^2\right) + \nabla \cdot \mathbf{Q}_\alpha = (n_\alpha q_\alpha \mathbf{E} + \mathbf{F}_{1\alpha}) \cdot \mathbf{V}_\alpha + Q_\alpha. \tag{29}$$

Energy weighted momentum conservation:

$$\frac{\partial \mathbf{Q}_\alpha}{\partial t} = \frac{q_\alpha}{M_\alpha}\left\{\mathbf{E} \cdot \left[\mathbf{P}_\alpha + M_\alpha n_\alpha \mathbf{V}_\alpha \mathbf{V}_\alpha + \left(\frac{3}{2}P_\alpha + \frac{1}{2}M_\alpha n_\alpha V_\alpha^2\right)\mathbf{I}\right] + \mathbf{Q}_\alpha \times \mathbf{B}\right\}$$
$$- \nabla \cdot \mathbf{r}_\alpha + \frac{T_\alpha}{M_\alpha}\left(\frac{5}{2}\mathbf{F}_{1\alpha} + \mathbf{F}_{2\alpha}\right). \tag{30}$$

Time-dependent equations for the higher moments Π_α, $\mathbf{q}_\alpha$, $\mathbf{r}_\alpha$, etc. can also be derived, although the procedure is quite tedious (as are the derivations of the expressions above).

Equations (27, 28, 29, 30) are the plasma *transport* (or *moment*) equations. They describe how the global quantities mass, momentum, energy, etc., move about (i.e., are *transported*) in the plasma. Knowledge of each successive moment provides slightly more information about the distribution function f_α. Knowing the zeroth moment (the density, n_α) allows us to reconstruct the distribution function at a single point in the velocity space (for each point on configuration space); it gives the average of f_α. Similarly, knowing the zeroth and first moments simultaneously (n_α and $\mathbf{V}_\alpha$; four pieces of information) gives us information about four points in velocity space, etc., and so on for the higher-order moments. It is easy to see that a significant amount of information about the plasma has been lost in going from a kinetic to a purely fluid description. However, we have gained a substantial simplification in formulation, and transport models have proven useful in describing plasma dynamics. (But you already know this because you have become experts in MHD!). In the end, comparison of the predictions of theory with experiment are the only test of efficacy of this approach.

We note that the moment equations inherit from the PKE, the property that the evolution of any moment of the distribution function depends on the next higher moment. The equation for the density (27) involves the velocity. The equation for the velocity involves the pressure and the stress tensor, the equation for the pressure involves $\mathbf{Q}_\alpha$, and so on at each order. We have not escaped the problem of closure, and we must find expressions for the higher-order moments in terms of the lower-order moments.

In the following sections, we will assume the "classical" closure of Braginskii, and will make reference to the so-called "neo-classical" closures, which are low collisionality forms that account for toroidal geometry. This will serve our purpose of exhibiting the form of the fluid equations, and allowing an analysis to determine the specific equations should be used in specific parameter regimes. However, we remark that the questions of where (i.e., at what order) to close the fluid equations and what are the closure relations are presently at the forefront of theoretical plasma physics research, and there is no universal agreement on this issue.

Finally, we note that there is a separate set of fluid equations for each species α and that the electromagnetic fields are to be interpreted in the average sense of Sect. 2.

4 Fluid Models

4.1 Drift Velocities

The transport models of Sect. 3 represent the constituents of the plasma (ion species and electrons) as continuous, interacting fluids characterized in configuration space by the average values of various parameters. Some information about the velocity distribution of the individual particles is contained in the higher-order moments. Although the form of these equations is familiar, there are significant differences between them and the ideal MHD equations are given in Table 1. In the following, we shall try to understand how MHD fits into this picture, when it is valid, and how MHD might be extended to describe more physical phenomena.

For simplicity of presentation, we restrict ourselves to a single positively charged ion species ($\alpha = i$, $q_\alpha = e$) and electrons ($\alpha = e$, $q_\alpha = -e$), will neglect the electron mass, since $M_e/M_i \ll 1$, and will use $M_i \to M$. (We could have defined a center-of-mass velocity, but this is equal to the ion velocity to $O(M_e/M_i)$. To an excellent approximation, the ions carry all the momentum.) We will also leave out all the frictional forces (the moments of the collision operator), not because they are small or unimportant, but because they do not alter the conclusions that will be drawn. We also drop specific mention of the energy equation(s), although their presence is implicit in discussions.

Under these assumptions, the dynamical equations for ions and electrons are

$$\frac{\partial n_i}{\partial t} = -\nabla \cdot (n_i \mathbf{V}_i), \tag{31}$$

$$\frac{\partial n_e}{\partial t} = -\nabla \cdot (n_e \mathbf{V}_e), \tag{32}$$

and
$$Mn_i \frac{d\mathbf{V}_i}{dt} = n_i e \left(\mathbf{E} + \mathbf{V}_i \times \mathbf{B}\right) - \nabla P_i - \nabla \cdot \Pi_i, \tag{33}$$

$$0 = -n_e e \left(\mathbf{E} + \mathbf{V}_e \times \mathbf{B}\right) - \nabla P_e - \nabla \cdot \Pi_e. \tag{34}$$

As stated above, ion and electron energy equations exist but will not be discussed here.

In ideal MHD, Ohm's law (the middle entry in Table 1) states that the (momentum carrying) velocity perpendicular (to **B**) is related to the electric and magnetic fields by

$$\mathbf{V}_\perp = \frac{\mathbf{E} \times \mathbf{B}}{B^2}. \tag{35}$$

This is just the "E cross B" drift. It is sometimes called the *MHD velocity*, and is labeled $\mathbf{V}_E$. (In ideal MHD, this is the only flow experienced by both the electrons and ions.) Taking $\mathbf{B} \times$ Eqs. (33) and (34), we find that the ion and electron perpendicular velocities are

$$\mathbf{V}_{i\perp} = \frac{\mathbf{E} \times \mathbf{B}}{B^2} + \frac{M}{eB^2}\mathbf{B} \times \frac{d\mathbf{V}_i}{dt} + \frac{1}{n_i e B^2}\mathbf{B} \times \nabla P_i + \frac{1}{n_i e B^2}\mathbf{B} \times \nabla \cdot \Pi_i, \tag{36}$$

and

$$\mathbf{V}_{e\perp} = \frac{\mathbf{E} \times \mathbf{B}}{B^2} - \frac{1}{n_e e B^2}\mathbf{B} \times \nabla P_e - \frac{1}{n_e e B^2}\mathbf{B} \times \nabla \cdot \Pi_e. \tag{37}$$

Clearly, the motions of the ions and electrons in the moment model are more complicated than in MHD. We recognized the first term on the right-hand side of Eqs. (36) and (37) as the MHD velocity, Eq. (35). It is independent of the sign or magnitude of the electric charge and is experienced by both electrons and ions. The remaining terms depend on the sign and magnitude of the electric charge and introduce additional physics. It is common to identify these terms as additional "drift velocities," e.g.,

The *ion polarization drift*:

$$\mathbf{V}_{pi} \equiv \frac{M}{eB^2}\mathbf{B} \times \frac{d\mathbf{V}_i}{dt}. \tag{38}$$

The *diamagnetic drift*:

$$\mathbf{V}_{*i,e} \equiv \pm\frac{1}{n_i e B^2}\mathbf{B} \times \nabla P_{i,e}. \tag{39}$$

The *stress drift*:

$$\mathbf{V}_{\pi i,e} \equiv \pm\frac{1}{n_i e B^2}\mathbf{B} \times \nabla \cdot \Pi_{i,e}. \tag{40}$$

There is also an electron polarization drift, but it is proportional to the electron mass and has been neglected.

4.2 The Single-Fluid Formulation

Equations (31, 32, 33, 34) describe ions and electrons as separate fluids. It is convenient, and conventional, to transform to a frame moving with the center of mass of the ions and electrons. For our purposes, this is the ion frame (since we have neglected the electron mass). Adding Eqs. (33) and (34), we have

$$Mn_i\frac{d\mathbf{V}_i}{dt} = \rho_q\mathbf{E} + \mathbf{J} \times \mathbf{B} - \nabla P - \nabla \cdot \Pi, \tag{41}$$

where we have defined the *charge density* as

$$\rho_q = e\,(n_i - n_e) \tag{42}$$

and the *current density* as

$$\mathbf{J} = e\,(n_i \mathbf{V}_i - n_e \mathbf{V}_e), \tag{43}$$

and have further defined $P = P_i + P_e$ and $\Pi = \Pi_i + \Pi_e$. Equation (41) is the *equation of motion* for the combined system of ions and electrons, the so-called *single-fluid form*. Using Eqs. (43) in (34) to eliminate $\mathbf{V}_e$ in favor of $\mathbf{V}_i$ yields

$$\mathbf{E} = -\mathbf{V}_i \times \mathbf{B} + \frac{1}{n_e e}\,(\mathbf{J} \times \mathbf{B} - \nabla P_e - \nabla \cdot \Pi_e). \tag{44}$$

This is referred to as the *generalized Ohm's law*. The first term on the right-hand side is identical to the ideal MHD Ohm's law. The remaining terms represent new physics that is not captured by the MHD model. There is no a priori reason for neglecting them.

Equations (31), (32), (41), and (44), along with an appropriate energy equation and closure relations, must be solved simultaneously with Maxwell's equations, Eqs. (12, 13, 14), using Eqs. (42) and (43). Unfortunately, this system is inconsistent. The fluid equations are Gallilean invariant (accurate to $O(V/c)$), while Maxwell's equations are Lorentz invariant (accurate to all orders of V/c). Either we must use relativistic fluid equations or render Maxwell's equations' Gallilean invariant (and justify this step). This is dealt with in the next section. We will see that this step also eliminates the electric force from the equation of motion by introducing (in fact, requiring) the concept of *quasi-neutrality*.

4.3 Low Frequencies and Quasi-neutrality

We can understand how to obtain the low-frequency form of the equations by introducing dimensionless variables. We define $\mathbf{V}' = \mathbf{V}/V_0$, $n' = n/n_0$, $t' = \omega_0 t$, $\nabla' = L\nabla$, $\mathbf{E}' = \mathbf{E}/E_0$, $\mathbf{B}' = \mathbf{B}/B_0$, and $\mathbf{J}' = \mathbf{J}/J_0$. We will then ignore terms that are $O(V_0^2/c^2)$. Then, in terms of these variables, Faraday's law is

$$\frac{\partial \mathbf{B}'}{\partial t'} = -\frac{E_0}{\omega_0 L B_0}\nabla' \times \mathbf{E}', \tag{45}$$

so that it is natural to choose $E_0 = \omega_0 B_0 L$. The non-dimensional form of Ampére's law is

$$\nabla' \times \mathbf{B}' = \frac{\mu_0 J_0 L}{B_0}\mathbf{J}' + \frac{E_0 \omega_0 L}{B_0 c^2}\frac{\partial \mathbf{E}'}{\partial t'}. \tag{46}$$

Choosing $J_0 = B_0/\mu_0 L$ and $\omega_0 = V_0/L$ results in

$$\nabla' \times \mathbf{B}' = \mathbf{J}' + \frac{V_0^2}{c^2}\frac{\partial \mathbf{E}'}{\partial t'}, \tag{47}$$

so that the displacement current is $O(V_0^2/c^2)$ and should be dropped from Ampére's law. The current density is

$$\mathbf{J}' = \frac{n_0 e V_0}{J_0} \left(n_i' \mathbf{V}_i' - n_e' \mathbf{V}_e' \right), \tag{48}$$

and we choose $V_0 = J_0/n_0 e = V_A^2/\Omega_i L$, where $V_A^2 = B_0^2/\mu_0 M n_0$ is the square of the Alfvén velocity, and $\Omega_i = e B_0/M$ is the ion cyclotron frequency. The Poisson equation is then

$$\nabla' \cdot \mathbf{E}' = \frac{\rho_q L}{\epsilon_0 E_0}. \tag{49}$$

The left-hand side is $O(1)$, since we are using dimensionless variables. For the right-hand side to be $O(1)$ requires

$$\frac{\rho_q}{n_0 e} = \frac{\epsilon_0 E_0}{n_0 e L} = \frac{V_0^2}{c^2}, \tag{50}$$

where we have used the definitions above and the relationship $c^2 = 1/\epsilon_0 \mu_0$. This implies that

$$\frac{n_i - n_e}{n_0} \approx O\left(\frac{V_0^2}{c^2}\right), \tag{51}$$

which is the usual statement of *quasi-neutrality*. Quasi-neutrality is thus a requirement for consistency, rather than an independent assumption.

We can now estimate the size of the electrostatic forces. The ratio of the electrostatic force to the Lorentz force is, approximately,

$$\frac{|\rho_q \mathbf{E}|}{|\mathbf{J} \times \mathbf{B}|} \approx \left(\frac{\rho_q}{n_0 e}\right) \left(\frac{V_0 n_0 e}{J_0}\right) \approx O\left(\frac{V_0^2}{c^2}\right), \tag{52}$$

so that the electrostatic forces are to be neglected.

We remark that quasi-neutrality does not imply that the electrostatic field must be small. Even within MHD, situations can occur where $\nabla \times \mathbf{E} = -\nabla \times (\mathbf{V} \times \mathbf{B}) \approx 0$, but $\nabla \cdot \mathbf{E} = -\nabla \cdot (\mathbf{V} \times \mathbf{B}) \neq 0$. Because of the large factor of $1/\epsilon_0$ on the right-hand side of the Poisson equation, it does not take much charge imbalance to produce significant electric fields, and we have shown above that these charge imbalances are consistent with low-frequency models, such as MHD. Quasi-neutrality does *not* mean that the charge density vanishes. It only means that the electrostatic force is negligible compared with the Lorentz force, and therefore need not be included in the dynamics. One can always calculate the charge density a posteriori by computing the divergence of the electric field.

Returning to dimensional variables, and setting $n_e = n_i = n$, the dynamical equations to be investigated are then as follows:

The continuity equation:

$$\frac{\partial n}{\partial t} = -\nabla \cdot (n \mathbf{V}_i). \tag{53}$$

The equation of motion:

$$Mn\frac{d\mathbf{V}_i}{dt} = \mathbf{J} \times \mathbf{B} - \nabla P - \nabla \cdot \Pi. \tag{54}$$

The generalized Ohm's law:

$$\mathbf{E} = -\mathbf{V}_i \times \mathbf{B} + \frac{1}{ne}(\mathbf{J} \times \mathbf{B} - \nabla P_e - \nabla \cdot \Pi_e). \tag{55}$$

Faraday's law:

$$\frac{\partial \mathbf{B}}{\partial t} = -\nabla \times \mathbf{E}. \tag{56}$$

Ampére's law:

$$\mu_0 \mathbf{J} = \nabla \times \mathbf{B}, \tag{57}$$

along with the energy equation and closure expressions. These have been called the *extended MHD equations*.

4.4 Closures

We must now briefly discuss closures, which are expressions that relate high-order velocity moments of the distribution function to lower-order moments. They contain information about the velocity distribution of the individual plasma particles, and deriving closure expressions usually requires solving the kinetic equation with some approximations. These closures usually appear in the fluid equations acted upon by a divergence operator. Examples are $\nabla \cdot \mathbf{q}$, the divergence of the heat flux, and $\nabla \cdot \Pi$, the divergence of the stress tensor. The heat flux thus represents the flux (or flow) of heat in some direction, the stress tensor represents the flux (or flow) of some component of momentum in some direction. They are inherent in the concept of transport.

We know that thermal equilibrium is characterized by the velocity distribution that is Maxwellian, and is parameterized by a density, a velocity, and a temperature. In cases where the macroscopic variables (e.g., temperature, density, etc.) vary in space, we can concieve of local Maxwellians that depend on the local values of these parameters. The system tends toward such states on what is called the *relaxation time*, which is usually related to the interaction of nearest neighbor particles (e.g., collisions). The process of relaxation is described macroscopically by transport, It is not surprising, then, that the fluxes encapsulated by the closures arise from deviations of the distribution function from local Maxwellian.

When the deviations from local Maxwellian are small, the kinetic equations may be linearized and solved, and expressions for the fluxes may be obtained. This procedure usually requires some further assumptions, usually regarding the existence of one or more small parameters. These parameters usually (but not always) appear as the ratio of some quantity characterizing microscopic processes to a corresponding macroscopic parameter. For example, it might be the ratio of the mean-free path to the macroscopic gradient scale length, the inverse of the collision frequency to the macroscopic time scale, or the particle gyro-radius to the macroscopic length scale. An exception is the ratio of the collision frequency to the gyro-frequency, which relates two microscopic parameters but may be small nonetheless.

In spite of these simplifying assumptions, these calculations are extremely complicated and have been carried out only in certain parameter regimes. There is no general solution for the closures of a strongly magnetized plasma. Here we will merely summarize some closure relations that are valid under the assumption that the mean-free path is small compared with the macroscopic scale length. These are called *classical*, or *collisional*, closures. We will also briefly present some results that are valid in lower collisionality regimes, and account for toroidal effects in fusion confinement devices.

It is customary to express the fluxes (flows) in terms of three orthogonal components: one parallel to $\mathbf{B}$, one perpendicular to $\mathbf{B}$, and one in the "cross" direction that is mutually perpendicular to the others. The parallel component is generally inversely proportional to the collision frequency (so that it decreases as the collision frequency increases), the perpendicular component is proportional to the collision frequency, and the "cross" component is independent of the collision frequency. The "cross" component is therefore *not dissipative*; it does not increase the entropy. It is often called the "gyro-" component because it arises from energy transport due to the gyro-motions of individual particles about the magnetic field, which are completely reversible.

Assuming that the mean-free path is small compared with the macroscopic scale length, and that the collision frequency is small compared with the gyro-frequency (i.e., the plasma is strongly magnetized), the classical closure for the heat flux for species α is

$$\mathbf{q}_\alpha = \mathbf{q}_{\alpha_\parallel} + \mathbf{q}_{\alpha_\wedge} + \mathbf{q}_{\alpha_\perp}$$
$$= -\kappa_\parallel^\alpha \nabla_\parallel T_\alpha - \kappa_\wedge^\alpha \mathbf{b} \times \nabla T_\alpha - \kappa_\perp^\alpha \nabla_\perp T_\alpha, \tag{58}$$

where $\nabla_\parallel = \mathbf{b}\mathbf{b} \cdot \nabla$, $\nabla_\perp = -\mathbf{b} \times (\mathbf{b} \times \nabla) = (\mathbf{I} - \mathbf{b}\mathbf{b}) \cdot \nabla$, and $\mathbf{b} = \mathbf{B}/B$. Note that this relates the the third velocity moment ($\mathbf{q}$) to the gradient of the second moment (T). It is the gradient in quantities that drive transport. The coefficients, called the *thermal conductivities*, of ions and electrons are

$$\kappa_\parallel^\alpha = A_\alpha \frac{n_\alpha T_\alpha}{M_\alpha \nu_\alpha}, \tag{59}$$

$$\kappa_\wedge^\alpha = B_\alpha \frac{n_\alpha T_\alpha}{M_\alpha \Omega_\alpha}, \tag{60}$$

and

$$\kappa_\perp^\alpha = C_\alpha \frac{n_\alpha T_\alpha \nu_\alpha}{M_\alpha \Omega_\alpha^2}, \tag{61}$$

where $A_i = 3.9$, $A_e = 3.2$, $B_i = B_e = 5/2$, $C_i = 2$, $C_e = 4.7$, and ν_α is the collision frequency. Note that, since $M_e \ll M_i$, the parallel heat flux is dominated by the electrons, but the perpendicular heat flux is dominated by the ions (since $\Omega_i \ll \Omega_e$). The "cross" components are comparable.

From the theory of elasticity, we know that in any material, the linear relation ship (if it exists) between the stress tensor and the rate of strain tensor is given by the *generalized Hookes' law*

$$\Pi_{ij} = E_{ijkl} W_{kl}, \tag{62}$$

where

$$W_{kl} = \frac{\partial V_l}{\partial x_k} + \frac{\partial V_k}{\partial x_l} - \frac{2}{3}\frac{\partial V_n}{\partial x_n}\delta_{kl} \tag{63}$$

is the rate of strain tensor and E_{ijkl} is the elastic constant tensor. In isotropic materials, it is a constant called Young's Modulus. In an anisotropic medium, like a magnetized plasma, it is defined by

$$E_{ijkl} \equiv \frac{\partial \Pi_{ij}}{\partial W_{kl}}. \tag{64}$$

Equation (63) can be written in a dyadic form as

$$\mathbf{W} = \nabla\mathbf{V} + \nabla\mathbf{V}^T - \frac{2}{3}\nabla\mathbf{V}\mathbf{I}, \tag{65}$$

where $(\dots)^T$ denotes the transpose.

The stress tensor can be calculated from the linearized kinetic equation in a manner analogous to the heat flux and under the same conditions. The classical result for the ion stress is

$$\Pi = \Pi_{\parallel} + \Pi_{\wedge} + \Pi_{\perp}, \tag{66}$$

with

$$\Pi_{\parallel} = -\frac{3}{2}\eta_0 \left(\mathbf{b} \cdot \mathbf{W} \cdot \mathbf{b}\right)\left(\mathbf{bb} - \frac{1}{3}\mathbf{I}\right), \tag{67}$$

and

$$\Pi_{\wedge} = -\frac{\eta_3}{2}\left[\mathbf{b} \times \mathbf{W} \cdot (\mathbf{I} + 3\mathbf{bb}) + transpose\right]. \tag{68}$$

Here, $\eta_0 = 0.96nT_i/\nu_i$ and $\eta_3 = nT_i/2\Omega_i$. The classical expression for $\Pi_{\perp}$ is even more complicated and will not be given here. In any case, it is proportional to the collision frequency and can often be ignored in high-temperature plasmas. The classical electron stress is smaller by a factor of the mass ratio and is usually excluded from extended fluid models.

Equation (68) is often called the *gyro-viscous stress*, and the coefficient η_3 is called the *gyro-viscosity*, although it is independent of the collision frequency and therefore is not dissipative. Like the "cross" component of the heat flux, it represents momentum transport due to the gyro-motion of individual particles. The gyro-viscous stress plays an important role in extended fluid models.

We remark that, at least formally, the gyroviscous stress should contain an additional term proportional to the gradient of the heat flux,

$$\Pi_q = \frac{2}{5\Omega_i} \left[\mathbf{b} \times W_q \cdot (\mathbf{I} + 3\mathbf{bb}) + transpose \right], \tag{69}$$

where

$$W_q = \nabla \mathbf{q} + \nabla \mathbf{q}^T - \frac{2}{3} \nabla \mathbf{q} \mathbf{I}. \tag{70}$$

As these terms are generally not included in the extended fluid models, their full effect on plasma dynamics is unknown at this time.

Finally, we mention that other expressions valid for a low collisionality plasma in a torus (where trapped particle effects can be important) have been derived. These are called *neo-classical closures*. Most of these are expressions for the flux surface average of some component of the viscous force, rather than expressions that are locally valid. For example, a neo-classical expression for the flux surface average of the parallel ion viscous force is

$$\langle \mathbf{B} \cdot \nabla \cdot \Pi_i^{\text{neo}} \rangle = M n \langle B^2 \rangle \mu_i \frac{V_{\theta_i}}{B_\theta} \mathbf{e}_\theta, \tag{71}$$

where θ represents the poloidal direction in a torus and μ_i is the "neo-classical damping coefficient." More complicated expressions for both ions and electrons have been given. They play an important role in extended fluid models of well-confined plasmas.

4.5 Ordered Fluid Equations

We are now in a position to identify different fluid models that are subsets of extended MHD. Again, this will be facilitated by introducing non-dimensional variables. This will lead to the identification of a small parameter, and the remaining non-dimensional constants can then be ordered as large or small with respect to this parameter. We will only retain terms in the equations that are of lowest order in this parameter.

As in Sect. 4.3, we measure the density in units of n_0, the velocity in units of V_0, the magnetic field in units of B_0, the electric field in units of $E_0 = V_0 B_0$, the current density in units of $J_0 = n_0 e V_0$, the time in units of ω_0^{-1}, and the length in units of $L = |\nabla^{-1}|$. We measure the pressure in units of $M n V_{\text{thi}}^2$, where $V_{\text{thi}} = \sqrt{2 T_i / M}$ is the ion thermal speed. Note that neither the characteristic velocity V_0 nor the characteristic frequency ω_0 have been specified. They will be used to "order" the equations.

We define the non-dimensional parameters

$$\epsilon = \frac{\omega_0}{\Omega_i}, \tag{72}$$

$$\xi = \frac{V_0}{V_{\text{thi}}}, \tag{73}$$

and

$$\delta = \frac{\rho_i}{L}, \tag{74}$$

where $\rho_i = V_{\text{thi}}/\Omega_i$ is the ion gyro-radius. In a strongly magnetized plasma, δ is always a small parameter. We will order ϵ and ξ as large or small compared with δ to obtain different fluid models.

With these normalizations and definitions, the extended MHD equations can be written as

$$\epsilon \frac{\partial n}{\partial t} = -\xi \delta \nabla \cdot (n \mathbf{V}_i), \tag{75}$$

$$n \left(\epsilon \xi \frac{\partial \mathbf{V}_i}{\partial t} + \xi^2 \delta \mathbf{V}_i \cdot \nabla \mathbf{v}_i \right) = \xi \mathbf{J} \times \mathbf{B} - \delta \nabla P - \xi \frac{\mu}{\Omega_i} \mathbf{bb} \cdot \nabla \Pi_i^{\text{neo}}$$
$$- \xi \delta^2 \left(\frac{1}{\nu/\Omega_i} \nabla \cdot \Pi_\parallel + \nabla \cdot \Pi_\wedge + \frac{\nu}{\Omega_i} \nabla \cdot \Pi_\perp \right), \tag{76}$$

$$\xi \mathbf{E} = -\xi \mathbf{V}_i \times \mathbf{B} + \xi \frac{1}{n} \mathbf{J} \times \mathbf{B} - \delta \frac{1}{n} \left(\nabla P_e + \nabla \cdot \Pi_e^{\text{neo}} \right), \tag{77}$$

along with the Maxwell equations

$$\epsilon \frac{\partial \mathbf{B}}{\partial t} = -\xi \delta \nabla \times \mathbf{E}, \tag{78}$$

$$\mathbf{J} = \xi \nabla \times \mathbf{B}, \tag{79}$$

and the constitutive relation

$$\mathbf{J} = n (\mathbf{V}_i - \mathbf{V}_e). \tag{80}$$

We are now ready to order the equations. In what follows we will drop the neoclassical stresses, not because they are not important, but because they apply to specialized geometry (generally a tokamak), and they do not provide further insight into the various fluid models. Further, the factor of $1/(\nu/\Omega_i)$ multiplying the parallel ion viscosity is an indication that the classical formalism is breaking down in the limit of $\nu/\Omega_i \to 0$, so we also disregard this term as unphysical. In practical calculations, it is often replaced by a heuristic viscosity proportional to $\nabla^2 \mathbf{V}_i$ (although this is also unphysical).

4.5.1 The Hall MHD Ordering ($\xi \sim 1/\delta$, $\epsilon \sim 1$)

We begin by considering fast flows and relatively high frequencies. We thus order $\xi \sim 1/\delta$ and $\epsilon \sim 1$. We are allowing flows faster than the ion thermal velocity (supersonic) and frequencies on the order of the ion gyro-frequency. We also assume $\nu/\Omega_i \sim \delta$. Then the extended MHD equations become

$$\frac{\partial n}{\partial t} = -\nabla \cdot (n \mathbf{V}_i), \tag{81}$$

$$\mathbf{J} \times \mathbf{B} = n \frac{d \mathbf{V}_i}{dt} + O\left(\delta^2\right), \tag{82}$$

and

$$E = -V_i \times B + \frac{1}{n}J \times B + O\left(\delta^2\right). \tag{83}$$

The second term on the right-hand side of Eq. (83) is called the *Hall term*. Since, from Eq. (82), $J \times B \sim O(1)$, it must be included when studying this parameter regime.

Equation (82) indicates that in this regime, there are unbalanced electromagnetic forces that are $O(1)$ and pressure forces do not play a significant role in the dynamics. These plasmas are far from confinement, as is consistent with the fast flows and high frequencies allowed by the ordering.

4.5.2 The MHD Ordering ($\xi \sim 1, \epsilon \sim \delta$)

We now consider flows that are of the order of the ion thermal speed and frequencies that are less than the ion gyro-frequency. We thus order $\xi \sim 1$ and $\epsilon \sim \delta$. The result is

$$\frac{\partial n}{\partial t} = -\nabla \cdot (nV_i), \tag{84}$$

$$J \times B = \delta\left(n\frac{dV_i}{dt} + \nabla P\right) + O\left(\delta^2\right), \tag{85}$$

and

$$E = -V_i \times B + \frac{1}{n}J \times B - \delta \nabla P_e. \tag{86}$$

However, from Eq. (85), the second term on the right-hand side of Eq. (86) is $O(\delta)$. Therefore, to the lowest order in δ, Ohm's law is

$$E = -V_i \times B. \tag{87}$$

This is just the ideal MHD Ohm's law.

Plasmas where the MHD model is valid are "force-free" to $O(\delta)$. Small imbalances in the Lorentz force appear as accelerations and pressure gradients. These plasmas are better confined than those considered in Sect. 4.5.1, but there can still be considerable dynamics (e.g., flows on the order of the ion sound speed). The ion velocity is just the MHD (or $E \times B$) velocity defined in Sect. 4.1.

4.5.3 The Drift Ordering ($\xi \sim \delta, \epsilon \sim \delta^2$)

We now consider a model that applies slow flows and very low frequencies, and we set $\xi \sim \delta$ and $\epsilon \sim \delta^2$. This is the so-called drift ordering, because the relative drifts of the different particle species (see Sect. 4.1) play an important role. We will see that it applies to plasmas that only deviate slightly from force balance. These plasmas are therefore well confined. It is not surprising that this ordering has been successfully applied to the onset of instabilities in tokamak plasmas.

The equations of the drift ordering are

$$\frac{\partial n}{\partial t} = -\nabla \cdot (n\mathbf{V}_i), \tag{88}$$

$$-\nabla P + \mathbf{J} \times \mathbf{B} = \delta^2 \left(n\frac{d\mathbf{V}_i}{dt} + \nabla \cdot \mathbf{\Pi}_\wedge \right), \tag{89}$$

and

$$\mathbf{E} = -\mathbf{V}_i \times \mathbf{B} + \frac{1}{n}(\mathbf{J} \times \mathbf{B} - \nabla P_e). \tag{90}$$

The pressure and Lorentz forces are in balance to $O(\delta^2)$, so that plasmas that are well described by this model can deviate only slightly from force balance. Further, note that the gyro-viscosity now enters at the same order as the ion inertia, so that it cannot be ignored.

4.5.4 The Drift Model

While the drift ordering introduces the lowest order finite-Larmor radius (FLR) corrections to the ideal MHD model (these corrections are the terms proportional to δ^2), it leads to more complicated equations and provides no special insights. In contrast, the *drift model* makes use of the velocity decomposition given in Sect. 4.1, along with a remarkable result called the *gyro-viscous cancellation*, to provide a simplified set of equations that have provided significant insight into the properties of tokamak plasmas.

Essentially, the drift model makes a velocity transformation to a frame moving with the MHD velocity $\mathbf{V}_E$ (see Sect. 4.1):

$$\mathbf{V}_i = \mathbf{V}_{\|_i} + \mathbf{V}_E + \mathbf{V}_{*_i} + O\left(\delta^2\right), \tag{91}$$

and

$$\mathbf{V}_e = \mathbf{V}_i - \frac{1}{n}\mathbf{J} = \mathbf{V}_{\|_i} + \mathbf{V}_E + \mathbf{V}_{*_i} - \frac{1}{n}\mathbf{J} + O\left(\delta^2\right). \tag{92}$$

The idea is to arrive at a set of equations written *in terms of the MHD velocity* $\mathbf{V}_E$ that look like the MHD equations, plus corrections.

Substituting directly into the generalized Ohm's law, Eq. (90), we find

$$\mathbf{E} = -\mathbf{V}_E \times \mathbf{B} - \frac{1}{n}\nabla_\| P_e + \frac{1}{n}\left(-\nabla_\| P + \mathbf{J} \times \mathbf{B}\right) + O\left(\delta^2\right). \tag{93}$$

But, from Eq. (89), the third term on the right-hand side of this equation is $O(\delta^2)$, so that Ohm's law in the drift model is, to lowest order in δ, is

$$\mathbf{E} = -\mathbf{V}_E \times \mathbf{B} - \frac{1}{n}\nabla_\| P_e. \tag{94}$$

The momentum equation becomes, to the second order in δ,

$$\delta^2 \left[n\frac{d}{dt}\left(\mathbf{V}_{\|_i} + \mathbf{V}_E\right) + n\frac{d\mathbf{V}_{*_i}}{dt} + \nabla \cdot \mathbf{\Pi}_\wedge \right] = -\nabla P + \mathbf{J} \times \mathbf{B}. \tag{95}$$

One reason for the utility of the drift model is an enormous simplification of the equation of motion that occurs because, in the proper reference frame, the $\nabla \cdot \Pi_\wedge$ algebraically cancels a significant fraction of the advective acceleration $n\mathbf{V}_i \cdot \nabla\mathbf{V}_i$. This *gyro-viscous cancellation* is usually written

$$n \left(\frac{\partial \mathbf{V}_{*_i}}{\partial t} + \mathbf{V}_i \cdot \nabla\mathbf{V}_{*_i} \right) + \nabla \cdot \Pi_\wedge \approx \nabla\chi - \mathbf{b}\mathbf{V}_{*_i} \cdot \nabla V_{\|_i}, \tag{96}$$

where $\chi = -P_i \mathbf{b} \cdot \nabla \times \mathbf{V}_{\perp_i}$. (We remark that the gyro-viscous cancellation has only been quantitatively derived in simple geometry, such as a slab, and is otherwise to be considered approximate. Nonetheless, it is widely used in the theory of tokamak plasmas.)

Using the gyro-viscous cancellation, and after some algebra, the equations of the drift model can be written in terms of parallel and perpendicular (to $\mathbf{B}$) components as

$$\frac{\partial n}{\partial t} + \nabla \cdot n\mathbf{V}_E = -\nabla \cdot n \left(\mathbf{V}_{*_i} + \mathbf{V}_{\|_i} \right), \tag{97}$$

$$n\delta^2 \left(\frac{d\mathbf{V}_E}{dt} \right)_{\text{MHD}} = -n\delta^2 \left(\mathbf{V}_{*_i} + \mathbf{V}_{\|_i} \right) \cdot \nabla\mathbf{V}_E - \nabla_\perp \left[P \left(1 + \delta^2\chi \right) \right] + \mathbf{J} \times \mathbf{B}, \tag{98}$$

and

$$\mathbf{E} = -\mathbf{V}_E \times \mathbf{B} - \frac{1}{n}\nabla_\|P_e, \tag{99}$$

where $(d/dt)_{\text{MHD}} = \partial/\partial t + \mathbf{V}_E \cdot \nabla$.

We have achieved the goal of obtaining a set of equations that look very much like the ideal MHD equations with lowest-order corrections.

When interpreting these equations, and their solutions, one must be aware that the dependent velocity variable, $\mathbf{V}_E$, is not the true flow velocity of the fluid, but the $\mathbf{E} \times \mathbf{B}$ velocity. There are several further caveats. In the first place, the derivation formally admits only slow flows, which is consistent with the result of force balance through first order in δ. Second, the assumption of very low frequency may limit the validity of the model to phenomena that evolve much slower than the Alfvén transit time L/V_A. Acceptable frequencies are of the order of the diamagnetic drift frequency $\omega_{*_i} \sim V_{*_i}/L \ll \omega_A$. Finally, the form of the gyro-viscous cancellation used here assumes a uniform magnetic field or at least a sheared slab. There is no generally accepted form that is much less restrictive.

4.5.5 The Transport Model

We briefly mention the *tansport model*, which is a special case of the drift ordering (see Sect. 4.5.3) that retains only corrections that are $O(\delta)$. At this order, there is complete force balance and the only corrections are to Ohm's law. The equations of the model are

$$\frac{\partial n}{\partial t} = -\nabla \cdot (n\mathbf{V}_i), \tag{100}$$

$$\nabla P = \mathbf{J} \times \mathbf{B}, \tag{101}$$

$$\mathbf{E} = -\mathbf{V} \times \mathbf{B} + \frac{1}{n}\nabla_\perp P_i, \tag{102}$$

and

$$\frac{\partial \mathbf{B}}{\partial t} = -\nabla \times \mathbf{E}. \tag{103}$$

Inertia has been ordered out of the system and with it all waves. In an axisymmetric equilibrium, the flows will be only perpendicular to the flux surfaces. Eq. (101) becomes the Grad–Shafranov equation, while, from Eq. (102), the flux of particles is

$$n\mathbf{V}_{\perp_i} = n\left(\mathbf{V}_E + \mathbf{V}_{*_i}\right). \tag{104}$$

When this expression is used with an equation of state, such as $P = nT$, Eq. (100) becomes a diffusion equation for the density. Field diffusion results from Eq. (103) if the resistivity is included in Ohm's law, Eq. (102).

The transport model is widely used in studies of the very long time scale evolution of tokamaks.

5 Properties of the Extended MHD Model

We have now derived a set of fluid equations that encapsulate the ideal MHD model, and extend it to include the lowest-order effects of relative particle drifts and finite ion Larmor radius (FLR). These are Eqs. (53, 54, 55, 56, 57). You are all now experts in the dispersion (wave propagation) and stability properties of the MHD equations. Here we investigate how the new correction terms affect these properties.

With regard to wave propagation, the most striking effect of the new terms is to introduce a new class of waves that are *dispersive*. In MHD, all waves (sound waves, and shear and compressible Alfvén waves) have dispersion relations of the form $\omega^2 \sim k^2$, so that all wavelengths propagate with the same phase velocity. In extended MHD, there are now modes of propagation with the property $\omega^2 \sim k^4$, so that shorter wavelengths have faster phase velocity. These are called *whistler*, *kinetic Alfvén* (KAWs), and *gyro-viscous* (for lack of a better term) waves.

With regard to stability, the most striking feature is that force operator ceases to be self-adjoint, so that all the nice properties associated with the ideal MHD energy principle are lost. However, the implications of the new terms for stability are easy to see heuristically. Ignoring pressure and stress forces, the linearized momentum equation and Ohm's law in the extended MHD model are

$$Mn_0 \frac{\partial \mathbf{V}}{\partial t} = \frac{1}{\mu_0}\left(\nabla \times \mathbf{B}\right) \times \mathbf{B}_0, \tag{105}$$

and

$$\mathbf{E} = -\mathbf{V} \times \mathbf{B}_0 + \frac{M}{e}\frac{\partial \mathbf{V}}{\partial t}, \tag{106}$$

where we have used the momentum equation to eliminate $\mathbf{J} \times \mathbf{B}_0$ in favor of $\partial \mathbf{V}/\partial t$ in Ohm's law. As in ideal MHD, we can introduce the displacement $\partial \xi/\partial t = \mathbf{V}$, and assuming a time dependence $\sim \exp i\omega t$, we have

$$-\omega^2 M n_0 \xi = \mathbf{F}_{\text{MHD}}(\xi) + \omega \mathbf{F}_{2F}(\xi), \tag{107}$$

where $\mathbf{F}_{\text{MHD}}(\xi) = \nabla \times [\nabla \times (\xi \times \mathbf{B}_0)] \times \mathbf{B}_0/\mu_0$ is the (self-adjoint) ideal MHD force operator, and $\mathbf{F}_{2F} = (iM/e\mu_0)\nabla \times (\nabla \times \xi) \times \mathbf{B}_0$ is the two-fluid force operator. This comes from the additional terms in Ohm's law. However, it is not self-adjoint, so that generally the eigenvalues ω^2 will be complex.

Dotting Eq. (107) with ξ^* and integrating over all space yields, at least heuristically,

$$\omega^2 - \omega_* \omega + \gamma_{\text{MHD}}^2 = 0, \tag{108}$$

where γ_{MHD} is the ideal MHD growth rate (for an unstable configuration) and we identify $\omega_* = (1/\mu_0 n_0 e) \int \xi^* \cdot \mathbf{F}_{2F}(\xi) d^3x$ as the *drift frequency*. Equation (108) appears often in plasma physics and is a common form for the extended MHD dispersion relation. The solution is

$$\omega = \frac{1}{2}\omega_* \pm \frac{1}{2}\sqrt{\omega_*^2 - 4\gamma_{\text{MHD}}^2}, \tag{109}$$

which is generally complex. Note, however, that the roots become purely real (i.e., stable) when

$$\omega_* > 2\gamma_{\text{MHD}}, \tag{110}$$

so that unstable MHD modes are stabilized if the drift frequency is large enough. This is a fairly general conclusion that holds under a variety of more complex circumstances. (However, this stabilization may be lost if compressibility is taken into account.)

5.1 Dispersive Waves

Consider first the effect of the Hall term (the $\mathbf{J} \times \mathbf{B}$ term in the generalized Ohm's law). The linearized electric field for the case of a uniform, straight magnetic field is then $\mathbf{E} = (1/n_0 e)\mathbf{J} \times \mathbf{B}_0$, and Faraday's law becomes

$$\frac{\partial \mathbf{B}}{\partial t} = -\frac{B_0}{\mu_0 n_0 e}\mathbf{b} \cdot \nabla(\nabla \times \mathbf{B}). \tag{111}$$

A wave equation can be found by taking the time derivative of Eq. (111). The result is

$$\frac{\partial^2 \mathbf{B}}{\partial t^2} = -\left(\frac{B_0}{\mu_0 n_0 e}\right)^2 (\mathbf{b} \cdot \nabla)^2 \nabla^2 \mathbf{B}. \tag{112}$$

This differs from the usual wave equation in that it has two time derivatives on the left, but *four* space derivatives on the right. Assuming spatial dependence $\sim \exp i(\mathbf{k} \cdot \mathbf{x} + \omega t)$, the algebraic dispersion relation is

$$\omega^2 = -\left(\frac{V_A^2}{\Omega_i}\right)^2 k_\parallel^2 k^2. \tag{113}$$

These are called *whistler waves*. They propagate parallel to the magnetic field, and the shorter wavelengths have larger phase velocity. They were first discovered (and named) during the early days of radio, when "whistles" proceeding from high to low frequencies were heard in the noise. They were just these waves propagating down the earth's magnetic field lines.

Whistler waves follow directly from Ohm's law, which is the electron equation of motion. They are thus an electron phenomenon, and occur at zero pressure. When the pressure is finite, other dispersive waves arise when the equation of motion and the (unspecified here) energy equation are taken into account. These are called *kinetic Alfvén waves* or KAWs. They are also dispersive. We will not discuss them further here.

Recall that the gyro-viscosity is non-dissipative. In fact, it also introduces a new family of dispersive waves. The gyro-viscosity is given in Eq. (68). For the case of a uniform plasma in a straight, uniform magnetic field in the z-direction, the expressions for the gyro-viscous force are

$$(\nabla \cdot \Pi)_x = -\eta_3 \nabla_\perp^2 V_y - \eta_4 \frac{\partial}{\partial z}\left(\frac{\partial V_y}{\partial z} + \frac{\partial V_z}{\partial y}\right), \tag{114}$$

$$(\nabla \cdot \Pi)_y = \eta_3 \nabla_\perp^2 V_x + \eta_4 \frac{\partial}{\partial z}\left(\frac{\partial V_x}{\partial z} + \frac{\partial V_z}{\partial x}\right), \tag{115}$$

and

$$(\nabla \cdot \Pi)_z = -\eta_4 \frac{\partial}{\partial z}\left(\frac{\partial V_x}{\partial y} - \frac{\partial V_y}{\partial x}\right), \tag{116}$$

where $\eta_3 = nT_i/2\Omega_i$ and $\eta_4 = 2\eta_3$. With constant density, and again assuming $\exp i\,(\mathbf{k} \cdot \mathbf{x} + \omega t)$ dependence for all quantities, the linearized extended MHD equations become

$$i\omega M n_0 V_x = -ik_x P + \frac{iB_0}{\mu_0}(k_z B_x - k_x B_z) - \left(\eta_3 k_\perp^2 + \eta_4 k_z^2\right) V_y - \eta_4 k_y k_z V_z, \tag{117}$$

$$i\omega M n_0 V_y = -ik_y P + \frac{iB_0}{\mu_0}(k_z B_y - k_y B_z) + \left(\eta_3 k_\perp^2 + \eta_4 k_z^2\right) V_x - \eta_4 k_x k_z V_z, \tag{118}$$

$$i\omega M n_0 V_z = -ik_z P - \eta_4 k_y k_z V_x + \eta_4 k_x k_z V_y, \tag{119}$$

$$i\omega B_x = ik_z B_0 V_x + \frac{B_0}{\mu_0 n_0 e}k_z\left(k_y B_z - k_z B_y\right), \tag{120}$$

$$i\omega B_y = ik_z B_0 V_y + \frac{B_0}{\mu_0 n_0 e}k_z(k_z B_x - k_x B_z), \tag{121}$$

$$i\omega B_z = -ik_y B_0 V_y - ik_x B_0 V_x + \frac{B_0}{\mu_0 n_0 e}k_z\left(k_x B_y - k_y B_x\right), \tag{122}$$

and

$$i\omega P = -i\Gamma P_0 k_x V_z, \tag{123}$$

where $k_\perp^2 = k_x^2 + k_y^2$ and Γ is the adiabatic index. (We have assumed an adiabatic energy equation.) The terms proportional to $B_0/\mu_0 n_0 e$, η_3, and η_4 are the new terms from extended MHD.

The dispersion relation can be obtained by setting the determinant of Eq. (117–123) to zero. This is relatively easy for the special cases of parallel and perpendicular propagation. For the case of parallel propagation ($k_x = k_y = 0$), we have, after some algebra,

$$(\omega \pm \omega_4)(\omega \pm \omega_W) = \omega_A^2, \tag{124}$$

where $\omega_A = |k_z| V_A$ is the Alfvén frequency, $\omega_W = (\omega_A/\Omega_i)^2 \Omega = (\rho_i k_z)^2/\beta$ is the whistler frequency, and $\omega_4 = (1/2)(\rho_i k_z)^2$ is the gyro-viscous frequency (for lack of a better term). Note that Eq. (124) is of the same general form as Eq. (108). The two $\pm$ signs represent left- and right-propagating waves that have left and right polarization. For $\rho_i k_z \ll 1$, as is required for the validity of the extended fluid model, the dispersion relations for these waves are of the form

$$\omega = k_z V_A \left[1 + \frac{1+\beta}{2\sqrt{\beta}} (\rho_i k_z) \right]. \tag{125}$$

The gyro-viscosity introduces an $O(\beta)$ correction to the whistler wave.

For the case of perpendicular propagation ($k_y = k_z = 0$), the vanishing of the determinant of Equations (117–123) leads to the relationship

$$\frac{\omega^2}{\omega_s^2 + \omega_A^2} = 1 + \frac{\omega_3^2}{\omega_s^2 + \omega_A^2}, \tag{126}$$

where $\omega_s^2 = c_s^2 k_x^2$ is the sound frequency and $\omega_3 = \omega_4/2$. For $\rho_i k_x \ll 1$, the solution is

$$\omega^2 = k_x^2 V_A^2 \left[1 + \frac{\Gamma\beta}{2} + \frac{\beta}{16} (\rho_i k_x)^2 \right]. \tag{127}$$

This is clearly a dispersive modification to the fast magneto-acoustic wave.

5.2 Stability

From Eqs. (108) and (109), we expect extended MHD to have a stabilizing effect on unstable MHD modes. This can be illustrated in the specific case of the gravitational interchange mode (or g-mode) in a uniform magnetic field.

We consider slab geometry, with the magnetic field in the z-direction and gravity in the y-direction. We assume an exponential density profile $\rho(y) = \rho_0 \exp(y/L)$, so that $d\rho/dy = \rho/L$. The equilibrium condition is

$$\frac{dP_0}{dy} = -M n_0 g. \tag{128}$$

The dynamical equations are

$$\frac{\partial n}{\partial t} = -\nabla \cdot n\mathbf{V}, \tag{129}$$

$$Mn\frac{d\mathbf{V}}{dt} = -\nabla \left(P + \frac{B^2}{2\mu_0} \right) + Mn\mathbf{g} - \nabla \cdot \Pi, \tag{130}$$

and

$$\mathbf{E} = -\mathbf{V} \times \mathbf{B} + \frac{1}{ne} \left(Mn\frac{d\mathbf{V}}{dt} + \nabla P_i - Mn\mathbf{g} + \nabla \cdot \Pi \right), \tag{131}$$

where, again, we have used the momentum equation to eliminate $\mathbf{J} \times \mathbf{B}$ in Ohm's law. The magnetic field only explicitly enters the dynamics through the *total pressure* $P_T = P + B^2/2\mu_0$. Therefore, as far as the dynamics are concerned, perturbations to the magnetic field can be ignored and all perturbed pressure forces can be viewed as entering through the pressure P. It is then a significant simplification to assume that all perturbations are *electrostatic*, so that we can set $\nabla \times \mathbf{E} = 0$. Assuming that the ions are isothermal, the result is

$$\nabla \cdot \mathbf{V} = -\frac{1}{\Omega} \mathbf{e}_z \cdot \left(\frac{d\mathbf{V}}{dt} - \frac{1}{\Omega Mn^2} \nabla n \times \nabla \cdot \Pi \right), \tag{132}$$

where Ω is the ion gyro-frequency. This condition closes the system of equations. In ideal MHD, $\nabla \cdot \mathbf{V} = 0$. In extended MHD, the velocity is non-solenoidal.

Equations (129), (130), and (132) can then be linearized and solved. The ideal MHD result is well known. The system is always unstable, and the growth rate is independent of the wave number:

$$\gamma_{\text{MHD}}^2 = \frac{g}{L}. \tag{133}$$

For extended MHD, the result of a more tedious calculation yields the equation

$$\omega^2 - \left(\frac{gk}{\Omega} + \frac{\eta_3 k}{Mn_0 L} \right) \omega + \gamma_{\text{MHD}}^2 = 0, \tag{134}$$

The first term in the coefficient of ω results from the additional two-fluid terms in Ohm's law. The second term in the coefficient represents the effect of gyro-viscosity. This equation is of the form of Eq. (108). We therefore expect the g-mode to be stabilized for wave numbers such that

$$k > \frac{g/L}{g/\Omega + \eta_3/Mn_0 L}. \tag{135}$$

This high-k stabilization by extended MHD effects is an important aspect of tokamak operation and is completely missed by the MHD model. The g-mode is therefore an important paradigm for understanding this effect in a situation that is simple enough to study analytically. (In a toroidal plasma, the field line curvature force $\mathbf{B} \cdot \nabla \mathbf{B}$ roughly corresponds to $\mathbf{g}$ in the preceding discussion.)

6 Current Research Problems

Problems requiring the application of the extended MHD model are at the forefront of current theoretical research in strongly magnetized plasmas. Several of these are described below.

6.1 Magnetic Reconnection

In ideal MHD, Ohm's law prevents the separate motion of the fluid and the magnetic field. This implies that the topology, or connectivity, of the magnetic field must remain invariant. This constraint is broken in the presence of even the smallest amount of resistivity, and the motions of the fluid and the field become decoupled. This allows new motions that can change magnetic topology. This process is called *magnetic reconnection*.

In ideal MHD, there is only one characteristic length scale in the problem, the macroscopic scale length. In resistive MHD (the ideal MHD Ohm's law with the addition of $\eta \mathbf{J}$), there are two scale lengths: one set by the ideal MHD dynamics and the other by resistive diffusion. This allows the possibility of coupling long wavelength ideal MHD motions that remain "frozen in" the plasma with magnetic field diffusion on small length scales that allows for changes in topology. The result is a magnetic reconnection process whose rate scales as some fractional power of the resistivity. The limiting physical factor is the rate at which plasma can be expelled from the reconnection region. Since the ions and electron motions are coupled in single-fluid models, this is limited to approximately the local Alfvén speed. This rate is much smaller than is commonly observed in both laboratory and astrophysical plasmas.

Extended MHD introduces additional symmetry breaking terms (the two-fluid terms) into Ohm's law that allow for an additional characteristic length scale. On this length scale, the electron and ion motions become decoupled. This allows for the fluid to exit the reconnection region more rapidly, thus increasing the reconnection rate. While there is no universal agreement on these issues, researchers hope that this process can help to explain the rapid reconnection rates observed in tokamaks, solar flares, and the earth's magnetosphere.

6.2 Magnetic Island Evolution

Tearing modes are linear instabilities that involve magnetic reconnection. They involve a coupling between large-scale Alfvén waves and small-scale resistive diffusion. The associated topology changes produce *magnetic islands*. They grow slowly (as a fractional power of the resistivity) and saturate (stop growing) at relatively small amplitude, often much smaller than is observed in experiments. Sometimes, also, magnetic islands are observed to grow in situations that are predicted to be stable within the single-fluid resistive MHD model.

In resistive MHD, the nonlinear evolution of the width of a magnetic island, W, is governed by a first-order differential equation of the form $dW/dt = AW + B$, where A is related to the bending of the field lines outside the inner (resistive) layer and B is a constant that is related to the instability drive. (The equation with $A = 0$ is known as the *Rutherford equation* and with $A \neq 0$ as the *modified Rutherford equation*.) The final island width W_s is found by setting $dW/dt = 0$. Since the resulting equation is linear in W, there is only one solution. (The Rutherford equation itself says nothing about the saturated width.)

When extended MHD effects, including the neo-classical terms, are taken into account, the Rutherford equation is further modified and becomes $dW/dt = A(W) + B$, where $A(W)$ is a nonlinear function of the island width. This allows for the possibility of multiple island widths at which $dW/dt = 0$. An interesting case involves two solutions, W_1 and W_2, say. When $W < W_1$ or $W > W_2$, $dW/dt < 0$, and when $W_1 < W < W_2$, $dW/dt > 0$. Thus an island with width $W < W_1$ will *decay* to zero (i.e., it is stable) and an island with $W > W_2$ will decay to W_1. However, an island in the range $W_1 < W < W_2$ will *grow* to have a width $W = W_2$. As in the single-fluid case, the growth is algebraic ($\sim t^n$) rather than exponential. These modes thus require a *finite initial island width*, or *seed island*, in order to grow. They are purely a nonlinear effect; they are linearly stable. These are called *neo-classical tearing modes* or NTMs.

This discussion raises two questions: How can you get a finite width seed island in a system that is linearly stable? and What is the driving force for the instability?

With regard to the initial seed island, there are at least two possibilities. The first is that it is nonlinearly driven by the interaction of some *other* instable modes. For example, neo-classical tearing modes in tokamaks are sometimes observed to grow after fast sawtooth crashes. The large amplitude kink modes that are responsible for the sawtooth may nonlinearly generate a neighboring perturbation with finite amplitude that forms the seed island. The second possibility is that a slowly growing unstable resistive mode simply attains large enough amplitude to trigger the neo-classical mode. Not all NTMs are associated with sawtooth crashes; some seem to appear spontaneously.

With regard to the drive for the instability, recall that, in the extended MHD model, Ohm's law has terms proportional to the electron pressure gradient and the electron stress tensor. In toroidal geometry, these contributions to the electric field can drive a toroidal current called the *bootstrap current*, so called because it is self-generated and not dependent on an applied voltage. A substantial fraction of the toroidal current in a tokamak is bootstrap. Also recall that anisotropic electron heat flux tends to equilibrate the temperature, and hence the pressure on a flux surface. Now consider the case of a magnetic island of finite width. It has its own "internal" flux surfaces. There is an equilibrium pressure gradient across the island.

Perpendicular heat flux tends to maintain this gradient. However, parallel heat flux tends to equilibrate the pressure within the island. The pressure distribution within the island is determined by the balance of these two competing processes. (These enter into the function $A(W)$, above.) If the island is large enough, equilibration within the island will win out over the equilibrium perpendicular heat flux, and the pressure gradient within the island will decrease. This lowers the drive for the bootstrap current. It turns out that the resulting perturbation to the total current is such as to cause the island to grow. Thus the presence of a critical seed island width is determined by competition between equilibrium perpendicular heat flux and parallel heat flux within the island, and the growth if the island is driven by the resulting perturbation to the bootstrap current.

The quantitative theory of the processes just described requires substantial mathematical fortitude and assumptions that limit the theoretical model to islands that are very small (but nonetheless finite). Since they invoke neo-classical closures of the fluid equations, they almost always involve flux surface averages, rather than local quantities. The details of the dynamics of these magnetic islands when variations *within* the island are accounted for are unknown and are a topic of considerable research.

6.3 Relaxation and Dynamo

Many systems naturally evolve toward states that exhibit some form of order on long length scales. Examples are the formation of isolated vortices in two-dimensional Navier–Stokes flow, the appearance of zonal flows in rotating fluids, the evolution of solitons in fluid and optical systems, and the characteristic structure of the magnetic field in laboratory plasma experiments. In all cases, long range order in one quantity is accompanied by short range disorder in another quantity, so that overall entropy increase is assured. The ordered states are robust in that their detailed structure remains relatively invariant across experimental realizations: the properties of these preferred states are independent of the way the system is initially prepared. This phenomenon is generally called self-organization.

Systems exhibiting self-organization have several common features. Their dynamics are described by nonlinear partial differential equations with dissipation; these equations admit quadratic (or higher order) quantities that are conserved in the absence of dissipation; and these conserved quantities decay at different rates when dissipation is taken into account. In the absence of dissipation (or on short enough time scales) the conserved quantities place severe constraints on the evolution of this ideal system. With the inclusion of dissipation, the ideal invariants decay. However, one (or a few) of these invariants decay slowly relative to the rest of the ideal invariants. Due to the robust nature of these privileged invariants, a variational principle is often invoked to predict the resulting global state with the slowly decaying invariants treated as a constraint condition.

In the resistive MHD model, the plasma is assumed to be a single, electrically conducting, charge neutral fluid that experiences both pressure and electromagnetic

(Lorentz) body forces. The electromagnetic field is described by the pre-Maxwell equations, in which the displacement current has been dropped from Amperes law. The dynamics of the fluid and the electromagnetic field are coupled by Ohms law, which relates the total electric field as seen by the moving fluid to the Ohmic (resistive) electric field, $\mathbf{E} + \mathbf{V} \times \mathbf{B} = \eta \mathbf{J}$. For the case of ideal MHD ($\eta = 0$), these equations admit an infinite number of conserved quantities. These are the total energy (assuming zero mean flow) and the infinite number of integrals $K_l = \int_{V_l} \mathbf{A} \cdot \mathbf{B} dV, l = 1, 2, \ldots,$ called the *Wöltjer invariants*. The integrals are to be taken over the volume of each and every flux tube in the plasma, and $\mathbf{B} = \nabla \times \mathbf{A}$. (These integrals are related geometrically to the local linkage of flux tubes. This topological property is preserved in ideal MHD where flux tubes retain their integrity. The invariance of these integrals is determined completely by the assumption of ideal MHD [Ohms law with $\eta = 0$]). The variational problem is then to minimize the energy W with the constraint that each of the infinite number of Wöltjer integrals remains constant. (Clearly, the unconstrained minimization of the energy leads to the trivial state $B = P = 0$.) The result of the constrained minimization is that the magnetic induction must satisfy the equation $\nabla \times \mathbf{B} = \mu_0 \lambda(\mathbf{r})\mathbf{B}$, with $\mathbf{B} \cdot \nabla \lambda = 0$. Magnetic fields that satisfy these conditions are called *force-free*, since the Lorentz force vanishes. The second condition, which comes from the requirement that $\nabla \cdot \mathbf{B} = 0$, is a statement that λ is constant along field lines.

In 1974, J. B. Taylor recognized a flaw in this variational approach. In an ideal, perfectly conducting fluid, flux tubes retain their integrity for all times. Thus, when the system is prepared (e.g., during the gas breakdown phase of an experiment), there is a value of λ associated with each flux tube in the initial state, and the details of the corresponding spatial distribution of are uncontrollable; they will vary greatly between different realizations of the experiment. Since λ must remain constant along flux tubes, the function $\lambda(\mathbf{r})$, and hence the magnetic field, will depend in a detailed manner on the way the system was prepared. The final state is thus not independent of the initial conditions, in contradiction to the observed properties of relaxed (or self-organized) states. The infinite number of constraints implied by ideal MHD limits the evolution of the system.

Taylor hypothesized that in a real plasma with large but finite electrical conductivity, the resulting breaking and merging of the individual flux tubes would render almost all of the Wöltjer invariants invalid. However, if the system is bounded by a perfectly conducting boundary, then the only flux tube that will retain its integrity is the one tangent to the conducting boundary. Under these conditions, only the single invariant $K_0 = \int_{V_0} \mathbf{A} \cdot \mathbf{B} dV$ would remain, where the integral is to be taken over the entire plasma volume. (This is called Taylors conjecture; K_0 is called the *magnetic helicity*.) The variational problem is thus to minimize the functional $I = W - \lambda K_0$, where λ is a Lagrange multiplier. The result is $\nabla \times \mathbf{B} = \mu_0 \lambda \mathbf{B}$, where λ is now a *constant*. (The constant λ can be related to the ratio of the total current to the total flux.) Solutions of this equation are thus independent of the initial conditions and can describe physically interesting states.

The above discussion applies to single-fluid-resistive MHD theory. When the extended MHD Ohm's law is used, the ions and electrons are placed on a more equal basis, and we might expect the predictions of relaxation theory to be modified.

Indeed, when the dynamical equations are (31, 32, 33, 34) are used, they can be combined with the expression $\mathbf{E} = -\partial\mathbf{A}/\partial t - \nabla\phi$, where ϕ is the scalar potential, into the useful form $-\partial\mathbf{A}_\alpha/\partial t - \nabla\phi_\alpha + \mathbf{V}_\alpha \times \mathbf{B}_\alpha = \mathbf{F}_\alpha$ for each species, where $\mathbf{A}_\alpha = \mathbf{A} + M_\alpha\mathbf{V}_\alpha/q_\alpha$ is related to the canonical momentum for each species, $\mathbf{F}_\alpha$ contains the pressure and stress forces, and $\mathbf{B}_\alpha = \nabla \times \mathbf{A}_\alpha = \mathbf{B} + M_\alpha\nabla \times \mathbf{V}_\alpha/q_\alpha$ is a generalized magnetic field related to the vector potential and the vorticity for each species, and $\phi_\alpha = \phi + M_\alpha V_\alpha^2/2q_\alpha$. From these equations it can be shown directly that, in the presence of fluctuations and dissipation, the generalized helicity integrals $K_\alpha = \int_{V_0} \mathbf{A}_\alpha \cdot \mathbf{B}_\alpha dV$, $\alpha = e, i$, are relatively invariant with respect to the total energy (magnetic W_M plus kinetic W_K). The relaxed state is now determined by minimizing the functional $I = W_M + W_K - \lambda_e K_e - \lambda_i K_i$ (where the λ_α are Lagrange multipliers) with respect to independent variations of $\mathbf{A}$, $\mathbf{V}_e$, and $\mathbf{V}_i$. With $M_e \to 0$ and $\mathbf{V} \approx \mathbf{V}_i$, the result of the variational calculation (the relaxed state) is given by

$$\mathbf{J} = 2(\lambda_e + \lambda_i)\mathbf{B} + \frac{2\lambda_i M_i}{e}\nabla \times \mathbf{V}, \tag{136}$$

$$\mathbf{V} = \frac{2\lambda_i}{ne}\left(\mathbf{B} + \frac{M_i}{e}\nabla \times \mathbf{V}\right), \tag{137}$$

and, with $\mathbf{V}_e = \mathbf{V}_i - \mathbf{J}/ne$,

$$\mathbf{V}_e = -\frac{2\lambda_e}{ne}\mathbf{B}. \tag{138}$$

The first term in Eq. (136) is similar to the single-fluid Taylor state, $\mathbf{J} = \lambda\mathbf{B}$. However, the current density is no longer parallel to the magnetic field, but is affected by the ion vorticity, and there is a net ion flow. The electrons continue to flow along the field lines.

With $M_e = 0$, the electron invariant K_e reduces to Taylor's invariant helicity K_0. By identifying a separate invariant for each fluid, two-fluid theory places ions and electrons on equal footing. The ion self-helicity, K_i, introduces flows, and the variational problem must then consider both the magnetic and kinetic energies. The relaxed state will therefore have both characteristic magnetic fields and zero-order flows. These flows are inevitably sheared, and sheared flows are ubiquitous in laboratory and, presumably, astrophysical plasmas. Their presence may be explained by this theory. The equivalent state obtained from resistive MHD theory does not contain these flows.

Finally, we comment on the relationship between plasma relaxation and the so-called dynamo. One definition of a dynamo is a solution to the Faraday–Ampère–Ohm equations, including resistivity, in which the total magnetic energy remains finite as $t \to \infty$. Since the Taylor relaxed state is a preferred state that maintains itself in the presence of finite resistivity, it has historically been associated with dynamo activity. This cannot occur in a state with too much symmetry, since, from Ohm's law, the parallel resistive electric field $\mathbf{E}_\parallel = \eta\mathbf{bb} \cdot \mathbf{J}$ will cause the magnetic field to decay. However, if there are fluctuations $\delta\mathbf{V}$ and $\delta\mathbf{B}$ superimposed on the symmetric fields, then the *mean* symmetric parallel electric field is $\langle\mathbf{E}\rangle_\parallel = -\mathbf{bb} \cdot \langle\delta\mathbf{V} \times \delta\mathbf{B}\rangle + \mathbf{bb} \cdot \langle\eta \cdot \mathbf{J}\rangle$. It can vanish, and the configuration can be preserved in the presence of resistivity. Thus, naturally occurring relaxed states are necessarily associated with fluctuations. The essential ingredient is the non-vanishing of the average $\langle \delta \mathbf{V} \times \delta \mathbf{B} \rangle$ quadratic term in Ohm's law. Extended MHD introduces the additional quadratic (and higher) nonlinearities $\mathbf{J} \times \mathbf{B}/n$ and $\nabla P_e/n$. Their mean values can also contribute to dynamo activity and sustainment. The role of these terms in plasma relaxation and discharge sustainment in different parameter ranges (and, in fact, in different regions of the same discharge) is also a topic of current research interest.

7 Summary

We began with a description of a plasma as N interacting particles obeying the laws of Hamiltonian dynamics. We naively hoped that, by integrating these equations forward in time, we could obtain a complete description of plasma dynamics. We found this to be impractical for two reasons. First, there are simply too many equations to integrate. Second, even if they could be integrated, it is not possible to know their initial conditions with sufficient accuracy to make the calculation meaningful. However, the fact that $N \gg 1$ allowed us to take advantage of the "fuzziness" in our knowledge of the initial conditions and instead apply a statistical methodology. The success of this approach depends on the ergodic hypothesis, which roughly states that the averaging a single system trajectory over time gives the same result as averaging an ensemble of identical systems over phase space. This allowed us to derive a time-dependent equation for the probability distribution function. Further averaging allowed us to reduce the number of degrees of freedom needed to describe the system from $6N$ to 6 and arrive at the so-called kinetic equation. This procedure required that we make a closure assumption, thereby relating certain high-order correlations of particle interactions to products of low-order correlations.

The physical quantities that commonly describe a macroscopic system can be defined as velocity space averages (or moments) of the distribution function. By applying these averages directly to the kinetic equation, we derived a set of time-dependent equations for these average quantities. These are called the moment equations, and they express the laws of conservation of mass, momentum, and energy. This reduced the number of degrees of freedom from 6 to 3 (for each species). These equations have the property that each successive equation for a moment contains the next higher-order moment, so we could not escape the problem of closure. In this case, closures express the high-order velocity moments in terms of combinations of the lower-order moments. Finding these moment expressions requires solving the kinetic equation in some approximation, and we noted that there was no universal agreement on either the order or form of the closures for strongly magnetized plasmas.

The moment equations needed to be solved simultaneously with Maxwell's equations for the electromagnetic field. Maxwell's equations had to be reduced to

Gallilean invariance in order to obtain a consistent model. We found that a direct consequence of this was that the electric force was always small compared with the magnetic force and could be neglected. This result is sometimes called quasi-neutrality.

There was a separate set of moment equations for each plasma species, ions, and electrons. By judicious combination of these, we were able to obtain a set of equations that looked like the familiar ideal MHD equations, but with extra terms. These new terms describe important effects of separate ion and electron drifts (two-fluid effects) and finite ion Larmor radius (FLR effects), which are not captured in the ideal MHD model. The resulting set of equations is called the extended MHD model.

We then introduced a set of dimensionless parameters, one of which was always small. By writing the extended MHD equations in terms of these parameters, and then ordering the parameters large or small with respect to the one small parameter, we were able to identify four different regimes in which four different sets of fluid equations are valid. These are fast flow and high frequencies (Hall MHD), moderate flows and frequencies (ideal MHD), very slow flows and frequencies (drift MHD), and extremely low flows and zero frequency (transport models). Thus, ideal MHD is only one of several fluid models that can be used to describe the dynamics of magnetized plasmas.

We then examined the wave and stability properties of the extended MHD model. We found that the correction terms introduced new families of dispersive waves. We also found that these terms generally have a stabilizing effect on ideal MHD instabilities.

Finally, we described some problems in extended MHD that are at the forefront of present theoretical plasma physics research.

So, we have come a long way in a short time. In trying to understand where ideal MHD came from, we shot completely through our target and arrived at the realization that ideal MHD is just a small subset of more complicated and physically richer model, the extended MHD model. Nonetheless, ideal MHD is a very useful model for understanding the basic behavior of magnetized plasmas. The goal now is to understand the extended MHD model, and its accompanying closures, to the same level that we understand ideal MHD. This is the focus of much modern research in theoretical plasma physics.

Bibliography

Because of the variety of topics to be emphasized, there is no appropriate single text for the planned course material. Instead I have drawn freely from the excellent sources listed below. Further references are given in the text as appropriate.

1. George Arfkin, *Mathematical Methods for Physicists*, 2nd Ed., Academic Press, New York (1970).
2. G. K. Batchelor, *The Theory of Homogenous Turbulence*, Cambridge University Press, Cambridge, UK (1953).
3. Dieter Biskamp, *Nonlinear Magnetohydrodynamics*, Cambridge University Press, Cambridge, UK (1993).
4. Dieter Biskamp, *Magnetic Reconnection in Plasmas*, Cambridge University Press, Cambridge, UK (2000).
5. Dieter Biskamp, *Magnetohydrodynamic Turbulence*, Cambridge University Press, Cambridge, UK (2003).
6. S. Chandrasekhar, *Hydromagnetic and Hydrodynamic Stability*, Dover Publications, New York (1981).
7. R. Courant and D. Hilbert, *Methods of Mathematic Physics*, Vol. 1, Interscience, New York (1953).
8. Jeffrey P. Freidberg, *Ideal Magnetohydrodynamics*, Plenum Press, New York (1987).
9. B. B. Kadomtsev, "Hydromagnetic Stability of a Plasma", in *Reviews of Plasma Physics*, Vol. 2, p. 153, Consultants Bureau, New York (1966).
10. L. D. Landau and E. M. Lifschitz, *Fluid Mechanics*, Pergamon Press, London, UK (1959).
11. L. D. Landau and E. M. Lifschitz, *Electrodynamics of Continuous Media*, Pergamon Press, Oxford, UK (1960).
12. Wallace M. Manheimer and Chris Lashmore-Davies, *MHD Instabilities in Simple Plasma Configurations*, Naval Research Laboratory, Washington, DC (1984).
13. Donald H. Menzel, *Mathematic Physics*, Dover Publications, New York (1961).
14. H. K. Moffatt, *Magnetic Field Generation in Electrically Conducting Fluids*, Cambridge University Press, Cambridge, UK (1978).
15. Sergio Ortolani and Dalton D. Schnack, *Magnetohydrodynamics of Plasma Relaxation*, World Scientific, Singapore (1993).
16. Eric R. Priest, *Solar Magnetohydrodynamics*, D. Reidel, Dortrecht (1982).
17. V. D. Shafranov, "Plasma Equilibrium in a Magnetic Field", in *Reviews of Plasma Physics*, Vol. 2, p. 103, Consultants Bureau, New York (1966).
18. J. B. Taylor, Rev. Mod. Phys. **58**, 741 (1986).

Index

A

Adiabatic law, 33, 76, 188
Adjoint matrix, 8
Adjoint operator, 126, 167
Advective derivative, 23, 33
Alfvén effect, 236
Alfvén, H., 3, 143
Alfvén speed, 37, 60, 141, 309
Alfvén velocity, 56, 67, 152, 203, 295
Alignment, 176, 236
Ampére's law, 35, 36, 38, 39, 46, 92, 93, 95, 96, 97, 113, 114, 294, 295, 296
Angular momentum, 26, 27, 28, 45, 231
Anisotropic medium, 1, 41, 298
Axial vector, 18
Axisymmetry/Axisymmetric, 108, 110, 116, 182, 255, 256, 269, 270, 271, 279, 304

B

Back reaction, 266, 280
Bad curvature, 162, 165
Basis vector, 6, 7, 14
Beltrami, 103
Bernoulli's theorem, 86
Bessel function model (BFM), 104, 252, 253, 256
Bessel's equation, 104, 251
Beta, 99–101, 188
BFM (Bessel function model), 104, 252, 253, 256
Bifurcation, 117, 120
Boundary conditions, 52, 53, 58, 60, 75, 105, 116, 119, 125, 133–136, 137, 138, 151, 154, 155, 156, 158, 163, 176, 191, 193, 195, 214, 215, 216, 217, 241, 242, 254, 256
Boundary layer, 58, 64, 216

Boundary-value problem, 136, 140, 215, 216, 217
Buoyancy, 23, 174, 271
Burger's equation, 230, 231
Butterfly diagram, 263, 281

C

Calculus of variations, 1, 140, 153–158, 244, 246
Cascade, 223
Christoffel symbols, 15
Closure, 21, 29, 33, 38, 39–42, 49, 225, 226, 227, 234, 237, 272, 288, 291, 292, 294, 296–299, 311, 314, 315
Collisionality, 41, 42, 292, 297, 299
Collision frequency, 41, 296, 297, 298
Column vector, 7
Complete set, 127, 140
Compressional waves, 3
Conductivity, 42, 49, 55, 62, 273, 312
Conservation of angular momentum, 45
Conservation of energy, 31, 46, 47, 140
Conservation law, 43–47
Conservation of mass, 21, 22, 23, 24, 43, 314
Conservation of momentum, 44
Continuity equation, 21, 24, 32, 33, 43, 44, 46, 57, 130, 266, 295
Contraction, 11
Contravariant, 15
Coordinate transformation, 5, 7, 17
Corona, 77, 204, 262, 263
Correlation tensor, 276
Covariant, 15
   derivative, 15
Cowling's theorem, 270, 271, 279
Cross-helicity, 71–76, 152, 236
Curl, 12, 13, 82, 103, 121, 207, 209, 232, 234, 235, 251

319